Mastering Flask Web and API Development

Build and deploy production-ready Flask apps seamlessly across web, APIs, and mobile platforms

Sherwin John C. Tragura

Mastering Flask Web and API Development

Group Product Manager: Kaustubh Manglurkar

Publishing Product Manager: Bhavya Rao

Book Project Manager: Sonam Pandey

Senior Editor: Anuradha Joglekar

Technical Editor: Reenish Kulshrestha

Copy Editor: Safis Editing

Proofreader: Anuradha Joglekar

Indexer: Rekha Nair

Production Designer: Prashant Ghare

DevRel Marketing Coordinator: Anamika Singh

Publication date: August 2024

Production reference: 2220724

Published by Packt Publishing Ltd.

Grosvenor House

11 St Paul's Square

Birmingham

B3 1RB, UK

ISBN 978-1-83763-322-7

www.packtpub.com

To my lola, Lila Calleja, my mama, Lorna Calleja, and the loving memory of my tatay, Cesar, for showing me what life should be to become successful. To Owen Estabillo, for being a partner from the start, through the thick and thin of life.

– Sherwin John C. Tragura

Contributors

About the author

Sherwin John C. Tragura is currently a subject matter expert and technical consultant in a company in the Philippines. He has been part of many development teams customizing Alfresco DMS/RMS and building Java and Python standalone and web projects. He is also a certified professional and bootcamp technical trainer who has delivered technical training on Java, Jakarta EE, C#, .NET, Python, Node frameworks, and other customized training courses since 2011. He was also associated as a lecturer with the Dela Salle University-Manila, Colegio de San Juan de Letran-Calamba, and the University of the Philippines-Los Banos. He is the author of other Packt books, including *Spring MVC Blueprints*, *Spring 5 Cookbook*, and *Building Python Microservices with FastAPI*.

First and foremost, I want to thank Packt and my team for the patience and understanding given to me for this project. A huge thanks to my cousin, Rhonalyn, for the road trips; Shon-Shon, my cute and puffy nephew, for making me laugh; and Owen, for being there during stressful days and coffee breaks. And to the Force for the blessings.

For my readers, to master something, you need to first learn how to share that something with others.

About the reviewers

Zubair Khan is a software engineer and the visionary behind Tech Marquee, a company dedicated to reimagining web services. His expertise in Python, Flask, FastAPI, Django, React, Vue, AWS, MySQL, Postgres, MongoDB, CI/CD, and BI analysis is globally recognized. An IGNITE Pakistan awardee for his driver drowsiness detection system using deep learning, Zubair is dedicated to quality and innovation, using SonarQube and Jenkins to ensure excellence. Inspired by the principles of decentralization, he strongly advocates for the open source community and regularly contributes with articles on AI, web development, and data science.

Vicente Marçal is a highly motivated self-taught Python developer with over 7 years of experience and is passionate about crafting clean, efficient, and scalable software solutions. Skilled in web development using popular frameworks such as Django, Flask, and FastAPI, he has a proven track record of delivering successful projects of varying sizes and complexities.

Table of Contents

2

Adding Advanced Core Features   35

3

Creating REST Web Services   73

4

Utilizing Flask Extensions   93

Part 2: Building Advanced Flask 3.x Applications

5

Building Asynchronous Transactions   123

6

Developing Computational and Scientific Applications   159

7

Using Non-Relational Data Storage   209

8

Building Workflows with Flask   263

9

Securing Flask Applications   307

Part 3: Testing, Deploying, and Building Enterprise-Grade Applications

10

Creating Test Cases for Flask   347

11

Deploying Flask Applications   381

12

Integrating Flask with Other Tools and Frameworks   421

Index   455

Other Books You May Enjoy   468

Preface

Since 2009, the time I started using the framework for software project development, Flask remains a powerful, lightweight, seamless, and easy-to-use Python framework for API and web application development. The non-boilerplate WSGI framework has grown its support, and now it has several utilities to support different features, even implementing asynchronous components.

The flexibility of Flask makes it the best tool for building various applications, from small-scale e-commerce to middle-scale enterprise applications, as per my experience. The framework is also an ideal choice for implementing scientific applications, such as laboratory calibration and monitoring systems, weather forecasting systems, and many other applications that require XLSX and CSV automation and report and graph generation.

This book showcases Flask 3 and how it translates and upgrades all my previous software development specifications with previous Flask versions using up-to-date features. I hope the book can help you understand Flask 3 and apply its components to create workarounds and provide solutions to challenging real-world problems.

Who this book is for

This book is for proficient Python developers seeking a deeper understanding of the Flask framework as a solution for tackling enterprise challenges. It is also a great resource for Flask-savvy readers eager to learn more about the framework's advanced capabilities and new features.

What this book covers

Chapter 1, A Deep Dive into the Flask Framework, introduces Flask as a simple and lightweight Python framework for web applications and showcases the installation of Flask 3 to jumpstart web application development using a non-standard project directory structure with basic Flask features, such as the view functions, class-based views, database connectivity, built-in Werkzeug server and libraries, and custom environment variables.

Chapter 2, Adding Advanced Core Features, provides the Flask 3 core features of web applications, such as session management, data management using **Object Relational Mapping** (**ORM**), view rendition using Jinja2 templates, message flashing, error handling, software logging, adding static contents, and applying blueprint and application factory design to project structuring.

Chapter 3, Creating REST Web Services, introduces API development using Flask 3 with request and response handling, implementation of JSON encoders and decoders for parsing the incoming request body and outgoing response, access of request and application context using `@before_request` and `@after_request` events, exception handling, and implementation of client applications to consume the REST services.

Chapter 4, Utilizing Flask Extensions, discusses how to save development time and effort by using helpful and efficient Flask modules in replacement to their ground-up equivalents, such as Flask-Session for non-browser-based session handling, Bootstrap-Flask for providing the presentation layer, Flask-WTF for building model-based web forms, Flask-Caching for creating cache, Flask-Mail for sending emails, and Flask-Migrate for building the database schema from data models.

Chapter 5, Building Asynchronous Transactions, explains the asynchronous features of Flask 3, which includes creating asynchronous view and API endpoint functions, implementing an asynchronous repository layer using SQLAlchemy, building asynchronous background tasks using Celery and Redis, implementing WebSocket and **Server-Sent Events** (**SSE**) with `asyncio` utilities, applying asynchronous signals for triggering transactions, and applying reactive programming, and introduces Quart as the ASGI-variant of Flask 3.

Chapter 6, Developing Computational and Scientific Applications, discusses the use of Flask in building scientific applications with XLSX and CSV uploading and tabular and graphical reports using popular Python libraries such as `numpy`, `pandas`, `matplotlib`, `seaborn`, `scipy`, and `sympy`, JavaScript libraries such as Chart.js, Bokeh, and Plotly, LaTeX tools for PDF generation, Celery and Redis for time-expensive background computations, and other scientific tools, such as Julia.

Chapter 7, Using Non-Relational Data Storage, explains how Flask can manage non-relational and big data using popular NoSQL databases, such as Apache HBase/Hadoop, Apache Cassandra, Redis, MongoDB, Couchbase, and Neo4J.

Chapter 8, Building Workflows with Flask, discusses how to implement non-BPMN and BPMN workflows with Flask 3, using Celery and Redis, SpiffWorkflow, Zeebe server from Camunda, Airflow 2, and Temporal.io.

Chapter 9, Securing Flask Applications, provides various ways to secure web-based and API-based Flask applications, such as implementing authentication and authorization mechanisms using HTTP Basic, Digest, and Bearer-token authentication schemes, OAuth2 authorization schemes, and Flask-Login; utilizing encoding/decoding and encryption/decryption libraries to protect user credentials; applying form validation and data sanitation to avoid different web attacks; replacing HTTP with the secured HTTPS to run applications; and controlling response headers to restrict or limit user access.

Chapter 10, Creating Test Cases for Flask, provides techniques for testing with or without mocking the Flask 3 components, such as model classes, repository transactions, native services, view and API endpoint functions, database connectivity, asynchronous processes, and WebSockets, using the PyTest framework.

Chapter 11, *Deploying Flask Applications*, discusses different options for deploying and running a web or API application, which includes using Gunicorn for standard and asynchronous Flask applications, uWSGI, the Docker platform through Docker Compose and Kubernetes deployments, and Apache HTTP Server.

Chapter 12, *Integrating Flask with Other Tools and Frameworks*, provides solutions for integrating Flask applications into different popular tools, such as GraphQL, React client applications, and Flutter mobile applications, and for building sub-applications built from Django, FastAPI, Tornado, and Flask frameworks within a microservice application using Flask's interoperability feature.

To get the most out of this book

To fully grasp the initial chapters of the book, you should have a background in Python web and API programming using any framework or a little background in Flask. But having a background in creating scripts using standard Python can also help you at least understand the first chapter, which is a jumpstart on how to use the Python language to build a web application using basic Flask concepts. Experienced developers can use the book to enrich their Flask experience further with the new utility classes and functions of the Flask 3.x framework.

Software/hardware covered in the book	Operating system requirements
Python 3.11.x	Windows 10, at least
React 18.3.1	Ubuntu (using PowerShell and WSL2)
Flutter 3.19.5	
PostgreSQL 13.4	
MongoDB Community Server 7.0.11	
Redis server 7.2.3 (Ubuntu)	
Redis server 3.0.504 (Windows)	
HBase/Hadoop 2.5.5	
Couchbase 7.2.0	
Cassandra 4.1.5	
Neo4J Desktop 1.5.8	
Julia 1.9.2	
Docker 25.0.3	
Kubernetes 5.0.4 (Docker-bundled)	
Zeebe 1.1.0 (Docker)	
Airflow 2.5	
Temporal.io server 1.22.0	

Software/hardware covered in the book	Operating system requirements
Apache HTTP Server 2.4	
Jaeger 1.5	
VS Code 1.88.0	

Optionally, it would help if you use a licensed Microsoft Excel to open XLSX and CSV files for *Chapter 6* and Foxit PDF Reader for the generated PDF files.

If you are using the digital version of this book, we advise you to type the code yourself or access the code from the book's GitHub repository (a link is available in the next section). Doing so will help you avoid any potential errors related to the copying and pasting of code.

We advise you to install the specified Python version to avoid unexpected errors related to version incompatibilities. Also, it will be advisable to download and read the code from GitHub while reading the chapters to understand the discussions. The code in GitHub is just a prototype and may serve as a guide to build your versions of the applications.

Download the example code files

You can download the example code files for this book from GitHub at https://github.com/PacktPublishing/Mastering-Flask-Web-Development. If there's an update to the code, it will be updated in the GitHub repository.

We also have other code bundles from our rich catalog of books and videos available at https://github.com/PacktPublishing/. Check them out!

Conventions used

There are a number of text conventions used throughout this book.

Code in text: Indicates code words in text, database table names, folder names, filenames, file extensions, pathnames, dummy URLs, user input, and Twitter handles. Here is an example: "However, to fully use this feature, install the flask[async] module using the following pip command:"

A block of code is set as follows:

```
from sqlalchemy import create_engine
from sqlalchemy.ext.declarative import declarative_base
from sqlalchemy.orm import sessionmaker, scoped_session

DB_URL = "postgresql://<username>:<password>@localhost:5433/sms"
```

When we wish to draw your attention to a particular part of a code block, the relevant lines or items are set in bold:

```
engine = create_engine(DB_URL)
db_session = scoped_session(sessionmaker(autocommit=False,
autoflush=False, bind=engine))

Base = declarative_base()

def init_db():
    import modules.model.db
```

Any command-line input or output is written as follows:

```
pip install flask[async]
```

Bold: Indicates a new term, an important word, or words that you see onscreen. For instance, words in menus or dialog boxes appear in **bold**. Here is an example: "Clicking this option will lead you to the **Enter interpreter path…** menu command and, eventually, to the **Find…** option."

> **Tips or important notes**
> Appear like this.

Get in touch

Feedback from our readers is always welcome.

General feedback: If you have questions about any aspect of this book, email us at customercare@packtpub.com and mention the book title in the subject of your message.

Errata: Although we have taken every care to ensure the accuracy of our content, mistakes do happen. If you have found a mistake in this book, we would be grateful if you would report this to us. Please visit www.packtpub.com/support/errata and fill in the form.

Piracy: If you come across any illegal copies of our works in any form on the internet, we would be grateful if you would provide us with the location address or website name. Please contact us at copyright@packtpub.com with a link to the material.

If you are interested in becoming an author: If there is a topic that you have expertise in and you are interested in either writing or contributing to a book, please visit authors.packtpub.com.

Share Your Thoughts

Once you've read *Mastering Flask Web and API Development*, we'd love to hear your thoughts! Scan the QR code below to go straight to the Amazon review page for this book and share your feedback.

https://packt.link/r/1-837-63322-3

Your review is important to us and the tech community and will help us make sure we're delivering excellent quality content.

Download a free PDF copy of this book

Thanks for purchasing this book!

Do you like to read on the go but are unable to carry your print books everywhere?

Is your e-book purchase not compatible with the device of your choice?

Don't worry!, Now with every Packt book, you get a DRM-free PDF version of that book at no cost.

Read anywhere, any place, on any device. Search, copy, and paste code from your favorite technical books directly into your application.

The perks don't stop there, you can get exclusive access to discounts, newsletters, and great free content in your inbox daily

Follow these simple steps to get the benefits:

1. Scan the QR code or visit the following link:

https://packt.link/free-ebook/9781837633227

2. Submit your proof of purchase.
3. That's it! We'll send your free PDF and other benefits to your email directly.

Part 1:
Learning the Flask 3.x
Framework

In this part, you will learn and understand the basic and core components to implement Flask-accepted web and API applications, including using the Blueprints and application factory functions to build the proper Flask project structure. This part will also teach you how to integrate Flask into the PostgreSQL databases using the psycopg2 and asyncpg drivers and implement the repository layers of your application using the native database driver or the **object-relational mapping (ORM)** tool. Moreover, you will learn how to use external Flask modules to implement the features of your applications without spending much time and effort.

This part includes the following chapters:

- *Chapter 1, A Deep Dive into the Flask Framework*
- *Chapter 2, Adding Advanced Core Features*
- *Chapter 3, Creating REST Web Services*
- *Chapter 4, Utilizing Flask Extensions*

A Deep Dive into the Flask Framework

Flask is a **Python** web framework that was created by Armin Ronacher to solve both web-based and API-related requirements that need a rapid development approach. It is a lightweight framework with helper classes and methods, a built-in server, a debugger, and a reloader, all of which are required for building scalable web applications and web services.

Unlike the Django framework, Flask is minimalistic and slimmer in that it requires more experience in using Python to craft various coding techniques and workarounds to implement its components. It is more open-ended and extensible than the full-stack Django, which is more friendly to newbies because of its easy-to-build projects and reusable components.

This first chapter will showcase the essential task itineraries that cover the initial components and base features of Flask 3.x that are essential in initiating our web development.

In this chapter, we will cover the following development tasks:

- Setting up the project baseline
- Creating routes and navigations
- Managing the requests and response data
- Implementing view templates
- Creating web forms
- Building the data layer with PostgreSQL
- Managing the project structure

Technical requirements

The first chapter will focus on building a prototype for an *Online Personal Counseling System* that simulates a face-to-face consultation between a patient and a counselor while highlighting the base components of *Flask 3.x*. The application will cover modules such as managing users, questionnaires, and some reports. The code for this chapter can be found at `https://github.com/PacktPublishing/Mastering-Flask-Web-Development/tree/main/ch01`.

Setting up the project baseline

Gathering and studying the system requirements for the development environment for the proposed project is essential. Some of these requirements include the correct versions of the installers and libraries, the appropriate servers, and the inclusion of other essential dependencies. We have to perform various setups before kicking off our projects.

Installing the latest Python version

All our applications will run on the *Python 11* environment for faster performance. The updated Python installer for all operating systems is available at `https://www.python.org/downloads/`.

Installing the Visual Studio (VS) Code editor

The Django framework has a `django-admin` command that generates a project structure, but Flask does not have that. We can use a terminal console or a tool such as the **Visual Studio (VS) Code** editor that can help developers create a Flask project. The VS Code installer is available at `https://code.visualstudio.com/download`.

After installing the VS Code editor, we can create a filesystem folder through it and start a Flask project. To create the folder, we should go to the **Open Folder** option under **File** or use the *Ctrl + K + O* shortcut to open the **Open Folder** mini-window. *Figure 1.1* shows a sample process of creating a Flask project using the editor:

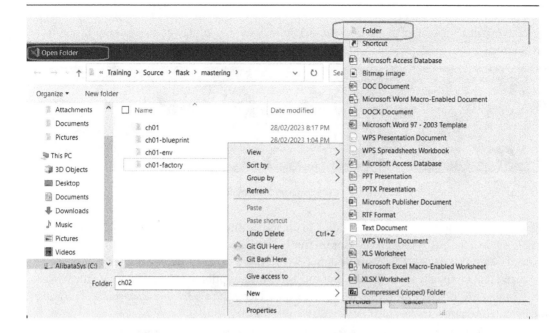

Figure 1.1 – Creating a Flask project folder using the VS Code editor

Creating the virtual environment

Another aspect of developing a Flask project is having a repository called a **virtual environment** that can hold its libraries. It is a mechanism or a tool that can manage all dependencies of a project by isolating these dependencies from the global repository and other project dependencies. The following are the advantages of using this tool in developing Flask-based applications:

- It can avoid broken module versions and collisions with other existing similar global repository libraries.

- It can help build a dependency tree for the project.

- It can help ease the deployment of applications with libraries to both physical and cloud-based servers.

A Python extension named `virtualenv` is required to set up these virtual environments. To install the extension, run the following command in the terminal:

```
pip install virtualenv
```

After this installation, we need to run `python virtualenv -m ch01-01` to create our first virtual environment for our Flask project. *Figure 1.2* shows a snapshot of creating our `ch01-env` repository:

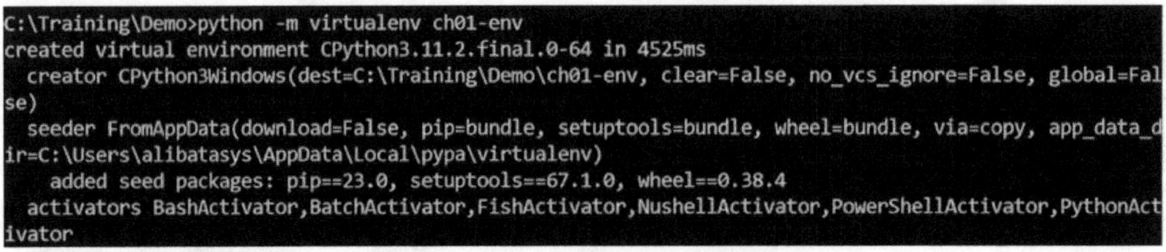

```
C:\Training\Demo>python -m virtualenv ch01-env
created virtual environment CPython3.11.2.final.0-64 in 4525ms
  creator CPython3Windows(dest=C:\Training\Demo\ch01-env, clear=False, no_vcs_ignore=False, global=Fal
se)
  seeder FromAppData(download=False, pip=bundle, setuptools=bundle, wheel=bundle, via=copy, app_data_d
ir=C:\Users\alibatasys\AppData\Local\pypa\virtualenv)
    added seed packages: pip==23.0, setuptools==67.1.0, wheel==0.38.4
  activators BashActivator,BatchActivator,FishActivator,NushellActivator,PowerShellActivator,PythonAct
ivator
```

Figure 1.2 – Creating a virtual environment

The next step is to open the project and link it to the virtual environment created for it. Pressing *Ctrl + Shift + P* in VS Code will open the **Command Palette** area. Here, we can search for `Python: Select Interpreter`. Clicking this option will lead you to the **Enter interpreter path...** menu command and eventually to the **Find...** option. This **Find...** option will help you locate the virtual environment's `Python.exe` file in the `/Scripts` folder. *Figure 1.3* shows a snapshot of locating the Python interpreter in the repository's `/Scripts` folder:

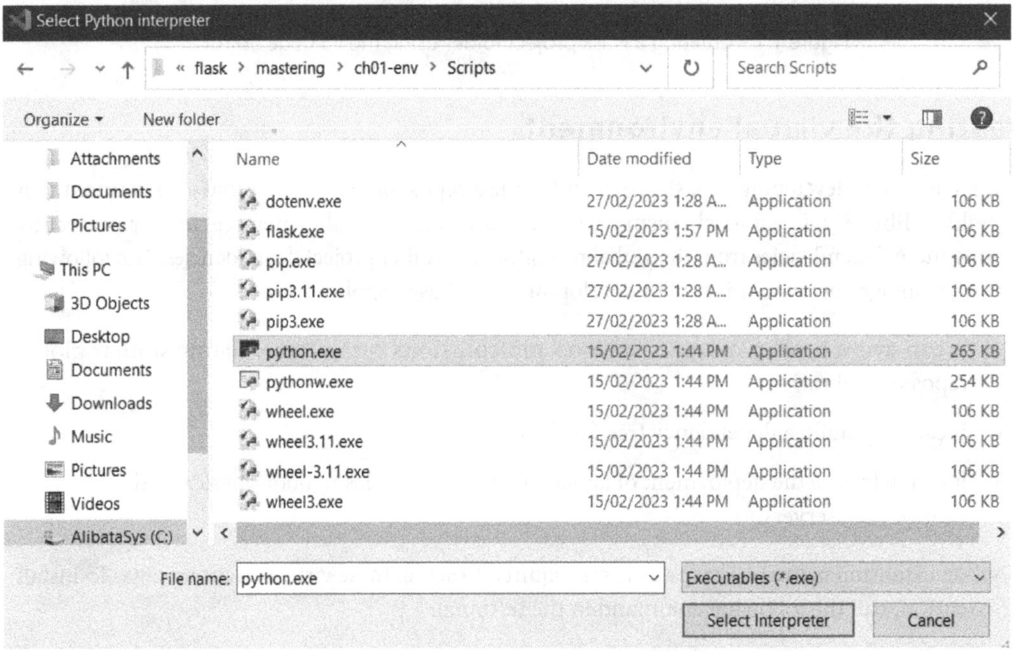

Figure 1.3 – Locating the Python interpreter of the virtual environment

Afterward, the virtual environment must be activated for the project to utilize it. You must run `/Scripts/activate.bat` in Windows or `/bin/activate` in Linux through the editor's internal console. Upon activation, the terminal should show the name of the virtual environment in its prompt (for example, `(ch01-env) C:\`).

Installing the Flask 3.x libraries

The integrated terminal of VS Code will appear after right-clicking the explorer portion of the editor, which leads to the **Open in Integrated Terminal** option. Once it appears on the lower right-hand side, activate the virtual environment first, then install all Flask dependencies into the repository by running `pip install flask`.

Once all the requirements are in place, we are ready to create our baseline application.

Creating the Flask project

The first component that must be implemented in the main project folder (that is, `ch01`) is the application file, which can be `main.py` or sometimes `app.py`. This component will become the top-level module the Flask will recognize when the server starts. Here is the baseline application file for our *Online Personal Counseling System* prototype:

```python
from flask import Flask
app = Flask(__name__)
@app.route('/', methods = ['GET'])
def index():
    return "This is an online … counseling system (OPCS)"
if __name__ == '__main__':
    app.run(debug=True)
```

Let's dissect and scrutinize the essential parts of the given `main.py` file:

- An imported `Flask` class from the `flask` package plays a considerable role in building the application. This class provides all the utilities that implement the **Werkzeug** specifications, which include features such as managing the requests and the responses of every route, redirecting pages, handling form data, accessing and creating cookies, parsing custom and built-in headers, and even providing debuggers for the development environment. In other words, the `Flask` instance is the main element in building a **Web Server Gateway Interface (WSGI)**-compliant application.

Werkzeug

`Werkzeug` is a WSGI-based library or module that provides Flask with the necessary utilities, including a built-in server, for running WSGI-based applications.

- The imported `Flask` instance must be instantiated once per application. The `__name__` argument must be passed to its constructor to provide `Flask` with a reference to the main module without explicitly setting its actual package. Its purpose is to provide Flask with the reach it needs in providing the utilities across the application and to register the components of the project to the framework.

- The `if` statement tells the Python interpreter to run Werkzeug's built-in development server if the module is `main.py`. This line validates the `main.py` module as the top-level module of the project.

- `app.run()` calls and starts the built-in development server of Werkzeug. Setting its `debug` parameter to `True` sets development or debug mode and enables Werkzeug's debugger tool and automatic reloading. Another way is to create a configuration file that will set `FLASK_DEBUG` to `True`. We can also set development mode by running `main.py` using the `flask run` command with the `--debug` option. Other configuration approaches before Flask 3.0, such as using `FLASK_ENV`, are already deprecated.

Running the `python main.py` command on the VS Code terminal will start the built-in development server and run our application. A server log will be displayed on the console with details that include the development mode, the debugger ID, and the **URL** address. The default port is `5000`, while the host is `localhost`.

Now, it is time to explore the view functions of our Flask application. These are the components that manage the incoming requests and outgoing responses.

Creating routes and navigations

Routing is a mapping of URL pattern(s) and other related details to a view function that's done using Flask's route decorators. On the other hand, the view function is a transaction that processes an incoming request from the clients and, at the same time, returns the necessary response to them. It follows a life cycle and returns an HTTP status as part of its response.

There are different approaches to assigning URL patterns to view functions. These include creating static and dynamic URL patterns, mapping URLs externally, and mapping multiple URLs to a view function.

Creating static URLs

Flask has several built-in route decorators that implement some of its components, and `@route` decorator is one of these. `@route` directly maps the URL address to the view function seamlessly. For instance, `@route` maps the `index()` view function presented in the project's `main.py` file to the root URL or `/`, which makes `index()` the view function of the root URL.

But @route can map any valid URL pattern to any view function. A URL pattern is accepted if it follows the following best practices:

- All characters must be in lowercase.

- Use only forward slashes to establish site hierarchy.

- URL names must be concise, clear, and within the business context.

- Avoid spaces and special symbols and characters as much as possible.

The following home() view function renders an introductory page of our ch01 application and uses the URL pattern of /home for its access:

```
@app.route('/home')
def home():
    return '''
        <html><head><title>Online Personal … System</title>
            </head><body>
            <h1>Online … Counseling System (OPCS)</h1>
            <p>This is a template of a web-based counseling
                application where counselors can … … …</em>
            </body></html>
    '''
```

Now, Flask accepts simple URLs such as /home or complex ones with slashes and path-like hierarchy, including these multiple URLs.

Assigning multiple URLs

A view function can have a stack of @route decorators annotated on it. Flask allows us to map these valid multiple URLs if there is no conflict with other view functions and within that stack of @route mappings. The following version of the home() view function now has three URLs, which means any of these addresses can render the home page:

```
@app.route('/home')
@app.route('/information')
@app.route('/introduction')
def home():
    return '''<html><head>
                <title>Online Personal … System</title>
            </head><body>
                <h1>Online … Counseling System (OPCS)</h1>

                … … … … …
            </body></html>
    '''
```

Aside from complex URLs, Flask is also capable of creating *dynamic routes*.

Applying path variables

Adding path variables makes a URL dynamic and changeable depending on the variations of the values passed to it. Although some SEO experts may disagree with having dynamic URLs, the Flask framework can allow view functions with changeable URL patterns to be implemented.

In Flask, a path variable is declared inside a diamond operator (<>) and placed within the URL path. The following view function has a dynamic URL with several path variables:

```
@app.route('/exam/passers/list/<float:rate>/<uuid:docId>')
def report_exam_passers(rating:float, docId:uuid4 = None):
    exams = list_passing_scores(rating)
    response = make_response(
      render_template('exam/list_exam_passers.html',
          exams=exams, docId=docId), 200)
    return response
```

As we can see, path variables are identified with data types inside the diamond operator (<>) using the <type:variable> pattern. These parameters are set to None if the path variables are optional. The path variable is considered a string type by default if it has no associated type hint. *Flask 3.x* offers these built-in data types for path variables:

- **string**: Allows all valid characters except for slashes.
- **int**: Takes integer values.
- **float**: Accepts real numbers.
- **uuid**: Takes unique 32 hexadecimal digits that are used to identify or represent records, documents, hardware gadgets, software licenses, and other information.
- **path**: Fetches characters, including slashes.

These path variables can't function without the corresponding parameters of the same name and type declared in the view function's parameter list. In the previous report_exam_passers() view function, the local rating and docId parameters are the variables that will hold the values of the path variables, respectively.

But there are particular or rare cases where path variables should be of a type different than the supported ones. View functions with path variables declared as `list`, `set`, `date`, or `time` will throw `Status Code 500` in Flask. As a workaround, the Werkzeug bundle of libraries offers a `BaseConverter` utility class that can help customize a variable type for paths that allows other types to be part of the type hints. The following view function requires a `date` type hint to generate a certificate in HTML format:

```
@app.route('/certificate/
accomp/<string:name>/ <string:course>/<date:accomplished_date>')
def show_certification(name:str, course:str, accomplished_date:date):
    certificate = """<html><head>
        <title>Certificate of Accomplishment</title>
      </head><body>
       <h1>Certificate of Accomplishment</h1>
        <p>The participant {} is, hereby awarded this certificate
of accomplishment, in {} course on {} date for passing all exams. He/
she proved to be ready for any of his/her future endeavors.</em>
        </body></html>
    """.format(name, course, accomplished_date)
    return certificate, 200
```

`accomplished_date` in `show_certification()` is a `date` hint type and will not be valid until the following tasks are implemented:

- First, subclass `BaseConverter` from the `werkzeug.routing` module. In the `/converter` package of this project, there is a module called `date_converter.py` that implements our `date` hint type, as shown in the following code:

```
from werkzeug.routing import BaseConverter
from datetime import datetime
class DateConverter(BaseConverter):
    def to_python(self, value):
        date_value = datetime.strptime(value, "%Y-%m-%d")
        return date_value
```

The given `DateConverter` will custom-handle date variables within our Flask application.

- `BaseConverter` has a `to_python()` method that must be overridden to implement the necessary conversion process. In the case of `DateConverter`, we need `strptime()` so that we can convert the path variable value in the `yyyy-mm-dd` format into the datetime type.

- Lastly, declare our new custom converter in the Flask instance of the `main.py` module. The following snippet registers `DateConverter` to app:

```
app = Flask(__name__)
app.url_map.converters['date'] = DateConverter
```

After following all these steps, the custom path variable type – for instance, date – can now be utilized across the application.

Assigning URLs externally

There is also a way to implement a routing mechanism without using the @route decorator, and that's by utilizing Flask's add_url_rule() method to register views. This approach binds a valid request handler to a unique URL pattern for every call to add_url_rule() of the app instance in the main.py module, not in the handler's module scripts, thus making this approach an external way of building routes. The following arguments are needed by the add_url_rule() method to perform mapping:

- The URL pattern with or without the path variables.

- The URL name and, usually, the exact name of the view function.

- The view function itself.

The invocation of this method must be in the main.py file, anywhere after its @route implementations and view imports. The following main.py snippet shows the external route mapping of the show_ honor_dismissal() view function to its dynamic URL pattern. This view function generates a termination letter for the counseling and consultation agreement between a clinic and a patient:

```
app = Flask(__name__)
def show_honor_dissmisal(counselor:str, effective_date:date,
patient:str):
    letter = """
    ... ... ... ... ...
        </head><body>
            <h1> Termination of Consultation </h1>
            <p>From: {}
            <p>Head, Counselor
            <p>Date: {}
            <p>To: {}
            <p>Subject: Termination of consultation
                    <p>Dear {},
            ... ... ... ... ... ...
                    <p>Yours Sincerely,
                    <p>{}
                </body>
            </html>
    """.format(counselor, effective_date, patient, patient, counselor)
    return letter, 200
```

```
app.add_url_rule('/certificate/
terminate/<string:counselor>/<date:effective_date>/<string:patient>',
'show_honor_dissmisal', views.certificates.show_honor_dissmisal)
```

Binding URL mappings to views using `add_url_rule()` is not only confined to the decorated function views but is also necessary for *class-based views*.

Implementing class-based views

Another way to create the view layer is through Flask's class-based view approach. Unlike the Django framework, which uses mixin programming to implement its class-based views, Flask provides two API classes, namely `View` and `MethodView`, that can directly subclass any custom view implementations.

The most common and generic class to implement HTTP GET operations is the `View` class from the `flask.views` module. It has a `dispatch_request()` method that executes the request-response transactions like a typical view function. Thus, subclasses must override this core method to implement their view transactions. The following class, `ListUnpaidContractView`, renders a list of patients with payments due to the clinic:

```
from flask.views import View
class ListUnpaidContractView(View):
    def dispatch_request(self):
        contracts = select_all_unpaid_patient()
        return render_template("contract/ list_patient_contract.html",
contracts=contracts)
```

`select_all_unpaid_patient()` will provide the patient records from the database. All these records will be rendered to the `list_patient_contract.html` template. Now, aside from overriding the `dispatch_request()` method, `ListUnpaidContractView` also inherits all the attributes and helper methods from the `View` class, including the `as_view()` static method, which creates a view name for the view. During view registration, this view name will serve as the `view_func` name of the custom `View` class in the `add_url_rule()` method with its mapped URL pattern. The following `main.py` snippet shows how to register `ListUnpaidContractView`:

```
app.add_url_rule('/contract/unpaid/patients', view_
func=ListUnpaidContractView.as_view('list-unpaid-view'))
```

If a `View` subclass needs an HTTP POST transaction, it has a built-class class attribute called `methods` that accepts a list of HTTP methods the class needs to support. Without it, the default is the `[ "GET" ]` value. Here is another custom `View` class of our *Online Personal Counselling System* app that deletes existing patient contracts of the clinic:

```
class DeleteContractByPIDView(View):
    methods = ['GET', 'POST']
    ... ... ... ... ... ...
```

```
def dispatch_request(self):
    if request.method == "GET":
        pids = list_pid()
        return render_template("contract/ delete_patient_contract.
html", pids=pids)
    else:
        pid = int(request.form['pid'])
        result = delete_patient_contract_pid(pid)
        if result == False:
            pids = list_pid()
            return render_template("contract/ delete_patient_
contract.html", pids=pids)
        contracts = select_all_patient_contract()
        return render_template("contract/ list_patient_contract.
html", contracts=contracts)
```

`DeleteContractByPIDView` handles a typical form-handling transaction, which has both a `GET` operation for loading the form page and a `POST` operation to manage the submitted form data. The `POST` operation will verify if the patient ID submitted by the form page exists, and it will eventually delete the contract(s) of the patient using the patient ID and render an updated list of contracts.

Other than the `View` class, an alternative API that can also build view transactions is the `MethodView` class. This class is suitable for web forms since it has the built-in `GET` and `POST` hints or templates that subclasses need to define but without the need to identify the `GET` transactions from `POST`, like in a view function. Here is a view that uses `MethodView` to manage the contracts of the patients in the clinic:

```
from flask.views import MethodView
class ContractView(MethodView):
    ... ... ... ... ... ...
    def get(self):
        return render_template("contract/ add_patient_contract.html")

    def post(self):
        pid = request.form['pid']
        approver = request.form['approver']

        ... ... ... ... ... ...
        result = insert_patient_contract(pid=int(pid), approved_
by=approver, approved_date=approved_date, hcp=hcp, payment_
mode=payment_mode, amount_paid=float(amount_paid), amount_
due=float(amount_due))
        if result == False:
            return render_template("contract/ add_patient_contract.
html")
        contracts = select_all_patient_contract()
```

```
        return render_template("contract/ list_patient_contract.html",
contracts=contracts)
```

The `MethodView` class does not have a `methods` class variable to indicate the HTTP methods supported by the view. Instead, the subclass can select the appropriate HTTP hints from `MethodView`, which will then implement the required HTTP transactions of the custom view class.

Since `MethodView` is a subclass of the `View` class, it also has an `as_view()` class method that creates a `view_func` name of the view. This is also necessary for `add_url_rule()` registration.

Aside from `GET` and `POST`, the `MethodView` class also provides the `PUT`, `PATCH`, and `DELETE` method hints for API-based applications. `MethodView` is better than the `View` API because it organizes the transactions according to HTTP methods and checks and executes these HTTP methods by itself at runtime. In general, between the decorated view function and the class-based ones, the latter approach provides a complete Flask view component because of the attributes and built-in methods inherited by the view implementation from these API classes. Although the decorated view function can support a flexible and open-ended strategy for scalable applications, it cannot provide an organized base functionality that can supply baseline view features to other related views, unlike in a class-based approach. However, the choice still depends on the scope and requirements of the application.

Now that we've created and registered the routes, let's scrutinize these view implementations and identify the essential Flask components that compose them.

Managing request and response data

At this point, we already know that routing is a mechanism for mapping view functions to their URLs. But besides that, routing declares any valid functions to be view implementations that can manage the incoming request and outgoing response.

Retrieving the request object

Flask uses its `request` object to carry cookies, headers, parameters, form data, form objects, authorization data, and other request-related details. But the view function doesn't need to declare a variable to auto-wire the request instance, just like in Django, because Flask has a built-in proxy object for it, the `request` object, which is part of the `flask` package. The following view function takes the `username` and `password` request parameters and checks if the credentials are in the database:

```
from __main__ import app
from flask import request, Response, render_template, redirect
from repository.user import validate_user
@app.route('/login/params')
def login_with_params():
    username = request.args['username']
    password = request.args['password']
```

```
      result = validate_user(username, password)
      if result:
        resp = Response(
          response=render_template('/main.html'), status=200, content_
type='text/html')
        return resp
      else:
          return redirect('/error')
```

For instance, running the URL pattern of the given view function, `http://localhost:5000/login/params?username=sjctrags&password=sjctrags2255`, will provide us with `sjctrags` and `sjctrags2255` as values when `request.args['username']` and `request.args['password']` are accessed, respectively.

Here is the complete list of objects and details that we can retrieve from the Request object through its request instance proxy:

- `request.args`: Returns a `MultiDict` class that carries URL arguments or request parameters from the query string.

- `request.form`: Returns a `MultiDict` class that contains parameters from an HTML form or JavaScript's `FormData` object.

- `request.data`: Returns request data in a byte stream that Flask couldn't parse to form parameters and values due to an unrecognizable mime type.

- `request.files`: Returns a `MultiDict` class containing all file objects from a form with `enctype=multipart/form-data`.

- `request.get_data()`: This function returns the request data in byte streams before calling `request.data`.

- `request.json`: Returns parsed JSON data when the incoming request has a `Content-Type` header of `application/json`.

- `request.method`: Returns the HTTP method name.

- `request.values`: Returns the combined parameters of `args` and `form` and encounters collision problems when both `args` and `form` carry the same parameter name.

- `request.headers`: Returns request headers included in the incoming request.

- `request.cookies`: Returns all the cookies that are part of the request.

The following view function utilizes some of the given request objects to perform an HTTP GET operation to fetch a user login application through an ID value and an HTTP POST operation to retrieve the user details, approve its preferred user role, and save the login details as new, valid user credentials:

```python
from __main__ import app
from flask import render_template
from model.candidates import AdminUser, CounselorUser, PatientUser
from urllib.parse import parse_qsl
@app.route('/signup/approve', methods = ['POST'])
@app.route('/signup/approve/<int:utype>',methods = ['GET'])
def signup_approve(utype:int=None):
    if (request.method == 'GET'):
        id = request.args['id']
        user = select_single_signup(id)
        … … … … … … …
    else:
        utype = int(utype)
        if int(utype) == 1:
            adm = request.get_data()
            adm_dict = dict(parse_qsl(adm.decode('utf-8')))
            adm_model = AdminUser(**adm_dict)
            user_approval_service(int(utype), adm_model)
        elif int(utype) == 2:
            cnsl = request.get_data()
            cnsl_dict = dict(parse_qsl(
                    cnsl.decode('utf-8')))
            cnsl_model = CounselorUser(**cnsl_dict)
            user_approval_service(int(utype), cnsl_model)
        elif int(utype) == 3:
            pat = request.get_data()
            pat_dict = dict(parse_qsl(pat.decode('utf-8')))
            pat_model = PatientUser(**pat_dict)
            user_approval_service(int(utype), pat_model)
        return render_template('approved_user.html',
message='approved'), 200
```

Our application has a listing view that renders hyperlinks that can redirect users to this `signup_approve()` form page with a context variable `id`, a code for a user type. The view function retrieves the variable `id` through `request.args`, checks what the user type `id` is, and renders the appropriate page based on the user type detected. The function also uses `request.method` to check if the user request will pursue either the `GET` or `POST` transaction since the given view function caters to both HTTP methods, as defined in its *dual* route declaration. When clicking the **Submit** button on the form page, its `POST` transaction retrieves all the form parameters and values in a byte stream type via `request.get_data()`. It is decoded to a query string object and converted into a dictionary by `parse_sql` from the `urllib.parse` module.

Now, if Flask can handle the request, it can also manage the outgoing response from the view functions.

Creating the response object

Flask uses `Response` to generate a client response for every request. The following view function renders a form page using the `Response` object:

```python
from flask import render_template, request, Response
@app.route('/admin/users/list')
def generate_admin_users():
    users = select_admin_join_user()
    user_list = [list(rec) for rec in users]
    content = '''<html><head>
                    <title>User List</title>
            </head><body>
                    <h1>List of Users</h1>
                    <p>{}
            </body></html>
            '''.format(user_list)
    resp = Response(response=content, status=200, content_type='text/html')
    return resp
```

`Response` is instantiated with its required constructor parameters and returned by the view function as a response object. The following are the required parameters:

- `response`: Contains the content that needs to be rendered either in a string, byte stream, or iterable of either of the two types.

- `status`: Accepts the HTTP status code as an integer or string.

- `content_type`: Accepts the mime type of the response object that needs rendering.

- `headers`: A dictionary that contains the response header(s) that is/are necessary for the rendition process, such as `Access-Control-Allow-Origin`, `Content-Disposition`, `Origin`, and `Accept`.

But if the purpose is to render HTML pages, Flask has a `render_template()` method that references an HTML template file that needs rendering. The following route function, `signup_users_form()`, yields the content of a signup page – that is, `add_signup.html` from the `/pages` template folder – for new user applicants:

```
@app.route('/signup/form', methods= ['GET'])
def signup_users_form():
    resp = Response(  response=render_template('add_signup.html'),
status=200, content_type="text/html")
    return resp
```

`render_template()` returns HTML content with its context data, if there is any, as a string. To simplify the syntax, Flask allows us to return the method's result and the *status code* instead of the `Response` instance since the framework can automatically create a `Response` instance from these details. Like the previous examples, the following `signup_list_users()` uses `render_template()` to show the list of new user applications subject to admin approval:

```
@app.route('/signup/list', methods = ['GET'])
def signup_list_users():
    candidates = select_all_signup()
    return render_template('reports/list_candidates.html',
records=candidates), 200
```

The given code emphasizes that `render_template()` can accept and pass context data to the template page. The `candidates` variable in this snippet handles an extracted list of records from the database needed by the template for content generation using the **Jinja2** engine.

> **Jinja2**
>
> Jinja2 is Python's fast, flexible, robust, expressive, and extensive templating engine for creating HTML, XML, LaTeX, and other supported formats for Flask's rendition purposes.

On the other hand, Flask has a utility called `make_response()` that can modify the response by changing headers and cookies before sending them to the client. This method is suitable when the base response frequently undergoes some changes in its response headers and cookies. The following code modifies the content type of the original response to XLS with a given filename – in this case, `question.xls`:

```
@app.route('/exam/details/list')
def report_exam_list():
    exams = list_exam_details()
    response = make_response( render_template('exam/list_exams.
html', exams=exams), 200)
    headers = dict()
    headers['Content-Type'] = 'application/vnd.ms-excel'
```

```
    headers['Content-Disposition'] = 'attachment;filename=questions.
xls'
    response.headers = headers
    return response
```

Flask will require additional Python extensions when serializing and yielding PDF, XLSX, DOCX, RTF, and other complex content types. But for old and simple mime type values such as `application/msword` and `application/vnd.ms-excel`, Flask can easily and seamlessly serialize the content since Python has a built-in serializer for them. Other than mime types, Flask also supports adding web cookies for route functions. The following `assign_exam()` route shows how to add cookies to the `response` value that renders a form for scheduling and assigning counseling exams for patients with their respective counselors:

```
@app.route('/exam/assign', methods=['GET', 'POST'])
def assign_exam():
    if request.method == 'GET':
        cids = list_cid()
        pids = list_pid()
        response = make_response( render_template('exam/assign_exam_
form.html', pids=pids, cids=cids), 200)
        response.set_cookie('exam_token', str(uuid4()))
        return response, 200
    else:
        id = int(request.form['id'])
        cid = request.form['cid']
        pid = int(request.form['pid'])
        exam_date = request.form['exam_date']
        duration = int(request.form['duration'])
        result = insert_question_details(id=id, cid=cid, pid=pid,
exam_date=exam_date, duration=duration)
        if result:
            task_token = request.cookies.get('exam_token')
            task = "exam assignment (task id {})".format(task_token)
            return redirect(url_for('redirect_success_
exam',        message=task ))
        else:
            return redirect('/exam/task/error')
```

The Response instance has a `set_cookie()` method that creates cookies before the view dispatches the response to the client. It also has `delete_cookie()`, which deletes a particular cookie before yielding the response. To retrieve the cookies, `request.cookies` has a `get()` method that can retrieve the cookie value through its cookie name. The given `assign_exam()` route shows how the `get()` method retrieves exam_cookie in its POST transaction.

Implementing page redirection

Sometimes, it is ideal for the route transaction to redirect the user to another view page using the redirect() utility method instead of building its own Response instance. Flask redirection requires a URL pattern of the destination to where the view function will redirect. For instance, in the previous assign_exam() route, the output of its POST transaction is not a Response instance but a redirect() method:

```
@app.route('/exam/assign', methods=['GET', 'POST'])
def assign_exam():
        ... ... ... ... ... ...
        if result:
            task_token = request.cookies.get('exam_token')
            task = "exam assignment (task id {})".format(task_token)
            return redirect(url_for('redirect_success_
exam', message=task ))
        else:
            return redirect('/exam/task/error')
```

When the result variable is False, redirection to an error view called /exam/task/error will occur. Otherwise, the route will redirect to an endpoint or view name called redirect_success_exam. Every @route has an endpoint equivalent, by default, to its view function name. So, redirect_success_exam is the function name of a route with the following implementation:

```
@app.route('/exam/success', methods=['GET'])
def redirect_success_exam():
    message = request.args['message']
    return render_template('exam/redirect_success_view.html',
message=message)
```

url_for(), which is used in the assign_exam() view, is a route handler that allows us to pass the endpoint name of the destination view to redirect() instead of passing the actual URL pattern of the destination. It can also pass context data to the Jinja2 template of the redirected page or values to path variables if the view uses a dynamic URL pattern. The redirect_success_exam() function shows a perfect scenario of context data passing, where it uses request.args to access a message context passed from assign_exam(), which is where the redirection call originated.

More content negotiations and how to serialize various mime types for responses will be showcased in the succeeding chapters, but in the meantime, let's scrutinize the view templates of our route functions. View templates are essential for web-based applications because all form-handling transactions, report generation, and page generation depend on effective dynamic templates.

Implementing view templates

Jinja2 is the default templating engine of the Flask framework and is used to create HTML, XML, LaTeX, and markup documents. It is a simple, extensive, fast, and easy-to-use templating approach with powerful features such as layout capabilities, built-in programming constructs, support for asynchronous operations, context data filtering, and utility for unit testing.

Firstly, Flask requires all template files to be in the `templates` directory of the main project. To change this setting, the `Flask()` constructor has a `template_folder` parameter that can set and replace the default directory with another one. Our prototype, for instance, has the following Flask instantiation that overrides the default templates directory with a more high-level directory name:

```
from flask import Flask
app = Flask(__name__, template_folder='pages')
```

In our given setup, the view functions always refer to the `pages` directory when calling the template files through the `render_template()` method.

When it comes to syntax, Jinja2 has a placeholder (`{{ }}`) that renders dynamic content passed by the view functions to its template file. It also has a Jinja block (`{% %}`) that supports control structures such as loops, conditional statements, macros, and template inheritance. In the previous route function, `assign_exam()`, the GET transaction retrieves a list of counselor IDs (`cids`) and patient IDs (`pids`) from the database and passes them to the `assign_exam_form.html` template found in the `exam` subfolder of the `pages` directory. The following snippet shows the implementation of the `assign_exam_form.html` view template:

```
<!DOCTYPE html>
<html lang="en"><head><title>Patient's Score Form</title>
    </head><body>
        <form action="/exam/score" method="POST">
            <h3>Exam Score</h3>
            <label for="qid">Enter Questionnaire ID:</label>
            <select name="qid">
                {% for id in qids %}
                    <option value="{{ id }}">{{ id }}</option>
                {% endfor %}
            </select><br/>
            <label for="pid">Enter patient ID:</label>
            <select name="pid">
                {% for id in pids %}
                    <option value="{{ id }}">{{ id }}</option>
                {% endfor %}
            </select><br/>

            ... ... ... ... ... ...
            <input type="submit" value="Assign Exam"/>
```

```
        </form></body>
</html>
```

This template uses the Jinja block to iterate all the IDs and embed each in the `<option>` tag of the `<select>` component with the placeholder operator.

More about Jinja2 and Flask 3.x will be covered in *Chapter 2*, but for now, let's delve into how Flask can implement the most common type of web-based transaction – that is, by capturing form data from the client.

Creating web forms

In Flask, we can choose from the following two approaches when implementing view functions for form data processing:

- Creating two separate routes, one for the GET operation and the other for the POST transaction, as shown for the following user signup transaction:

```python
@app.route('/signup/form', methods= ['GET'])
def signup_users_form():
    resp = Response(response= render_template('add_signup.
html'), status=200, content_type="text/html")
    return resp


@app.route('/signup/submit', methods= ['POST'])
def signup_users_submit():
    username = request.form['username']
    password = request.form['password']
    user_type = request.form['utype']
    firstname = request.form['firstname']
    lastname = request.form['lastname']
    cid = request.form['cid']
    insert_signup(user=username, passw=password, utype=user_
type, fname=firstname, lname=lastname, cid=cid)
    return render_template('add_signup_submit.html',
message='Added new user!'), 200
```

- Utilizing only one view function for both the GET and POST transactions, as shown in the previous `signup_approve()` route and in the following `assign_exam()` view:

```python
@app.route('/exam/assign', methods=['GET', 'POST'])
def assign_exam():
    if request.method == 'GET':
        cids = list_cid()
        pids = list_pid()
        response = make_response(render_template('exam/assign_
```

```
exam_form.html', pids=pids, cids=cids), 200)
        response.set_cookie('exam_token', str(uuid4()))
        return response, 200
    else:
        id = int(request.form['id'])
        … … … … … …
        duration = int(request.form['duration'])
        result = insert_question_details(id=id, cid=cid, pid=pid,
exam_date=exam_date, duration=duration)
        if result:
            exam_token = request.cookies.get('exam_token')
            return redirect(url_for('introduce_exam',
message=str(exam_token)))
        else:
            return redirect('/error')
```

Compared to the first, the second approach needs `request.method` to separate GET from the POST transaction.

In setting up the form template, binding context data to the form components through `render_template()` is a fast way to provide the form with parameters with default values. The form model must derive the names of its attributes from the form parameters to establish a successful mapping, such as in the `signup_approve()` route. When it comes to retrieving the form data, the `request` proxy has a `form` dictionary object that can store form parameters and their data while its `get_data()` function can access the entire query string in byte stream type. After a successful POST transaction, the view function can use `render_template()` to load a success page or go back to the form page. It may also apply redirection to bring the client to another view.

But what happens to the form data after form submission? Usually, form parameter values are rendered as request attributes, stored as values of the session scope, or saved into a data store using a data persistency mechanism. Let's explore how Flask can manage data from user requests using a relational database such as PostgreSQL.

Building the data layer with PostgreSQL

PostgreSQL is an object-relational database system, and Flask can utilize it as a data storage platform if the activated virtual environment has the `psycopg2-binary` extension module. To install this extension module into the `venv`, run the following command:

```
pip install psycopg2-binary
```

Now, we can write an approach to establish a connection to the PostgreSQL database.

Setting up database connectivity

There are multiple ways to create a connection to a database, but this chapter will showcase a Pythonic way to extract that connection using a custom decorator. In the project's /config directory, there is a connect_db decorator that uses psycopgy2.connect() to establish connectivity to the opcs database of our prototype. Here is the implementation of this custom decorator:

```python
import psycopg2
import functools
from os import environ


def connect_db(func):
    @functools.wraps(func)
    def repo_function(*args, **kwargs):
        conn = psycopg2.connect(
            host=environ.get('DB_HOST'),
            database=environ.get('DB_NAME'),
            port=environ.get('DB_PORT'),
            user = environ.get('DB_USER'),
            password = environ.get('DB_PASS'))
        resp = func(conn, *args, **kwargs)
        conn.commit()
        conn.close()
        return resp
    return repo_function
```

The given decorator provides the connection instance, conn, to a repository function and commits all the changes to the database after a transaction's successful execution. Also, it will close the database connection at the end of the process. All the database details, such as DB_HOST, DB_NAME, and DB_PORT, are stored as environment variables inside a .env file. To retrieve them using the environ dictionary of the os module, run the following command to install the required extension:

```
pip install python-dotenv
```

However, there are other ways to manage these custom and built-in configuration variables instead of storing them as .env variables. The next topic will expound on this, but first, let's apply @connect_db to our repository layer.

Implementing the repository layer

The following `insert_signup()` transaction adds a new user signup record to the database. It gets the conn instance from the `@connect_db` decorator. Our application has no **object-relational mapper** yet and solely depends on the `psycopg2` driver to perform the CRUD operation. The `cursor` instance created by conn executes the *INSERT* statement of the following transaction with form data provided by its view function:

```python
from config.db import connect_db
from typing import Dict, Any, List


@connect_db
def insert_signup(conn, user:str, passw:str, utype:str, fname:str,
lname:str, cid:str) -> bool:
    try:
        cur = conn.cursor()
        sql = 'INSERT INTO signup (username, password, user_type,
firstname, lastname, cid) VALUES (%s, %s, %s, %s, %s, %s)'
        values = (user, passw, utype, fname, lname, cid)
        cur.execute(sql, values)
        cur.close()
        return True
    except Exception as e:
        cur.close()
        print(e)
    return False
```

`cursor` is an object derived from conn that uses a database session to perform insert, update, delete, and fetch operations. So, just like `insert_signup()`, the following transaction uses cursor again to execute the *UPDATE* statement:

```python
@connect_db
def update_signup(conn, id:int, details:Dict[str, Any]) -> bool:
    try:
        cur = conn.cursor()
        params = ['{} = %s'.format(key) for key in details.keys()]
        values = tuple(details.values())
        sql = 'UPDATE signup SET {} where id = {}'.format(',
'.join(params), id);
        cur.execute(sql, values)
        cur.close()
        return True
    except Exception as e:
        cur.close()
```

```
            print(e)
        return False
```

To complete the CRUD operations for the `signup` table, here is the *DELETE* transaction from our application:

```
@connect_db
def delete_signup(conn, id) -> bool:
    try:
        cur = conn.cursor()
        sql = 'DELETE FROM signup WHERE id = %s'
        values = (id, )
        cur.execute(sql, values)
        cur.close()
        return True
    except Exception as e:
        cur.close()
        print(e)
    return False
```

The use of an ORM to build the model layer will be part of *Chapter 2*'s discussions. For now, the views and services of our application rely on a repository layer that manages PostgreSQL data directly through the `psycopg2` driver.

After creating the repository layer, many applications can build a service layer to provide loose coupling between the CRUD operations and the views.

Creating the service layer

The service layer of the application builds the business logic of the view functions and the repository. Instead of loading the view functions with transaction-related and business processes, we place all these implementations in the service layer by creating lists of all the counselor and patient IDs, validating where to persist the newly approved user, and creating a list of patients who excelled in the examinations. The following service function evaluates and records patients' exam scores:

```
def record_patient_exam(formdata:Dict[str, Any]) -> bool:
    try:
        pct = round((formdata['score'] / formdata['total']) * 100, 2)
        status = None
        if (pct >= 70):
            status = 'passed'
        elif (pct < 70) and (pct >= 55):
            status = 'conditional'
        else:
```

```
            status = 'failed'
        insert_patient_score(pid=formdata['pid'], qid=formdata['qid'],
    score=formdata['score'], total=formdata['total'], status=status,
    percentage=pct)
        return True
    except Exception as e:
        print(e)
    return False
```

Instead of directly accessing `insert_patient_score()` to save patient exam scores, `record_score()` accesses the `record_patient_exam()` service to compute some formulas before invoking `insert_patient_score()` from the repository layer for record insertion. The service lessens some friction between the database transactions and the view layer. The following snippet is the view function that accesses the `record_patient_exam()` service for record exam record insertion:

```
@app.route('/exam/score', methods=['GET', 'POST'])
def record_score():
    if request.method == 'GET':
        pids = list_pid()
        qids = list_qid()
        return render_template( 'exam/add_patient_score_form.
html', pids=pids, qids=qids), 200
    else:
        params = dict()
        params['pid'] = int(request.form['pid'])
        params['qid'] = int(request.form['qid'])
        params['score'] = float(request.form['score'])
        params['total'] = float(request.form['total'])
        result = record_patient_exam(params)
        ... ... ... ... ... ...
        else:
            return redirect('/exam/task/error')
```

Aside from calling `record_patient_exam()`, it also utilizes the `list_pid()` and `list_qid()` services to retrieve the IDs. The use of services can help separate the abstraction and use cases from the route functions, which has a beneficial impact on the scope, clean coding, and runtime performance of the routes. Moreover, the project structure can also contribute to clear business flow, maintainability, flexibility, and adaptability.

Managing the project structure

Flask provides developers with the convenience of building their desired project structure. It is open to any design patterns and architectural strategies for building a project directory because of its Pythonic characteristics. The focus of this discussion revolves around setting up our *Online Personal Counseling System* application using the simple and single-structured project approach while highlighting the different configuration variable setups.

Building the directory structure

The first aspect to consider in building the project structure is the level of complexity of the project scope. Since our project focuses only on small-scale clientele, a typical *single-structured* approach is enough to cater to a less scalable application. Second, we must ensure the proper layering or breakdown of various project components from the view layer down to the test modules so that the developers can identify what parts to prioritize, maintain, bug-fix, and test. The following is a screenshot of the directory structure of our prototype:

Figure 1.4 – The single-structured project directory

Chapter 2 will discuss other project structure techniques, especially when applications are scalable and complex.

Setting up a development environment

A Flask application, by default, is production-ready, even though its server, the Werkzeug's built-in server, is not. We need to replace it with an enterprise-grade server to be fully ready for production setup. However, our goal is to set up a Flask project with a development environment that we can sample and experiment on with various features and test cases. There are three ways to set up a Flask 3.x project for development and testing purposes:

- Running the server with `app.run(debug=True)` in `main.py`.
- Setting the `FLASK_DEBUG` and `TESTING` built-in configuration variables to `true` in the configuration file.
- Running the application with the `flask run --debug` command.

Setting the development environment will also enable automatic reloading and the default debugger of the framework. However, turn off debugging mode after deploying the application to production to avoid security risks for the applications and software logging problems. The following screenshot shows the server log when running a Flask project with a development environment setup:

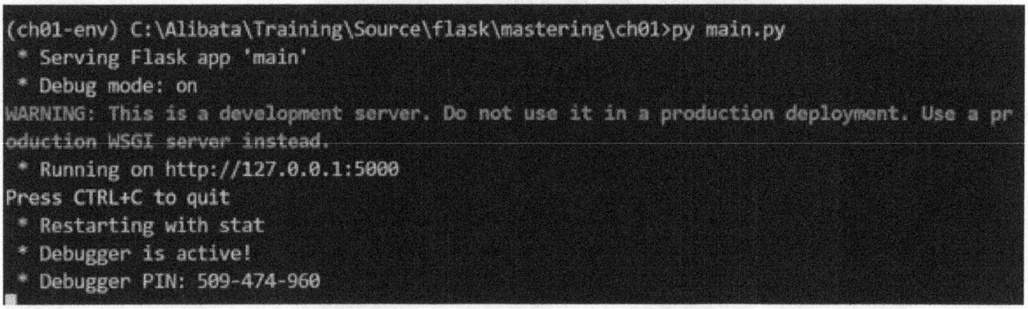

```
(ch01-env) C:\Alibata\Training\Source\flask\mastering\ch01>py main.py
 * Serving Flask app 'main'
 * Debug mode: on
WARNING: This is a development server. Do not use it in a production deployment. Use a pr
oduction WSGI server instead.
 * Running on http://127.0.0.1:5000
Press CTRL+C to quit
 * Restarting with stat
 * Debugger is active!
 * Debugger PIN: 509-474-960
```

Figure 1.5 – The server log of Flask's built-in server

Figure 1.5 shows that debug mode is set to ON and that the debugger is enabled and given a `PIN` value.

Implementing the main.py module

When creating a simple project like our specimen, the main module usually contains the Flask instantiation and some of its parameters (for example, `template_folder` for the new directory of the HTML templates) and the required imports of the views below it. The following is the complete code of our `main.py` file:

```python
from flask import Flask
from converter.date_converter import DateConverter

app = Flask(__name__, template_folder='pages')
```

```
app.url_map.converters['date'] = DateConverter

@app.route('/', methods = ['GET'])
def index():
    return "This is an online … counseling system (OPCS)"

import views.index
import views.certificates
import views.signup
import views.examination
import views.reports
import views.admin
import views.login
import views.profile

app.add_url_rule('/certificate/
terminate/<string:counselor>/<date:effective_date>/<string:patient>',
'show_honor_dissmisal', views.certificates.show_honor_dissmisal)

if __name__ == '__main__':
    app.run()
```

The imports to the views are placed below the Flask instantiation to avoid *circular dependency* problems. In this type of project structure, conflict always happens when a view module imports the app instance of the main module while the main module has the imports to the views declared at the beginning. This occurrence is called a circular dependency between two modules importing components from each other, which leads to some circular import issues. To avoid this problem with the main and view modules, the area below the Flask instantiation is where we place these view imports. The if statement at the bottom of main.py, on the other hand, verifies that only the main module can run the Flask server through the app.run() command.

The main module usually sets the configuration settings through its app instance to build the sessions and other context-based objects or integrate other custom components, such as the security and database modules. But the ideal setup doesn't recommend including them there; instead, you should place them separately from the code, say using a configuration file, to seamlessly manage the environment variables when configuration blunders arise, to avoid performance degradation or congestion when the Flask app instance has several variables to load at server startup, and to replicate and back up the environment settings with less effort during project migration or replication.

Creating environment variables

Configuration variables will always be part of any project setup, and how the frameworks or platforms manage them gives an impression of the kind of framework they are. A good framework should be able to decouple both built-in and custom configuration variables from the implementation area while maintaining their easy access across the application. It can support having a configuration file that can do the following:

- Contain the variables in a structured and readable manner.

- Easily integrate with the application.

- Allow comments to be part of its content.

- Work even when deployed to other servers or containers.

- Decouple the variables from the implementation area.

Aside from the .env file, Flask can also support configuration files in JSON, Python, and **Tom's Obvious Minimal Language** (**TOML**) format. Flask will not require an extension module if configuration files are in JSON and Python formats. The following is the application's config.json file, which contains the database and Flask development environment settings:

```
{
    «DB_USER»  :  «postgres»,
    «DB_PASS»  :  «admin2255»,
    «DB_PORT»  :  5433,
    "DB_HOST"  :  "localhost",
    "DB_NAME"  :  "opcs",
    "FLASK_DEBUG"  :  true,
    "TESTING":  true
}
```

This next is a Python config.py file with the same variable settings in config.json:

```
DB_USER = «postgres»
DB_PASS = «admin2255»
DB_PORT = 5433
DB_HOST = "localhost"
DB_NAME = "opcs"
FLASK_DEBUG = True
TESTING = True
```

The app instance has the `config` attribute with a `from_file()` method that can load the JSON file, as shown in the following snippet:

```
app.config.from_file("config.json", load=json.load)
```

On the other hand, `config` has a `from_pyfile()` method that can manage the Python config file when invoked, as shown in this snippet:

```
app.config.from_pyfile('myconfig.py')
```

The recent addition to the supported type, **TOML**, requires Flask to install the `toml` extension module before loading the `.toml` file into the platform. After running the `pip install toml` command, the `config` attribute's `from_file()` method can now load the following settings of the `config.toml` file:

```
DB_USER = «postgres»
DB_PASS = «admin2255»
DB_PORT = 5433
DB_HOST = "localhost"
DB_NAME = "opcs"
FLASK_DEBUG = true
TESTING = true
```

TOML, like JSON and Python, has data types. It supports arrays and tables and has structural patterns that may seem more complex than the JSON and Python configuration syntax. A TOML file will have the `.toml` extension.

When accessing variables from these file types, the Flask instance uses its `config` object to access each variable. This can be seen in the following version of our `db.py` module for database connectivity, which uses the `config.toml` file:

```
from __main__ import app
import psycopg2
import functools

def connect_db(func):
    @functools.wraps(func)
    def repo_function(*args, **kwargs):
        conn = psycopg2.connect(
            host=app.config['DB_HOST'],
            database=app.config['DB_NAME'],
            port=app.config['DB_PORT'],
            user=app.config['DB_USER'],
            password=app.config['DB_PASS'])
        resp = func(conn, *args, **kwargs)
```

```
        conn.commit()
        conn.close()
        return resp
    return repo_function
```

Summary

This chapter has presented the initial requirements to set up a development environment for a single-structured Flask project. It provided the basic elements that are essential to creating a simple Flask prototype, such as the `main.py` module, routes, database connectivity, repository, services, and configuration files. The nuts and bolts of every procedure in building every aspect of the project describe Flask as a web framework. The many ways to store the configuration settings, the possibility of using custom decorators for database connectivity, and the many options to capture the form data are indicators of Flask being so flexible, extensible, handy, and Pythonic in many ways. The next chapter will focus on the core components and advanced features that Flask can provide in building a more scalable application.

2

Adding Advanced
Core Features

After the setup, configuration, and initial development of a Flask web application in *Chapter 1*, it is time to include other essential components of the Flask framework that will complete a web application. These components, such as *session handling*, *flash messaging*, *error handling*, and *software logging*, can monitor and manage the interactions between the user and the internal transactions. Moreover, Flask can also provide an understanding of how the system will cope with critical issues such as running time, security, smooth performance, and changes to adapt to the ever-changing production environment.

These major web components for building enterprise-grade applications supported by Flask will be the focus of this chapter. We will also discuss the various approaches in designing the project structure once these core components become part of the application.

Here are the topics that we will cover in this chapter:

- Structuring huge and scalable projects
- Applying object-relational mapping (ORM)
- Configuring the logging mechanism
- Creating user sessions
- Applying flash messages
- Utilizing some advanced Jinja2 features
- Implementing error-handling solutions
- Adding static resources

Technical requirements

This chapter will focus on ordering and product management transactions. The application prototype for this chapter, an *Online Shipping Management System*, covers some generic product inventory, an ordering module, a basic shipping flow structure, and some parts of the delivery management module. This prototype comes in three different implementations suited for a complex and scalable Flask web application, namely, the following:

- The `ch02-factory` project that utilizes the application factory design pattern.
- The `ch02-blueprint` project that uses the Flask blueprint.
- The `ch02-blueprint-factory` project that uses both the application factory and blueprint structure.

Like in *Chapter 1*, the application uses *PostgreSQL* as a database, but this time with an ORM called **SQLAlchemy**. All these projects are uploaded at `https://github.com/PacktPublishing/Mastering-Flask-Web-Development/tree/main/ch02`.

Structuring huge and scalable projects

Creating a directory structure for a *simple* Flask web application is very handy and easy, especially if there is only one module to build with few software features. But for complex and scalable enterprise-grade applications with an overwhelming number of features to support, the most common issue is always the *circular import problem*.

> **Important note**
>
> A *circular import problem* happens when two or more modules import each other, creating a mutual dependency loop before the application's full-blown execution. This scenario always causes unexpected application loading errors, missing modules, and even weird runtime problems.

Flask as a framework is very Pythonic, which means developers can decide on their approaches to structuring their applications. Unfortunately, not all directory structure designs push through due to circular import problems. However, three design patterns can provide a baseline structure for Flask projects: the *application factory design*, the *Blueprint approach*, and the *combined application factory and Blueprint template*.

Using the application factory

In *Chapter 1*, the project structure used in our application was composed of modules and packages of various components such as the models, repository, services, templates, and the `main.py` file. The code organization was not within Flask's standard but was considered a clean directory structure.

One approach in building a Flask project is to use the *application factory*, a method consisting of the instantiation and configuration of the Flask instance. It loads the configuration file into the platform, sets up the necessary extension modules such as SQLAlchemy, and initializes the Flask constructor with parameters, such as `template_folder` and `static_folder`, before the app's instantiation. With factory application, there is flexibility in dealing with configurations. An application may have a separate factory method for *testing*, *development*, and *production*, depending on the stages the application will undergo.

But where to place this method definition? Before implementing this approach, separate all views and their related components from the general application components, such as exception classes and error pages. Place the files in sub-folders, but you may also add more folders underneath to organize the modules further. Afterward, create an `__init__.py` file anywhere inside these sub-folders to implement the application factory method. In our case, the `__init__.py` file of the app sub-folder is where we defined the application factory. *Figure 2.1* shows the directory structure for the `ch02-factory` project that contains the version of the *Online Shipping Management System* prototype with the application factory:

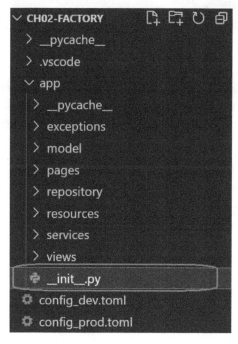

Figure 2.1 – Flask project directory with application factory

The `__init__.py` file converts any directory to a package with files and folders importable to other modules. Any module script imported inside the `__init__.py` file exposes the script for import outside the package directory. Likewise, it also allows exposing modules from other packages that are out of reach of the internal modules due to relative path problems. On the other hand, the application automatically loads all imported modules and executes method calls found inside the `__init__.py` file. Thus, placing our application factory inside the `__init__.py` file of the app package exposes the function anywhere within the Flask project. The following is the content of our `app/__init__.py` file:

```
from flask import Flask
from flask_sqlalchemy import SQLAlchemy
import logging
import logging.config
import logging.handlers
import sys
import toml


# Extension modules initialization
db = SQLAlchemy()
def configure_func_logging(log_path):
    logging.getLogger("werkzeug").disabled = True
    console_handler =  logging.StreamHandler(stream=sys.stdout)
    console_handler.setLevel(logging.DEBUG)
    logging.basicConfig(level=logging.DEBUG,  format='%(asctime)s
%(levelname)s %(module)s
        %(funcName)s %(message)s',
    datefmt='%Y-%m-%d %H:%M:%S',  handlers=[logging.handlers.
RotatingFileHandler(
        log_path, backupCount=3, maxBytes=1024 ),
        console_handler])


def create_app(config_file):
    app = Flask(__name__, template_folder='../app/pages', static_
folder='../app/resources')
    app.config.from_file(config_file, toml.load)
    db.init_app(app)
    configure_func_logging('log_msg.txt')


    with app.app_context():
        from app.views import login
        from app.views import menu
        from app.views import customer
        from app.views import admin
        from app.views import product
```

```
                from app.views import order
                from app.views import payment
                from app.views import shipping
        return app
```

The common and standard name given to an application factory function is `create_app()`, but anyone can replace it with something appropriate for their project. In the given snippet, our application factory creates the Flask's app instance, calls `db_init()` of the SQLAlchemy's db instance to define and configure the ORM with the `app` object, and sets up the logging mechanism. Since it is in `__init__.py`, the `main.py` file must import the factory method to eventually create the object and run the application by calling the app's `run()` method.

To make the application factory method flexible and configurable, add local parameters to it. For instance, it can take a filename as a string parameter to accept a configuration filename, such that when an application runs in development mode, it can take `config_dev.toml` as its configuration file. When shifting the deployment to the production server, it can accept a new filename and replace the existing config with a production configuration file, say `config_prod.toml`, to reload all the environment variables intended for the production server.

Utilizing the current_app proxy

Using the application factory design pattern in structuring the application makes it impossible to access the `app` instance from `main.py` in the views and other components requiring it, without encountering *circular import issues*. Instead of importing the `app` object in the modules of the app directory, we establish the application context in `create_app()` to utilize the proxy app object called `current_app`.

In Flask, the application context manages configuration variables, view data, loggers, database details, and other custom objects at the application level during a request. There are two ways of creating the application context:

- Explicitly pushing the application context using the `push()` method, allowing access to the `current_app` from anywhere in the application for every request:

```
    app_ctx = app.app_context()
    app_ctx.push()
```

- Automatically pushing the application context using the `with`-block, setting a limit on where to access the application-level components, such as allowing access to `current_app` from the `views` module only, as depicted in the following example:

```
    with app.app_context():
        from app.views import login
        from app.views import menu
        from app.views import customer
```

```
from app.views import admin
from app.views import product
from app.views import order
from app.views import payment
from app.views import shipping
```

Instead of accessing the app instance from the __init__.py file to implement views in the views.shipping module, which definitely can cause circular import problems due to the current_app()'s import to the views.shipping module, the application can now allow the use of the current_app proxy to build views.shipping because of the with-context block pushing the module script to the application context. The following code shows the use of the proxy object in creating the add_delivery_officer view function that inserts delivery officer profile details into the database:

```
from flask import current_app
@current_app.route('/delivery/officer/add', methods = ['GET', 'POST'])
def add_delivery_officer():
    if request.method == 'POST':
        current_app.logger.info('add_delivery_officer POST view
executed')
        repo = DeliveryOfficerRepository(db)
        officer = DeliveryOfficer(id=int(request.form['id']),
firstname=request.form['firstname'], middlename=request.
form['middlename'],

        … … … … … …
        result = repo.insert(officer)
        return render_template(
         'shipping/add_delivery_officer_form.html'), 200
    current_app.logger.info('add_delivery_officer GET view executed')
    return render_template(
        'shipping/add_delivery_officer_form.html'), 200
```

The current_app is part of the flask module that can provide all the necessary decorators and utilities for building the view functions and other components, as long as the module scripts are *within the bounds of the application context* pushed by create_app(). In the add_delivery_officer view function, the current_app provides the route() decorator and the logger instance configured by the application factory.

Storing data to the application context

Now, the application context in Flask is like a mini-layer created on top of the request context for every request-response transaction. There is always a new context in every request, so all application-level data are short-lived in Flask applications. During this span, we can store data using another application-level proxy object, the g component. The following snippet shows how to create and destroy application context data objects:

```python
def get_database():
    if 'db' not in g:
        g.db = db
        app.logger.info('storing … as context data')


@app.teardown_appcontext
def teardown_database(exception):
    db = g.pop('db', None)
    app.logger.info('removing … as context data')
```

The get_database() method stores the db instance instantiated by create_app() into the context through the g proxy. Before storing the data, it is always a good practice to first verify whether the object is already in g. On the other hand, the teardown_database() has a @app.teardown_appcontext decorator that allows the automatic calling of the method before the request context ends. The pop() method of g removes or deallocates the data from the context before Flask destroys the whole application context.

Accessing g for storing data will not always work in the application. The appropriate place to create the context data or call our get_database() method is in the @before_request method. This method automatically executes before any request transactions start. The context data in g will only be accessible to any view function after the execution of the @before_request event method. In other words, all resources shared through g will be accessible only within the request-response scope. Accessing g context data not set in @before_request can cause ValueError. Thus, we invoke our get_database() method in the following @before_request implementation:

```python
@app.before_request
def init_request():
    get_database()
```

This will allow the following `list_login` view function to access the database for query transactions:

```
@current_app.route('/login/list', methods=['GET'])
def list_login():
    repo = LoginRepository(g.db)
    users = repo.select_all()
    session['sample'] = 'trial'
    flash('List of user credentials')
    return render_template('login/login_list.html', users=users) ,
200
```

Although it is an advantage to verify the existence of a context attribute in g before accessing it, sometimes accessing the data right away is inevitable, such as in our `list_login` view function that directly passes the g.db context object into `LoginRepository`. Another approach to accessing the context data is through the g.set() method that allows a default value if the data is non-existent, such as using g.get('db', db) instead of g.db, where db in the second parameter is a backup connection.

> **Important note**
>
> Both application and request contexts exist only during a request-response life cycle. The application context provides the `current_app` proxy and the g variable, while the request context contains the request variables and the view function details. Unlike in other web frameworks, Flask's application context will not be valid for access after the life cycle destroys the request context.

Aside from the application factory, we can also use Flask's blueprints to establish a project directory.

Using the Blueprint

`Blueprints` are Flask's built-in components from its `flask` module. Their core purpose is to organize all related views with the repository, services, templates, and other associated features to form a solid and self-contained structure. The strength of the Blueprint is that it can break down a single huge application into independent business units that can be considered sub-applications but are still dependent on the main application. *Figure 2.2* shows the organization of the `ch02-blueprint` project, a version of our *Online Shipping Management System* that uses Blueprints in building its project structure:

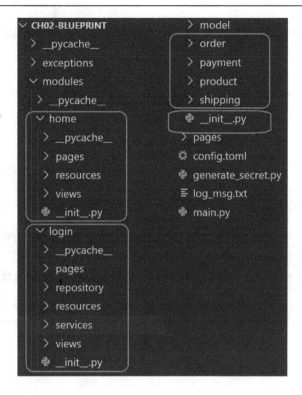

Figure 2.2 – Flask project directory with Blueprints

Defining the Blueprint

Instead of placing all view files in one folder, the project's related views are grouped based on business units, then assigned their respective Blueprint sub-projects, namely, home, login, order, payment, product, and shipping. Each Blueprint represents a section with the templates, static resources, repositories, services, and utilities needed to build a sub-application.

Now, the __init__.py file of each sub-application is very important because this is where the Blueprint object is created and instantiated. The blueprint home's __init__.py file, for instance, has the following Blueprint definition:

```
from flask import Blueprint
home_bp = Blueprint('home_bp', __name__,
    template_folder='pages',
    static_folder='resources', static_url_path='static')


import modules.home.views.menu
```

Meanwhile, the `login` section's Blueprint has the `__init__py` file that contains this:

```
from flask import Blueprint
login_bp = Blueprint('login_bp', __name__,
    template_folder='pages',
    static_folder='resources', static_url_path='static')


import modules.login.views.login
import modules.login.views.admin
import modules.login.views.customer
```

The constructor of the `Blueprint` class requires two parameters for it to instantiate the class:

- The first parameter is the *name of Blueprint*, usually the name of the reference variable of its instance.
- The second parameter is `__name__`, which depicts the *current package* of the section.

The name of the Blueprint must be unique to the sub-application because it is responsible for its internal routings, which must not have any collisions with the other Blueprints. The Blueprint package, on the other hand, will indicate the root path of the sub-application.

Implementing the Blueprint's routes

One purpose of using a Blueprint is to avoid circular import issues in implementing the routes. Rather than accessing the app instance from `main.py`, sub-applications can now directly access their respective Blueprint instance to build their routes. The following code shows how the login Blueprint implements its route using its Blueprint instance, the `login_bp` object:

```
from modules.login import login_bp


@login_bp.route('/admin/add', methods = ['GET', 'POST'])
def add_admin():
    if request.method == 'POST':
        app.logger.info('add_admin POST view executed')
        repo = AdminRepository(db_session)
        ... ... ... ... ... ...
        return render_template('admin_details_form.html',
logins=logins), 200
    app.logger.info('add_admin GET view executed')
    logins = get_login_id(1, db_session)
    return render_template('admin_details_form.html', logins=logins),
200
```

The `login_bp` object is instantiated from `__init__.py` of the `login` directory, thus importing it from there. But this route and the rest of the views will only work after registering these Blueprints with the `app` instance.

Registering the blueprints

Blueprint registration happens in the `main.py` file, which is the location of the Flask app. The following snippet is part of the `main.py` file of our `ch02-blueprint` project that shows how to establish the registration procedure correctly:

```
from flask import Flask
app = Flask(__name__, template_folder='pages')
app.config.from_file('config.toml', toml.load)

from modules.home import home_bp
from modules.login import login_bp
from modules.order import order_bp
from modules.payment import payment_bp
from modules.shipping import shipping_bp
from modules.product import product_bp

app.register_blueprint(home_bp, url_prefix='/ch02')
app.register_blueprint(login_bp, url_prefix='/ch02')
app.register_blueprint(order_bp, url_prefix='/ch02')
app.register_blueprint(payment_bp, url_prefix='/ch02')
app.register_blueprint(shipping_bp, url_prefix='/ch02')
app.register_blueprint(product_bp, url_prefix='/ch02')

from modules.model.db import *

if __name__ == '__main__':
    app.run()
```

The `register_blueprint()` method from the `app` instance has three parameters, namely, the following:

- The Blueprint object imported from the sub-application.
- The `url_prefix`, the assigned URL base route.
- The `url_defaults`, the dictionary of parameters required by the views linked to the Blueprint.

Registering the Blueprints can also be considered a workaround in providing our Flask applications with a context root. The context root defines the application, and it serves as the base URL that can be used to access the application. In the given snippet, our application was assigned the /ch02 context root through the url_prefix parameter of register_blueprint(). On the other hand, as shown from the given code, the imports to the Blueprints must be placed below the app instantiation to avoid circular import issues.

Another way of building a clean Flask project is by combining the application factory design technique with the Blueprint approach.

Utilizing both the application factory and the Blueprint

To make the Blueprint structures flexible when managing configuration variables and more organized by utilizing the application context proxies g and current_app, add an __init__.py file in the modules folder. *Figure 2.3* shows the project structure of ch02-blueprint-factory with the __init__.py file in place to implement the factory method definition:

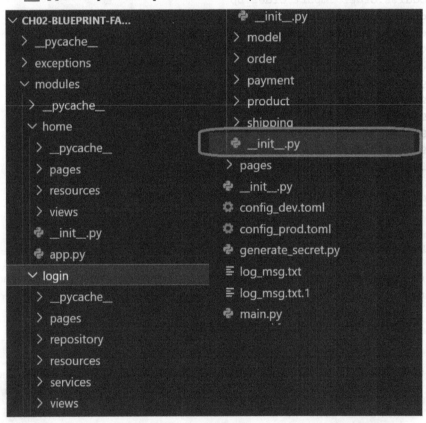

Figure 2.3 – Flask directory with Blueprints and application factory

The `create_app()` factory method can now include the import of the Blueprints and their registration to the app. The rest of its setup is the same as the `ch01` project. The following code shows its entire implementation:

```
import toml
from flask import Flask
from flask_sqlalchemy import SQLAlchemy


db = SQLAlchemy()


def create_app(config_file):
    app = Flask(__name__, template_folder='../pages', static_
folder='../resources')
    app.config.from_file(config_file, toml.load)
    ... ... ... ... ... ...

    ... ... ... ... ... ...

    with app.app_context():
        from modules.home import home_bp
        from modules.login import login_bp
        from modules.order import order_bp

        ... ... ... ... ... ...

        ... ... ... ... ... ...

        app.register_blueprint(home_bp, url_prefix='/ch02')
        app.register_blueprint(login_bp, url_prefix='/ch02')
        app.register_blueprint(order_bp, url_prefix='/ch02')

        ... ... ... ... ... ...
    return app
```

The application factory here uses the `with`-block to bind the application context only within the Blueprint components.

Depending on the scope of the software requirements and the appropriate architecture, any of the given approaches will be reliable and applicable in building organized, enterprise-grade Flask applications. Adding more packages and other design patterns is possible, but the core structure emphasized in the previous discussions must remain intact to avoid cyclic imports.

From project structuring, it is time to discuss the setup indicated in the application factory and `main.py` of the Blueprints, which is about the configuration of SQLAlchemy ORM and Flask's logging mechanism.

Applying object-relational mapping (ORM)

The most used **object-relational mapping** (**ORM**) that can work perfectly with the Flask framework is SQLAlchemy. This ORM is a boilerplated interface that aims to create a database-agnostic data layer to connect to any database engine. But compared to other ORMs, SQLAlchemy has support in optimizing native SQL statements, which makes it popular with many database administrators. When formulating its queries, it only requires Python functions and expressions to pursue CRUD operations.

Before using the ORM, the `flask-sqlalchemy` and `psycopg2-binary` extensions for the PostgreSQL database must be installed in the virtual environment using the following command:

```
pip install psycopg2-binary flask-sqlalchemy
```

What follows next is the setup of the database connectivity.

Setting up the database connectivity

Now, we are ready to implement the configuration file for our database setup. Flask 3.x supports the declarative extension of SQLAlchemy, which is the commonly used approach in implementing SQLAlchemy ORM in most frameworks such as FastAPI.

In this approach, the first step is to create the database connectivity by building the SQLAlchemy engine, which manages the connection pooling and the installed dialect. The `create_engine()` function from the `sqlalchemy` module derives the engine object with a **database URL** (**DB URL**) string as its main parameter. This URL string contains the database name, DB API driver, account credentials, IP address of the database server, and its port.

Now, the engine is required to create the session factory through the `sessionmaker()` method. And this session factory becomes the essential parameter to the `session_scoped()` method in extracting the session registry, which provides the session to SQLAlchemy's CRUD operations. The following is the database configuration found in the `/modules/model/config.py` module of the `ch02-blueprint` project:

```
from sqlalchemy import create_engine
from sqlalchemy.ext.declarative import declarative_base
from sqlalchemy.orm import sessionmaker, scoped_session

DB_URL = "postgresql://<username>:<password>@localhost:5433/sms"

engine = create_engine(DB_URL)
db_session = scoped_session(sessionmaker(autocommit=False,
autoflush=False, bind=engine))
```

```
Base = declarative_base()

def init_db():
    import modules.model.db
```

When the sessions are all set, the derivation of the base object from the `declarative_base()` method is the next focus for the model layer implementation. The instance returned by this method will subclass all the SQLAlchemy entity or model classes.

Building the model layer

Each entity class needs to extend the `Base` instance to derive the necessary properties and methods in mapping the schema table to the ORM platform. It will allow the classes to use the `Column` helper class to build the properties of the actual column metadata. There are support classes that the models can utilize such as `Integer`, `String`, `Date`, and `DateTime` to define the data types and other constraints of the columns, and `ForeignKey` to establish parent-child table relationships. The following are some model classes from the `/modules/model/db.py` module:

```python
from sqlalchemy import Time, Column, ForeignKey, Integer, String,
Float, Date, Sequence
from sqlalchemy.orm import relationship
from modules.model.config import Base
class Login(Base):
    __tablename__ = 'login'
    id = Column(Integer, Sequence('login_id_seq', increment=1),
primary_key = True)
    username = Column(String(45))
    password = Column(String(45))
    user_type = Column(Integer)

    admins = relationship('Admin', back_populates="login",
uselist=False)
    customer = relationship('Customer', back_populates="login",
uselist=False)

    def __init__(self, username, password, user_type, id = None):
        self.id = id
        self.username = username
        self.password = password
        self.user_type = user_type

    def __repr__(self):
        return f"<Login {self.id} {self.username} {self.password}
{self.user_type}>"
```

Now, the `relationship()` directive in the code links to model classes based on their actual reference and foreign keys. The model class invokes the method and configures it by setting up some parameters, beginning with the name of the entity it must establish a relationship with and the backreference specification. The `back_populates` parameter refers to the complementary attribute names of the related model classes, which express the rows needed to be queried based on some relationship loading technique, typically the lazy type. Using the `backref` parameter instead of `back_populates` is also acceptable. The following `Customer` model class shows its one-to-one relationship with the `Login` entity model as depicted in their respective calls to the `relationship()` directive:

```
class Customer(Base):
    __tablename__ = 'customer'
    id = Column(Integer, ForeignKey('login.id'), primary_key = True)
    firstname = Column(String(45))
    lastname = Column(String(45))
    middlename = Column(String(45))
    ... ... ... ... ... ...

    login = relationship('Login', back_populates="customer")
    orders = relationship('Orders', back_populates="customer")
    shippings = relationship('Shipping', back_populates="customer")
        ... ... ... ... ... ...
```

The return value of the `relationship()` call in `Login` is the scalar object of the filtered `Customer` record. Likewise, the `Customer` model has the joined `Login` instance because of the directive. On the other hand, the method can also return either a `List` collection or scalar value if the relationship is a *one-to-many* or *many-to-one* type. When setting this setup in the parent model class, the `useList` parameter must be omitted or set to `True` to indicate that it will return a filtered list of records from its child class. However, if `useList` is set to `False`, the indicated relationship is *one-to-one*.

The following `Orders` class creates a *many-to-one* relationship with the `Products` and `Customer` models but a *one-to-one* relationship with the `Payment` model:

```
class Orders(Base):
    __tablename__ = 'orders'
    id = Column(Integer, Sequence('orders_id_seq', increment=1),
primary_key = True)
    pid = Column(Integer, ForeignKey('products.id'), nullable = False)
    ... ... ... ... ... ...
    product = relationship('Products', back_populates="orders")
    customer = relationship('Customer', back_populates="orders")
    payment = relationship('Payment', back_populates="order",
uselist=False)
        ... ... ... ... ... ...
```

```
class Payment(Base):
    __tablename__ = 'payment'
    id = Column(Integer, Sequence('payment_id_seq', increment=1),
primary_key = True)
    order_no = Column(String, ForeignKey('orders.order_no'), nullable =
False)
    ... ... ... ... ... ...
    order = relationship('Orders', back_populates="payment")
    payment_types = relationship('PaymentType', back_
populates="payment")
    shipping = relationship('Shipping', back_populates="payment",
uselist=False)
```

The `main.py` module needs to call the custom method, the `init_db()` method found in the `config.py` module, to load, and register all these model classes for the repository classes.

Implementing the repository layer

Each repository class requires SQLAlchemy's `Session` instance to implement its CRUD transactions. The following `ProductRepository` code is a sample repository class that manages the `product` table:

```
from typing import List, Any, Dict
from modules.model.db import Products
from main import app
from sqlalchemy.orm import Session


class ProductRepository:
    def __init__(self, sess:Session):
        self.sess = sess
        app.logger.info('ProductRepository instance created')
```

`ProductRepository`'s constructor is essential in accepting the `Session` instance from the view or service functions and preparing it for internal processing. The first transaction is the `INSERT` product record transaction that uses the `add()` method of the `Session`. SQLAlchemy always imposes transaction management in every CRUD operation. Thus, invoking `commit()` of its `Session` object is required after successfully executing the `add()` method. The following `insert()` method shows the correct implementation of an `INSERT` transaction in SQLAlchemy:

```
    def insert(self, prod:Products) -> bool:
        try:
            self.sess.add(prod)
            self.sess.commit()
            app.logger.info('ProductRepository inserted record')
            return True
```

```
    except Exception as e:
        app.logger.info(f'ProductRepository insert error: {e}')
    return False
```

The `Session` object has an `update` method that can perform an `UPDATE` transaction. The following is an `update()` implementation that updates a `product` record based on its primary key ID:

```
def update(self, id:int, details:Dict[str, Any]) -> bool:
    try:
        self.sess.query(Products).filter(Products.id == id).
update(details)
        self.sess.commit()
        app.logger.info('ProductRepository updated record')
        return True
    except Exception as e:
        app.logger.info(f'ProductRepository update error: {e}')
    return False
```

The `Session` also has a `delete()` method that performs record deletion based on a constraint, usually by ID. The following is an SQLAlchemy way of deleting a `product` record based on its ID:

```
def delete(self, id:int) -> bool:
    try:
        login = self.sess.query(Products).filter( Products.id ==
id).delete()
        self.sess.commit()
        app.logger.info('ProductRepository deleted record')
        return True
    except Exception as e:
        app.logger.info(f'ProductRepository delete error: {e}')
    return False
```

And lastly, the `Session` supports a query transaction implementation through its `query()` method. It can allow the filtering of records using some constraints that will result in retrieving a list or a single one. The following snippet shows a snapshot of these query implementations:

```
def select_all(self) -> List[Any]:
    users = self.sess.query(Products).all()
    app.logger.info('ProductRepository retrieved all record')
    return users


def select_one(self, id:int) -> Any:
    users =  self.sess.query(Products).filter( Products.id == id).
one_or_none()
    app.logger.info('ProductRepository retrieved one record')
```

```
        return users

    def select_one_code(self, code:str) -> Any:
        users =  self.sess.query(Products).filter( Products.code ==
code).one_or_none()
        app.logger.info('ProductRepository retrieved one record by
product code')
        return users
```

Since the `selectall()` query transaction must return a list of `Product` records, it needs to call the `all()` method of the Query object. On the other hand, both `select_one()` and `select_one_code()` use Query's `one_to_many()` method because they need to return only a single `Product` record based on `select_one()`'s primary key or `select_one_code()`'s unique key filter.

In the `ch02-blueprint` project, each Blueprint module has its repository classes placed in their respective `/repository` directory. Whether these repository classes use the `SQLAlchemy` instance or `Session` of the declarative approach, Flask 3.x has no issues supporting either of these repository implementations.

Service and repository layers are among the components that require a logging mechanism to audit all the process flows occurring within these two layers. Let us now explore how to employ software logging in Flask web applications.

Configuring the logging mechanism

Flask utilizes the standard logging modules of Python. The app instance has a built-in `logger()` method, which is pre-configured and can log views, repositories, services, and events. The only problem is that this default logger cannot perform info logging because the default severity level of the configuration is `WARNING`. By the way, turn off debug mode when running applications with a logger to avoid logging errors.

The Python logging mechanism has the following severity levels:

- `Debug`: This level has a *severity value of 10* and can provide traces of results during the debugging process.

- `Info`: This level has a *severity value of 20* and can provide general details about execution flows.

- `Warning`: This level has a *severity value of 30* and can inform about areas of the application that may cause problems in the future due to some changes in the platform or API classes.

- `Error`: This level has a *severity value of 40* and can track down executions that encountered failures in performing the expected features.

- `Critical`: This level has a *severity value of 50* and can show audits of serious issues in the application.

> **Important note**
> The log level value or severity value provides a numerical weight on a logging level that signifies the importance of the audited log messages. Usually, the higher the value, the more critical the priority level or log message is.

The logger can only log events with a severity level greater than or equal to the severity level of its configuration. For instance, if the logger has a severity level of WARNING, it can only log transactions with warnings, errors, and critical events. Thus, Flask requires a custom configuration of its logging setup.

In all our three projects, we implemented the following ways to configure the Flask logger:

- **Approach 1**: Set up the logger, handlers, and formatter programmatically using the classes from Python's logging module, as shown in the following method:

```python
def configure_func_logging(log_path):
    logging.getLogger("werkzeug").disabled = True
    console_handler =
        logging.StreamHandler(stream=sys.stdout)
    console_handler.setLevel(logging.DEBUG)
    logging.basicConfig(level=logging.DEBUG,
        format='%(asctime)s %(levelname)s %(module)s
            %(funcName)s %(message)s',
        datefmt='%Y-%m-%d %H:%M:%S',
        handlers=[logging.handlers.RotatingFileHandler(  log_
path, backupCount=3, maxBytes=1024), console_handler])
```

- **Approach 2**: Set up dictConfig using the JSON format, as shown in the following snippet:

```python
def configure_logger(log_path):
        logging.config.dictConfig({
            'version': 1,
            'formatters': {
            'default': {'format': '%(asctime)s
                %(levelname)s %(module)s %(funcName)s
                %(message)s', 'datefmt': '%Y-%m-%d
                    %H:%M:%S'}
            },
        'handlers': {
            'console': {
                'level': 'DEBUG',
                'class': 'logging.StreamHandler',
                'formatter': 'default',
                'stream': 'ext://sys.stdout'
            },
            'file': {
```

```
                    'level': 'DEBUG',
                    'class':
                    'logging.handlers .RotatingFileHandler',
                    'formatter': 'default',
                    'filename': log_path,
                    'maxBytes': 1024,
                    'backupCount': 3
                }
            },
            'loggers': {
                'default': {
                    'level': 'DEBUG',
                    'handlers': ['console', 'file']
                }
            },
            'disable_existing_loggers': False
        })
```

In maintaining clean logs, it is always a good practice to disable all default loggers, such as the werkzeug logger. In applying *Approach 1*, disable the server logging by explicitly deselecting the werkzeug logger from the working loggers. When using *Approach 2*, on the other hand, setting the disable_existing_loggers key to False disables the werkzeug logger and other unwanted ones.

All in all, both of the given configurations produce a similar logging mechanism. The ch02-factory project of our *Online Shipping Management System* applied the programmatical approach, and its add_payment() view function has the following implementation with logging:

```
@current_app.route('/payment/add', methods = ['GET', 'POST'])
def add_payment():
    if request.method == 'POST':
        current_app.logger.info('add_payment POST view executed')
        repo_type = PaymentTypeRepository(db)
        ptypes = repo_type.select_all()
        orders = get_all_order_no(db)
        repo = PaymentRepository(db)
        payment = Payment(order_no=request.form['order_no'], mode_
payment=int(request.form['mode']),
            ref_no=request.form['ref_no'], date_payment=request.
form['date_payment'],
            amount=request.form['amount'])
        result = repo.insert(payment)
        if result == False:
            abort(500)
```

```
        return render_template('payment/add_payment_form.html',
orders=orders, ptypes=ptypes), 200
    current_app.logger.info('add_payment GET view executed')
    repo_type = PaymentTypeRepository(db)
    ptypes = repo_type.select_all()
    orders = get_all_order_no(db)
    return render_template('payment/add_payment_form.html',
orders=orders, ptypes=ptypes), 200
```

Regarding logging the repository layer, the following is a snapshot of `ShippingRepository` that manages shipment transactions and uses logging to audit all these transactions:

```
class ShippingRepository:
    def __init__(self, db):
        self.db = db
        current_app.logger.info('ShippingRepository instance created')

    def insert(self, ship:Shipping) -> bool:
        try:
            self.db.session.add(ship)
            self.db.session.commit()
            current_app.logger.info('ShippingRepository inserted
record')
            return True
        except Exception as e:
            current_app.logger.error(f'ShippingRepository insert
error: {e}')
        return False
        … … … … … …
```

The given `insert()` method of the repository uses the `info()` method to log the insert transactions found in the `try` block, while the `error()` method logs the `exception` block.

Now, every framework has its own way of managing session data, so let us learn how Flask enables its session handling mechanism.

Creating user sessions

Assigning an uncompromised value to Flask's `SECRET_KEY` built-in configuration variable pushes the `Session` context into the platform. Here are the ways to generate the secret key:

- Apply the `uuid4()` method from the `uuid` module.
- Utilize any `openssl` utility.

- Use the `token_urlsafe()` method from the `secrets` module.

- Apply encryption tools such as AES, RSA, and SHA.

Our three applications include a separate Python script that runs the `token_urlsafe()` method to generate a random key string with 16 random bytes for the `SECRET_KEY` environment variable. The following snippet shows how our applications set the secret key with the `app` instance:

```
(config_dev.toml)
SECRET_KEY = "SpOn1ZyV4KE2FT1AUrWRZ_h7o5s"


(main.py)
app = Flask(__name__, template_folder='../app/pages', static_
folder='../app/resources')
app.config.from_file("config_dev.toml", toml.load)
```

Since our application loads the `config_dev.toml` file using the `from_file()` method of the property config, adding the `SECRET_KEY` environment variable in TOML file with the random key string to the config file will enable the user session automatically. Generally, it is always best practice to set a `SECRET_KEY` for any Flask applications.

Managing session data

After successfully pushing the session context, our application can readily store data in the session by using the session object imported from the `flask` module. The following `login_db_ath()` view function stores the username in the session after a successful database validation of the user credentials:

```
@current_app.route('/login/auth', methods=['GET', 'POST'])
def login_db_auth():
    if request.method == 'POST':
        current_app.logger.info('add_db_auth POST view executed')
        repo = LoginRepository(db)
        username = request.form['username'].strip()
        password = request.form['password'].strip()
        user:Login = repo.select_one_username(username)
        if user == None:
            flash(f'User account { request.form["username"] } does not
exist.', 'error')
            return render_template('login/login.html') , 200
        elif not user.password == password:
            flash('Invalid password.', 'error')
            return render_template('login/login.html') , 200
        else:
            session['username'] = request.form['username']
            return redirect('/menu')
```

```
current_app.logger.info('add_db_auth GET view executed')
return render_template('login/login.html') , 200
```

Calling the `session` object with the name of the session attribute inside the brackets (e.g., `session["username"]`) retrieves the session data at runtime. On the other hand, removing the session requires calling the `pop()` method of the session object. For instance, removing the username requires executing the following code:

```
session.pop("username", None)
```

Validating the session attributes first before removing them or performing other transactions is always a recommendation, and the following snippet will show us how to validate session attributes:

```
@app.before_request
def init_request():
    get_database()
    if (( request.endpoint != 'login_db_auth' and  request.endpoint
!= 'index' and request.endpoint != 'static')  and 'username' not in
session):
        app.logger.info('a user is unauthenticated')
        return redirect('/login/auth')
    elif (( request.endpoint == 'login_db_auth' and  request.endpoint
!= 'index' and request.endpoint != 'static')  and 'username' in
session):
        app.logger.info('a user is already logged in')
        return redirect('/menu')
```

As previously discussed, the method with the `@before_request` decorator always executes first before any route function performs. It processes some pre-condition transactions before the request reaches the route. In the given snippet, `@before_request` executes the `get_database()` method and checks whether an authenticated user has already logged into the applications. If there is a logged user, access to any endpoint, except for index and static resources, will always redirect the user to the menu page. Otherwise, it will always redirect the user to the login page.

Clearing all session data

Instead of removing every session attribute, the `session` object has a `clear()` method that removes all session data in just one call. The following is a `logout` route that deletes all the session data before redirecting a user to the login page:

```
@current_app.route('/logout', methods=['GET'])
def logout():
    session.clear()
    current_app.logger.info('logout view executed')
    return redirect('/login/auth')
```

There is no easy way in Flask to invalidate the session, but `clear()` can help prepare the session for another user to access it.

Now, another component that depends much on session handling is flash messaging, which stores messages of string type on a session.

Applying flash messages

Flash messages are usually seen on validated forms rendering error messages for every text field with an invalid input value. Sometimes flash messages are headlines or important notifications printed in all caps on a web page.

Flask has a flash method that any view function can import to make flash messages. The following authentication process creates a flash message after validating the user credentials from the database:

```
@current_app.route('/login/add', methods=['GET', 'POST'])
def add_login():
    if request.method == 'POST':
        current_app.logger.info('add_login POST view executed')
        login = Login(username=request.form['username'],
password=request.form['password'], user_type=int(request.form['user_
type']) )
        repo = LoginRepository(db)
        result = repo.insert(login)
        if result == True:
            flash('Successully added a user', 'success')
        else:
            flash(f'Error adding { request.form["username"] }',
'error')
        return render_template('login/login_add.html') , 200
    current_app.logger.info('add_login GET view executed')
    return render_template('login/login_add.html') , 200
```

The given add_login() view function uses `flash()` to create an error message if the credentials accepted by the route are already in the database. But it also sends a notification through `flash()` if the `insert` transaction is successful.

> **Important note**
> The Flask flashing system records messages to the user session at the end of every request and retrieves them on the following immediate request transaction.

Figure 2.4 shows a sample screen result after adding an existing username and password:

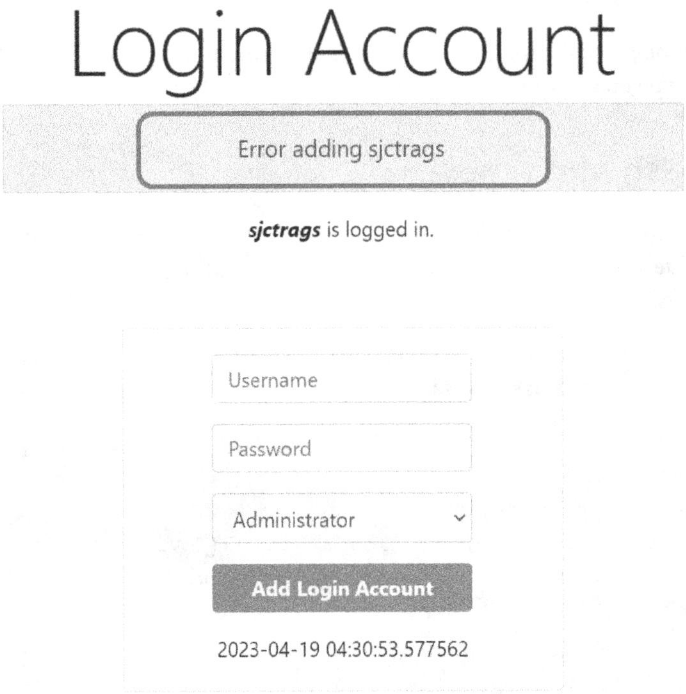

Login Account

Error adding sjctrags

sjctrags is logged in.

Username

Password

Administrator

Add Login Account

2023-04-19 04:30:53.577562

Figure 2.4 – A flash message for an invalid insert transaction

The Jinja2 template has access to Flask's get_flashed_messages() method that retrieves all the flash messages or just the categorized ones. The following Jinja2 macro of the /login/login_add. html template renders the error flash message in *Figure 2.4*:

```
{% macro render_error_flash(class_id) %}
    {% with errors = get_flashed_messages(category_filter=["error"])
%}
        {% if errors %}
            <p id="{{class_id}}" class="w-lg-50">
            {% for msg in errors %}
                {{ msg }}
            {% endfor %}
            </p>
        {% endif %}
    {% endwith %}
{% endmacro %}
```

The `with`-block provides the context for checking whether there are error-typed flash messages that need rendering. If there are, a `for`-block will retrieve all these retrieved flash messages.

On the other hand, Jinja2 can also retrieve uncategorized or generic flash messages from the view functions. The following macro retrieves a flash message from the `list_login()` route:

```
{%macro render_list_flash()%}
    {% with messages = get_flashed_messages() %}
        {% if messages %}
            <h1 class="display-4 ">
                {% for message in messages %}
                    {{ message }}
                {% endfor %}
            </h1>
        {% endif %}
    {% endwith %}
{%endmacro%}
```

Given the use of macros in rendering flash messages, let us explore other advanced features of Jinja2 templates of our applications that can provide better template implementation.

Utilizing some advanced Jinja2 features

Chapter 1 introduced the Jinja2 engine and templating, and some of these Jinja constructs were applied to render HTML contents:

- `{{ variable }}`: The placeholder expression that renders a single-valued object from view functions.
- `{% statement %}`: The expression that implements `if-else`-conditions, `for`-loops, block-expressions for calling layout fragments, `with`-blocks for managing context, and macro calls.

But some Jinja2 features, such as applying the `with`-statement, macros, filters, and comments, can help generate better views for our routes.

Applying with-blocks and macros

In the *Applying flash messages* section, templates used the `{% with %}` statement to extract the flash messages from the view functions and `{% macro %}` in optimizing our Jinja2 transactions. The `{% with %}` statement sets a context to limit the access or scope of some variables within the `with`-block. Access outside the block produces a Jinja2 error.

The {% macro %} block, on the other hand, pursues modular programming in Jinja2 templating. Every macro has a name and can have local parameters for reusability, and any templates can import and call them like typical methods. The following /login/login_list.html template renders the list of user credentials with a call on a macro that outputs an uncategorized flash message:

```
{% from "macros/flask_segment.html" import render_list_flash with
context %}
<!DOCTYPE html>
<html lang="en">
    <head>
        <title>List Login Accounts</title>
        <link rel="stylesheet" href="{{ url_for('static',
filename='css/styles.css') }}">
        <link rel="stylesheet" href="{{ url_for('static',
filename='css/bootstrap.min.css') }}">
        <script src="{{ url_for('static', filename='js/jquery-
3.6.4.js') }}"></script>
        <script src="{{ url_for('static', filename='js/bootstrap.
bundle.min.js') }}"></script>
    </head>
    <body>
        <section class="position-relative py-4 py-xl-5">
            <div class="container position-relative">
                <div class="row d-flex">
                    <div class="col-md-8 col-xl-6 text-center mx-
auto">
                        {{render_list_flash()}}
                        ... ... ... ... ... ...
                    </div>
                </div>
                ... ... ... ... ... ...
        </section>
    </body>
</html>
```

All macros are placed in a template file, like any Jinja2 expressions. In our application, the macros are found in /macros/flask_segment.html, and any template must import them from this file using the {% from ... import ... with context %} statement before utilizing them. In the given template, render_list_flash() is imported first before calling it like a method using the {{}} placeholder expression.

Applying filters

To improve the look and feel and clarity of the rendered data, Jinja2 has several filter operations that can provide additional aesthetics that can make the rendition more appealing to the users. This process is called **piping** because we use the pipe symbol (|) to pass the value to these operations. The following `product/list_product.html` page uses filter methods in rendering the list of products:

```html
<!DOCTYPE html>
<html lang="en">
    <head>
        <title>List of Products</title>
        <link rel="stylesheet" href="{{ url_for('static',
filename='css/styles.css')}}">
        <link rel="stylesheet" href="{{ url_for('static',
filename='css/bootstrap.min.css')}}">
        <script src="{{ url_for('static', filename='js/jquery-
3.6.4.js') }}"></script>
        <script src="{{ url_for('static', filename='js/bootstrap.
bundle.min.js') }}"></script>
    </head>
    <body>
        <table>
        {% for p in prods %}
            <tr>
                <td>{{p.id}}</td>
                <td>{{p.name|trim|upper}}</td>
                <td>{{"\u20B1%.2f"|format(p.price) }}</td>
                <td>{{p.code}}</td>
            </tr>
        {% endfor %}
        </table>
    </body>
</html>
```

The given template uses the trim filter to strip the name data with leading and trailing whitespaces and an upper filter to convert the names to uppercase. Through the format filter, all the price data now includes a Philippine peso currency sign with two decimal places. Jinja2 supports several built-in filters that can help derive other features from, compute, manipulate, modify, compress, expand, and sanitize the raw data from the view functions to render all these details in a more presentable outcome.

Adding comments

It is always best practice to add comments in every template using the {# comment #} expression for sectioning and internal documentation purposes. These comments are not part of the rendition provided by the Jinja2 template engine.

Jinja2 expressions are not only applied to route views but also to error pages. Let us now learn how to render error pages in the Flask 3.x framework.

Implementing error-handling solutions

Chapter 1 showcased the use of the redirect() method in rendering error pages given a status code, such as the status code 500. We will now discuss a better way of managing exceptions and status codes, including triggering error pages per status code.

Flask applications must always implement an error-handling mechanism using any of the following strategies:

- Registers a custom error function using the app's register_error_handler() method.
- Creates an error handler using the app's errorhandler decorator.
- Throws a custom Exception class.

Using the register_error_handler method

The declarative way to implement an error handler is to create a custom function and register it to the app's register_error_handler() method. The custom function must have a single local parameter that will accept the injected error message from the platform. It must also return its assigned error page using the make_response() and render_template() methods with the option of passing the error message as context data to the template for rendering. The following is a snippet that shows the steps:

```
def server_error(e):
    print(e)
    return make_response(render_template("error/500.html",
title="Internal server error"), 500)
app.register_error_handler(500, server_error)
```

The register_error_handler() method has two parameters:

- The status code that will trigger the error handling.
- The function name of the custom error handler.

There should only be one registered custom error handler per status code.

Applying the @errorhandler decorator

The easiest way to implement error handlers is to decorate customer error handlers with the app's errorhandler() decorator. The structure and behavior of the custom method are the same as the previous approach except that it has an errorhandler decorator with the assigned status code. The following shows the error handlers implemented using the decorator:

```
@app.errorhandler(404)
def not_found(e):
    return make_response(render_template("error/404.html", title="Page
not found"), 404)

@app.errorhandler(400)
def bad_request(e):
    return make_response(render_template("error/400.html", title="Bad
request"), 400)
```

Accessing an invalid URL path will auto-render the error page in *Figure 2.5* because of the given error handler for HTTP status code 404:

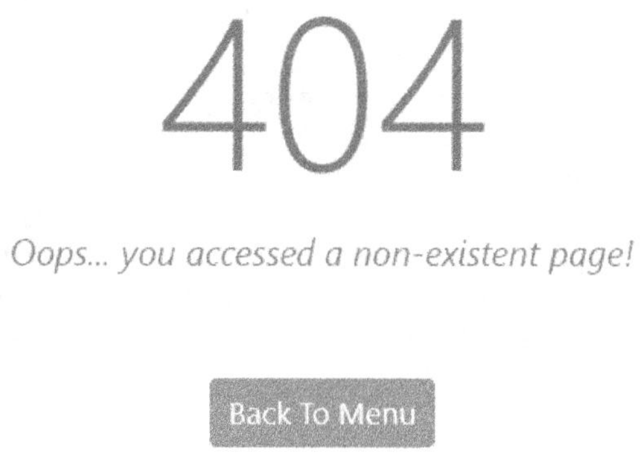

Oops... you accessed a non-existent page!

Back To Menu

Figure 2.5 – An error page rendered by the not_found() error handler

Creating custom exceptions

Another wise approach is assigning custom exceptions to error handlers. First, create a custom exception by subclassing HttpException from the werkzeug.exceptions module. The following shows how to create custom exceptions for Flask transactions:

```
from werkzeug.exceptions import HTTPException
from flask import render_template, Response

class DuplicateRecordException(HTTPException):
```

```
code = 500
description = 'Record already exists.'

def get_response(self, environ=None):
    resp = Response()
    resp.response = render_template('error/generic.html',
        ex_message=self.description)
    return resp
```

View functions and repository methods can throw this custom `DuplicateRecordException` class when an `INSERT` record transaction encounters a primary or unique key duplicate error. It requires setting the two inherited fields from the parent `HTTPException` class, namely the `code` and `description` fields. Once triggered, the exception class can auto-render its error page when it has an overridden `get_response()` method that creates a custom `Response` object to make way for the rendering of its error page with the exception message.

But overriding the `get_response()` instance method of the custom exception is just an option. Sometimes, assigning values to the code and description fields is enough, and then we map them to a custom error handler for the rendition of its error page, either through the `@errorhandler` decorator or `register_error_handler()`. The following code shows this kind of approach:

```
@app.errorhandler(DuplicateRecordException)
def insert_record_exception(e):
    return e.get_response()
```

When the raise command triggers the custom `DuplicateRecordException` class, the event handler will return its overridden `get_response()` method with the mapped Jinja2 error page and the HTTP status code 500. But how about if the exception triggered is a Python-based type?

Managing built-in exceptions

All the handlers previously presented manage only the Flask exceptions, not the Python-specific exceptions. To include handling of those exceptions generated by some Python runtime issues, create a dedicated custom method handler that listens to all these exceptions, such as in the following implementation:

```
from werkzeug.exceptions import HTTPException
@app.errorhandler(Exception)
def handle_built_exception(e):
    if isinstance(e, HTTPException):
        return e
    return render_template("error/generic.html", title="Internal
server error", e=e), 500
```

The given error handler filters out all Flask-related exceptions and throws them for Flask handlers to process, but it renders a custom error page for any Python runtime exception.

Triggering the error handlers

Sometimes it is recommended to explicitly trigger the error handler, especially in projects that utilize Blueprints as building blocks of their applications. A Blueprint module is not an independent sub-application that can own a URL context that listens to and calls the precise error handlers directly. So, to avoid some problems in calling the exact error handlers, transactions can invoke the `abort()` method with the proper HTTP status code, such as in the following snippet:

```
@current_app.route('/payment/add', methods = ['GET', 'POST'])
def add_payment():
    if request.method == 'POST':
        current_app.logger.info('add_payment POST view executed')
        ... ... ... ... ... ...
        result = repo.insert(payment)
        if result == False:
            abort(500)
        return render_template('payment/add_payment_form.html',
orders=orders, ptypes=ptypes), 200
    current_app.logger.info('add_payment GET view executed')
    ... ... ... ... ... ...
    return render_template('payment/add_payment_form.html',
orders=orders, ptypes=ptypes), 200
```

For custom or built-in exceptions, transactions can call the `raise()` method to trigger the error handler for the raised exception. The following view function raises the `DuplicateRecordException` class when an issue arises during order record insertion:

```
@current_app.route('/orders/add', methods=['GET', 'POST'])
def add_order():
    if request.method == 'POST':
        current_app.logger.info('add_order POST view executed')
        repo = OrderRepository(db)
        ... ... ... ... ... ...
        result = repo.insert(order)
        if result == False:
            raise DatabaseException()
        customers = get_all_cid(db)
        products = get_all_pid(db)
        return render_template('order/add_order_form.html',
customers=customers, products=products), 200
    current_app.logger.info('add_order GET view executed')
    customers = get_all_cid(db)
    products = get_all_pid(db)
    return render_template('order/add_order_form.html',
customers=customers, products=products), 200
```

All the generic error handlers are placed in the `main.py` module, while the custom and component-specific exception classes are in separate modules outside of the Blueprints for coding standard purposes and easy debugging.

Now, error pages and the rest of the Jinja2 templates can also use *CSS*, *JavaScript*, *images*, and other static resources to add look-and-feel features to their content.

Adding static resources

Static resources provide the user experience for Flask web applications. These static resources include the needed CSS, JavaScript, images, and video files to be used by some template pages. Now, Flask does not allow adding these files anywhere in the project. Generally, the Flask constructor has a `static_folder` parameter that accepts a relative path of a dedicated directory for these files.

In `ch02-factory`, `create_app()` configures the Flask instance to allow placing the resources in the `/resources` folder of the main project directory. The following snippet of `create_app()` shows the Flask instantiation with the `resource` folder setup:

```
def create_app(config_file):
    app = Flask(__name__, template_folder='../app/pages',
                   static_folder='../app/resources')
    app.config.from_file(config_file, toml.load)
    db.init_app(app)
    configure_func_logging('log_msg.txt')
    … … … … … …
```

Meanwhile, in the `ch02-blueprint` project, the main project and its Blueprints can have their respective `/resources` directory. The following snippet shows a Blueprint configuration with its own `resources` folder setup:

```
shipping_bp = Blueprint('shipping_bp', __name__,
    template_folder='pages',
    static_folder='resources', static_url_path='static')
```

Figure 2.6 shows the location and the content of the /resources folder of the main application, while *Figure 2.7* shows the /resources folder of the shipping Blueprint package:

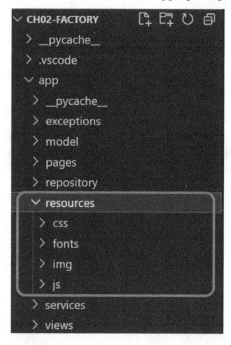

Figure 2.6 – The location of /resources in the main application

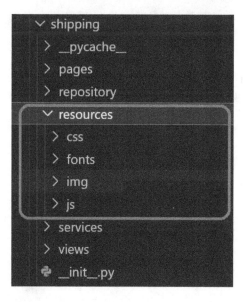

Figure 2.7 – The location of /resources in the shipping Blueprint

The `static` directory is the default and common folder name used to contain the Flask application's web assets. But in the succeeding chapters, we will use `/resources` instead of `/static` for naming convention purposes.

Accessing the assets in the templates

To avoid accessing relative paths, Flask manages, accesses, and loads static files or web assets in template pages using `static_url_path`, a logical path name used to access web resources from the `static` folder. Its default path value is `static`, but applications can set an appropriate value if needed.

Our application uses the Bootstrap 4 framework to apply responsive web design. All its assets are in the `/resources` folder, and the following `menu.html` template shows how to access these assets from the folder:

```
<!DOCTYPE html>
<html lang="en">
    <head>
        <title>Main Menu</title>
        <meta name="viewport" content="width=device-width, initial-
scale=1.0, shrink-to-fit=no">
        <link rel="stylesheet" href="{{ url_for('static',
filename='css/styles.css')}}">
        <link rel="stylesheet" href="{{ url_for('static',
filename='css/bootstrap.min.css')}}">
        <script src="{{ url_for('static', filename='js/jquery-
3.6.4.js') }}"></script>
        <script src="{{ url_for('static', filename='js/bootstrap.
bundle.min.js') }}"></script>
    </head>
    <body>
        <div class="container py-4 py-xl-5">
            <div class="row mb-5">
                <div class="col-md-8 col-xl-6 text-center mx-auto">
                    <h2 class="display-4">Supply Management System
Menu</h2>
                    <p class="w-lg-50"><strong><em>{{
session['username']}}</em></strong> is logged in.</p>
                </div>
            </div>
            <div class="row row-cols-1 row-cols-md-2 row-cols-xl-3">
                <div class="col">
                    <div class="d-flex p-3">

                        <div class="px-2">
                            <h5 class="mb-0 mt-1"><a href="#">Add
Delivery Officer</a></h5>
```

```
                            </div>
                        </div>
                    </div>
                    <div class="col">
                        <div class="d-flex p-3">
            ...  ...  ...  ...  ...  ...
        </body>
    </html>
```

The `url_for()` function, used to access view endpoints in the templates, is the way to access the static resources from the `/resources` folder using `static_url_path` as the directory name.

Summary

This chapter provided information on additional features of Flask that can support building complete, enterprise-grade, scalable, and complex but manageable Flask web applications. The details about adding error handling in many ways, integrating the Bootstrap framework to the application without using extensions, implementing SQLAlchemy using the declarative and standard approaches, and optimizing the Jinja2 templates using macros, indicate that the Flask framework is a lightweight but powerful solution to building web applications.

After learning about creating full-blown web applications with Flask, let us discuss and highlight, in the next chapter, the components and procedures for building API-based applications using the Flask 3.x framework.

3

Creating REST Web Services

Although Flask is a popular lightweight web framework, it can also support RESTful web service implementations. It has its own JSON encoders and decoders, built-in JSON support mechanisms for response generation and error handling, easy-to-manage RESTful request dispatching, and a lean configuration approach. Unlike the other API frameworks, Flask uses more modules and packages because of the required project structure it needs to maintain. However, after setting up the directory structure accordingly, the succeeding steps will be seamless, lightweight, and straightforward.

This chapter will introduce the part of the Flask framework that deals with building API endpoints to provide data and services to sub-modules or other applications. The goal is to understand how Flask manages the incoming requests and outgoing responses for REST endpoints that will run on its environment. Also, this chapter will discuss the various components that will comprise the Flask's API endpoint implementation.

Here are the topics that this chapter will cover to understand API development with Flask:

- Setting up a RESTful application
- Implementing API endpoints
- Managing requests and responses
- Utilizing response encoders and decoders
- Filtering API requests and responses
- Handling exceptions
- Consuming API endpoints

Technical requirements

This chapter utilizes a simple online pizza ordering system to showcase the capability of the Flask framework in developing REST web services. The ch03 application has a login, product inventory, and ordering and payment modules with the business scope to identify the necessary Flask components and utilities for the development. Moreover, a client application, ch03-client, is also included in the deliverables to showcase how to consume Flask API endpoints. Both applications use PostgreSQL as their database management system, with SQLAlchemy as their ORM. All these projects are uploaded at https://github.com/PacktPublishing/Mastering-Flask-Web-Development/tree/main/ch03.

Setting up a RESTful application

First, create the project's virtual environment, which will serve as the local repository of the needed module extensions. Next, open the VSCode editor to make the main project folder and install the flask extension module using the pip command through the VSCode's command line interpreter.

Afterward, manage the directory structure, such as the arrangement made for *Chapter 2* projects. From the three approaches, namely the application factory design, Blueprint, and the hybrid between these two, our online pizza ordering application will use the application factory approach to place its custom exception classes, models, repository, services, utilities, APIs, and database configuration in one app folder and register all these components using the create_app() method. *Figure 3.1* shows the project directory structure of our prototype application.

Figure 3.1 – Project directory structure for the RESTful application

The logging setup, the SQLAlchemy declarative configuration, and the sessions created in *Chapter 2* are all retained and used in this application. On the other hand, the `create_app()`, still placed in the `__init__.py` of the app package, is implemented as follows:

```
def create_app(config_file):
    app = Flask(__name__)
    app.config.from_file(config_file, toml.load)
    init_db()
    configure_logger('log_msg.txt')
    with app.app_context():
        from app.api import index
        ... ... ... ... ... ... ...
        from app.api import orders
    return app
```

The `main.py` still has the error handlers and the `app.run()` method for the server startup. The same command, `python main.py`, will run the application. However, the `ch03` application will not be web-based but API-based.

Let us dissect our application and identify the Flask components used to build REST services.

Implementing API endpoints

The implementation of API endpoints uses the same bolts and knots applied in creating web-based components in *Chapters 1* and *2*, such as declaring path variables, accessing the request through the `request` proxy object, returning the same `Response` object, and using the same `@route()` decorator. A GET API endpoint that returns a *JSON* response is as follows:

```
@current_app.route("/index", methods = ['GET'])
def index():
    response = make_response(jsonify(message='This is an Online Pizza
Ordering System.', today=date.today()), 200)
    return response
```

This `index()` function, found in the `app/api/index.py` module, is exactly similar to the web-based view function, except that `make_response()` requires the `jsonify()` instead of the `render_template()` method.

The `jsonify()` is a Flask utility method that serializes any data to produce an `application/json` response. It converts *multiple values* into an *array of data* and *key-value pairs* to a *dictionary*. It can also accept a *single-valued* entry. The `jsonify()` in the given `index()` function converts its arguments into a dictionary before calling Python's `json.dumps()` method. After `json.dumps()`'s JSON serialization, `jsonify()` will contain and render the result as part of `Response` with a mime-type of `application/json` instead of a plain JSON string. Thus, running the given `/index` endpoint with the `curl -i` command will generate the following request header result:

```
C:\Users\alibatasys>curl -i GET http://localhost:5000/index
HTTP/1.1 200 OK
Server: Werkzeug/2.2.3 Python/3.11.2
Date: Thu, 04 May 2023 02:07:13 GMT
Content-Type: application/json
Content-Length: 95
Connection: close

{"message":"This is an Online Pizza Ordering System.","today":"Thu, 04 May 2023 00:00:00 GMT"}
```

Figure 3.2 – Running the /index endpoint using cURL

The response body provided by running the curl command against `/index` has a message body and response headers composed of `Server`, `Date`, `Content-Type`, `Content-Length`, and `Connection`. `Content-Type` indicates the resource type the `/index` API will return to the client. Aside from strings, the `jsonify()` method can also serialize and render an array of objects like in the following API function that returns an array of string data and some single-valued objects:

```
@current_app.route("/introduction", methods = ['GET'])
def introduction():
    response = make_response(jsonify('This is an application that … … …
order requests, and provides payment receipts.'), 200)
    return response


@current_app.route("/company/trademarks", methods = ['GET'])
def list_goals():
    response = make_response(jsonify(['Eat', 'Live', 'Happy']), 200)
    return response
```

When the response data is not serializable, `jsonify()` can throw an exception, so it is advisable to enable error handlers. Now, it is customary to exclude `make_response` in returning the response data since `jsonify()` can already manage the `Response` generation alone for the endpoint function. Thus, the following versions of the `index()`, `introduction()`, and `list_goals()` endpoint functions are acceptable:

```
@current_app.route("/index", methods = ['GET'])
def index():
    response = jsonify(message='This is an Online Pizza Ordering
```

```
System.', today=date.today()), 200
   return response


@current_app.route("/introduction", methods = ['GET'])
def introduction():
   response = jsonify('This is an application that … … … order
requests, and provides payment receipts.'), 200
   return response


@current_app.route("/company/trademarks", methods = ['GET'])
def list_goals():
   response = jsonify(['Eat', 'Live', 'Happy']), 200
   return response
```

Using the `@app.route()` decorator to bind the URL pattern to the function and define the HTTP request is always valid. But Flask 3.x had released some decorator shortcuts that can assign one HTTP request per endpoint function, unlike the `@app.route()`, which can bind more than one HTTP request. These shortcuts are the following:

- `get()`: This defines an endpoint function that will listen to incoming *HTTP GET* requests, such as retrieving data from the database servers.

- `post()`: This defines an endpoint function to process an *HTTP POST* request, such as receiving a body of data for internal processing.

- `put()`: This defines an endpoint function to cater to any *HTTP PUT* requests, such as receiving a body of data containing updated details for the database server.

- `patch()`: This defines an endpoint to listen to an *HTTP PATCH* request that aims to modify some backend resources.

- `delete()`: This defines an *HTTP DELETE* endpoint function that will delete some server resources.

The following employee-related transactions of our `ch03` application are all implemented using the shortcut routing decorators:

```
@current_app.post('/employee/add')
def add_employee():
    emp_json = request.get_json()
    repo = EmployeeRepository(db_session)
    employee = Employee(**emp_json)
    result = repo.insert(employee)
    if result:
        content = jsonify(emp_json)
```

```
        current_app.logger.info('insert employee record successful')
        return make_response(content, 201)
    else:
        raise DuplicateRecordException("insert employee record
encountered a problem", status_code=500)
```

The given add_employee() endpoint function performs a database INSERT transaction of a record of employee details received from the client. The decorated @current_app.post() makes the API function an HTTP POST request method. On the other hand, the following is an API function that responds to an HTTP GET client request:

```
@current_app.get('/employee/list/all')
def list_all_employee():
    repo = EmployeeRepository(db_session)
    records = repo.select_all()
    emp_rec = [rec.to_json() for rec in records]
    current_app.logger.info('retrieved a list of employees
successfully')
    return jsonify(emp_rec)
```

The list_all_employee(), defined by the @current_app.get() decorator, processes the incoming HTTP GET requests for retrieving a list of employee records from the database server. For an HTTP PUT transaction, here is an API that updates employee details:

```
@current_app.put('/employee/update')
def update_employee():
    emp_json = request.get_json()
    repo = EmployeeRepository(db_session)
    result = repo.update(emp_json['empid'], emp_json)
    if result:
        content = jsonify(emp_json)
        current_app.logger.info('update employee record successful')
        return make_response(content, 201)
    else:
        raise NoRecordException("update employee record encountered a
problem", status_code=500)
```

The given API endpoint requires an `empid` path variable, which will serve as the key to search for the employee record that needs updating. Since this is an HTTP PUT request, the transaction requires all the new employee details to be replaced by their new values. But the following is another version of the update transaction that does not need a complete employee detail update:

```python
@current_app.patch('/employee/update/<string:empid>')
def update_employee_name(empid:str):
    emp_json = request.get_json()
    repo = EmployeeRepository(db_session)
    result = repo.update(empid, emp_json)
    if result:
        content = jsonify(emp_json)
        current_app.logger.info('update employee firstname,
middlename, and lastname successful')
        return make_response(content, 201)
    else:
        raise NoRecordException("update employee firstname,
middlename, and lastname encountered a problem", status_code=500)
```

`update_employee()`, decorated by `@current_app.patch()`, only updates the first name, middle name, and last name of the employee identified by the given employee ID using its path variable `empid`. Now, the following API function deletes an employee record based on the `empid` path variable:

```python
@current_app.delete('/employee/delete/<string:empid>')
def delete_employee(empid:str):
    repo = EmployeeRepository(db_session)
    result = repo.delete(empid)
    if result:
        content = jsonify(message=f'employee {empid} deleted')
        current_app.logger.info('delete employee record successful')
        return make_response(content, 201)
    else:
        raise NoRecordException("delete employee record encountered a
problem", status_code=500)
```

`delete_employee()`, decorated by `@current_app.delete()`, is an HTTP DELETE request method with the path variable `empid`, used for searching employee records for deletion.

These shortcuts of binding HTTP requests to their respective request handler methods are appropriate for implementing REST services because of their definite, simple, and straightforward one-route approach to managing incoming requests and serializing the required responses.

Let us now explore how Flask API captures the incoming body of data for POST, PUT, and PATCH requests and, aside from `make_response()`, what other ways the API can generate JSON responses.

Managing requests and responses

Unlike in other frameworks, it is easy to capture the request body of the incoming POST, PUT, and PATCH request in Flask, which is through the `get_json()` method from the `request` proxy object. This utility method receives the incoming JSON data, parses the data using `json.loads()`, and returns the data in a Python dictionary format. As seen in the following `add_customer()` API, the value of `get_json()` is converted into a `kwargs` argument by Python's `**` operator before passing the request data to the model class's constructor, an indication that the captured request data is a `dict` convertible into `kwargs`:

```python
@current_app.post('/customer/add')
def add_customer():
    cust_json = request.get_json()
    repo = CustomerRepository(db_session)
    customer = Customer(**cust_json)
    result = repo.insert(customer)
    if result:
        content = jsonify(cust_json)
        current_app.logger.info('insert customer record
successful')
        return make_response(content, 201)
    else:
        content = jsonify(message="insert customer record
encountered a problem")
        return make_response(content, 500)
```

Another common approach is to use the `request.json` property to capture the incoming message body, which is raw and with the mime-type `application/json`. The following endpoint function captures the incoming request through `request.json` and stores the data in the database as `category` information:

```python
@current_app.post('/category/add')
def add_category():
    if request.is_json:
        cat_json = request.json
        cat = Category(**cat_json)
        repo = CategoryRepository(db_session)
        result = repo.insert(cat)
        ... ... ... ... ... ...
    else:
        abort(500)
```

Unlike `request.get_json()`, which uses serialization, validation, and other utilities to transform and return incoming data to JSON, the `request.json` property has no validation support other than raising an `HTTP status 400` or `Bad Data` error if the data is not JSON serializable. The `request.get_json()` returns None if the request data is not parsable. That is why it is best to pair the `request.is_json` Boolean property with `request.json` to verify the incoming request and filter the non-JSON message body to avoid `HTTP Status Code 500`. Another option is to check if the `Content-Type` request header of the incoming request is `application/json`, as showcased by the following API function:

```
@current_app.post('/nonpizza/add')
def add_nonpizza():
    content_type = request.headers.get('Content-Type')
    if content_type == 'application/json':
        nonpizza_json = request.json
        nonpizza = NonPizza(**nonpizza_json)
        ... ... ... ... ... ...
    else:
        abort(500)
```

This `add_nonpizza()` function inserts a new record for the non-pizza menu options for the application, and it uses `request.json` to access the JSON-formatted input from the client. Both `request.json` and `request.get_json()` yield a dictionary object that makes the instantiation of model objects in the `add_category()` and `add_non_pizza()` API functions easier because `kwargs` transformation from these JSON data is straightforward.

On the other hand, validation of incoming requests using `request.is_json` and `Content-Type` headers is also applicable to the POST, PUT, and DELETE message body retrieval through `request.get_json()`. Now, another approach to accessing the message body that requires `request.is_json` validation is through `request.data`. This property captures POST, PUT, or PATCH message bodies regardless of any `Content-Type`, thus requiring a thorough validation mechanism. The following API function captures user credentials through `request.data` and inserts the login details in the database:

```
@current_app.route('/login/add', methods = ['POST'])
def add_login():
    if request.is_json:
        login_json = loads(request.data)
        login = Login(**login_json)
        ... ... ... ... ... ...
    else:
        abort(500)
```

It is always feasible to use `request.data` for HTTP POST transactions, such as in the given `add_login()` function, but the API needs to parse and serialize the `request.data` using Flask's built-in `loads()` decoder from the `flask.json` module extension because the request data is not yet JSON-formatted. Additionally, the process needs tight data type validation for each JSON object in the captured request data before using it in the transactions.

Aside from these variations of managing the incoming requests, Flask also has approaches to dealing with outgoing JSON responses. Instead of `jsonify()`, another way to render a JSON response is by instantiating and returning `Response` to the client. The following is a `list_login()` endpoint function that retrieves a list of `Login` records from the database using the `Response` class:

```python
@current_app.route('/login/list/all', methods = ['GET'])
def list_all_login():
    repo = LoginRepository(db_session)
    records = repo.select_all()
    login_rec = [rec.to_json() for rec in records]
    current_app.logger.info('retrieved a list of login successfully')
    resp = Response(response = dumps(login_rec), status=200,
mimetype="application/json" )
    return resp
```

When using `Response`, an encoder such as `dumps()` of the `flask.json` module can be used to create a JSONable object from an object, list, or dictionary. And the `mime-type` should always be `application/json` to force the object to become JSON.

Let us focus now on Flask's built-in support for JSON types and the serialization and de-serialization utilities it has to process JSON objects.

Utilizing response encoders and decoders

Flask framework supports the built Python `json` module by default. The built-in encoders, `dumps()`, and `loads()`, are found in the `flask.json` module. In the *Managing the requests and responses* section, the `add_login()` endpoint function uses the `flask.json.loads()` to de-serialize and transform the `request.data` into a JSONable dictionary. Meanwhile, the `flask.json.dumps()` provided the `Response` class with a JSONable object for some JSON response output, as previously highlighted in the `list_all_login()` endpoint.

But any application can override these default encoding and decoding processes to solve some custom requirements. Customizing an appropriate JSON provider by sub-classing Flask's JSONProvider, found in the `flask.json.provider`, can allow the overriding of these JSON processes. The following is a custom implementation of a JSONProvider with some modifications to the dumps () and loads () algorithms:

```python
from flask.json.provider import JSONProvider
import json
class ImprovedJsonProvider(JSONProvider):
    def __init__(self, *args, **kwargs):
        self.options = kwargs
        super().__init__(*args, **kwargs)

    def default(self, o):
        if isinstance(o, date):
            return o.strftime("%m/%d/%Y")
        elif isinstance(o, datetime):
            return o.strftime("%m/%d/%Y, %H:%M:%S")
        return super().default(self, o)

    def dumps(self, obj, **kwargs):
        kwargs.setdefault("default", self.default)
        kwargs.setdefault("ensure_ascii", True)
        kwargs.setdefault("sort_keys", True)
        return json.dumps(obj, **kwargs)

    def loads(self, s: str | bytes, **kwargs):
        s_dict:dict = json.loads(s.decode('utf-8'))
        s_sanitized = dict((k, v) for k, v in s_dict.items() if v)
        s_str = json.dumps(s_sanitized)
        return json.loads(s_str, **kwargs)
```

To apply this new JSON provider to the application, modify the `main.py` module and assign the app's json attribute with the instance of the custom provider with the app object as its constructor parameter. The following is the setup of our custom ImprovedJsonprovider in the online pizza ordering prototype:

```python
app = create_app('../config_dev.toml')
app.json = ImprovedJsonProvider(app)
```

Sub-classing the JSONProvider requires overriding its dump () and loads () methods. Additional custom features, such as formatting encoded dates, filtering empty JSON properties, and validating key and value types, can be helpful to custom implementation. For the serializer and de-serializer, the preferred JSON utility in customizing the JSONProvider is Python's built-in json module.

The `ImprovedJsonprovider` class includes a custom `default()` method that validates the property value types during encoding. It coerces the `date` or `datetime` objects to have a defined format. For the application to utilize this method during encoding, the overridden `dumps()` must pass this `default()` to Python's `json.dumps()` as the `kwargs["default"]` value. In addition, there are also other keyword arguments that can smoothen the encoding process, such as `ensure_scii`, which enables the replacement of non-ASCII characters with whitespaces, and `sort_keys`, which sorts the keys of the resulting dictionary in ascending order.

On the other hand, `ImprovedJsonprovider`'s overridden `loads()` method initially converts the string request data into a dictionary using Python's `json.loads()` before removing all the key-value pairs with empty values. Afterward, `json.dumps()` serializes the sanitized dictionary back to its string type before submitting it for JSON de-serialization. Thus, running the `add_category()` endpoint with a message body that has an empty description value will lead to *HTTP Status Code 500*, as shown in *Figure 3.3*:

```
C:\Users\alibatasys>curl -i -X POST http://localhost:5000/category/add -d "{\"name\": \"salad\",
\"description\": \"\"}" -H "Content-Type: application/json"
HTTP/1.1 500 INTERNAL SERVER ERROR
Server: Werkzeug/2.2.3 Python/3.11.2
Date: Mon, 08 May 2023 23:45:07 GMT
Content-Type: application/json
Content-Length: 77
Connection: close

{"error": "__init__() missing 1 required positional argument: 'description'"}
```

Figure 3.3 – Applying the overridden flask.json.loads() decoder

The removal of the `description` property by the custom `loads()` caused the constructor error flagged in the cURL command execution in *Figure 3.3*.

Now, the following are the deprecated features that will not work anymore in Flask 3.x and onwards:

- JSONEncoder and JSONDecoder APIs customize `flask.json.dumps()` and `flask.json.loads()`, respectively.

- `json_encoder` and `json_decoder` attributes set up JSONEncoder and JSONDecoder, respectively.

Also, the following setup applied in Python's `json` encoder and decoder during customization will not work here in the Flask framework:

- Specifying the `flask.json.loads()` encoder directly with the custom.

- Specifying the `flask.json.dumps()` decoder directly with the custom implementation class using the non-existent `cls` kwarg.

Since JSONEcoder and JSONDecoder will be obsolete soon, there will be no other means to customize these JSON utilities but through the JSONProvider.

However, there are instances where the incoming message body or the outgoing JSON responses are complex and huge, which cannot be handled optimally by the built-in JSON provider. In this case, Flask allows replacing the existing provider with a fast, accurate, and flexible provider, such as ujson and orjson. The following class is a sub-class of the JSONProvider that uses the orjson encoder and decoder.

```
from flask.json.provider import JSONProvider
import orjson
class OrjsonJsonProvider(JSONProvider):
    def __init__(self, *args, **kwargs):
        self.options = kwargs
        super().__init__(*args, **kwargs)

    def dumps(self, obj, **kwargs):
        return orjson.dumps(obj, option=orjson.OPT_NON_STR_KEYS).
decode('utf-8')

    def loads(self, s, **kwargs):
        return orjson.loads(s)
```

The OrjsonJsonProvider implements a custom JSON provider that uses orjson, one of the fastest JSON libraries that supports several types, such as datetime, dataclass, numpy types, and **Universally Unique Identifiers (UUID)**.

Another essential add-on that can further improve our RESTful application's validation and handling of incoming body requests and outgoing responses is *route filters*.

Filtering API requests and responses

In *Chapter 1*, the CRUD operations of every view function became possible without an ORM because of a custom decorator, @connect_db. The decorator was responsible for database connectivity and closure for every execution of the view function. Like in any Python decorator, the @connect_db executes first before the view function starts receiving the requests from the client and executes after the view generates the response.

On the other hand, *Chapter 2* introduced the use of @before_request and @after_request decorators in managing the application context of the view functions. Our applications used them to access the session db object for SQLAlchemy's database connectivity, impose user authentication, and perform software logging.

Using decorators to manage the requests and responses for a view or API function is called route filtering. The following are implementations of Flask's `before_request` and `after_request` methods used by the `ch03` application to filter the request–response handshake:

```python
from flask import request, abort, Response
@app.before_request
def before_request_func():
    api_method = request.method
    if api_method in ['POST', 'PUT', 'PATCH']:
        if request.json == '' or request.json == None:
            abort(500, description="request body is empty")
    api_endpoint_func = request.endpoint
    api_path = request.path
    app.logger.info(f'accessing URL endpoint: {api_path}, function
name: {api_endpoint_func} ')


@app.after_request
def after_request_func(response:Response):
    api_endpoint_func = request.endpoint
    api_path = request.path
    resp_allow_origin = response.access_control_allow_origin
    app.logger.info(f"access_control_allow_origin header: {resp_allow_
origin}")
    app.logger.info(f'exiting URL endpoint: {api_path}, function name:
{api_endpoint_func} ')
    return response
```

In this application, `before_request` checks if the incoming request body of HTTP POST, PUT, or PATCH transaction is not empty or None. Otherwise, it raises an `HTTP Status Code 500` with the error message `request body is empty`. It also performs logging for audit purposes. The `after_request` method, on the other hand, logs the basic details of the API for tracing purposes and checks the `access_control_allow_origin` response header. The mandatory parameter response allows us to access the response headers for modification if given by the software requirement. Also, this is the best spot to create cookies and execute the last database commits since this is the last moment of access to the response object before the `after_request` method sends it to the client.

Like FastAPI, the Flask framework has its version of creating middleware-like components, which can act as a global route filter. Our application has the following implementation, which serves as a middleware to the API endpoints:

```python
import werkzeug.wrappers
import werkzeug.wsgi
class AppMiddleware:
    def __init__(self, app):
        self.app = app
```

```
    def __call__(self, environ, start_response):
        request = werkzeug.wrappers.Request(environ)
        api_path = request.url
        app.logger.info(f'accessing URL endpoint: {api_path} ')

        iterator:werkzeug.wsgi.ClosingIterator = self.app(environ,
start_response)
        app.logger.info(f'exiting URL …: {api_path} ')
        return iterator
```

When implementing Flask middleware components, such as our AppleMiddleware, the involved Request API class is from the werkzeug module or the core platform itself since the implementation is server-level. Instantiating the werkzeug.wrappers.Request with the environ parameter as its constructor argument will give us access to the details of the incoming request of the API endpoint. Unfortunately, there is no direct way of accessing the response object within the filter class. Some implementations require the creation of hook methods by registering custom decorators to Flask through the custom middleware, and others use external modules to implement a middleware that acts like a URL dispatcher. Now, our custom middleware must be a callable class type, so all the implementations must be in its overridden __call__() method.

Moreover, we can also associate Blueprint modules with their respective custom before and after filter methods, if required. The following app configuration assigns filter methods to the order_client_bp and pizza_client_bp Blueprints of the ch03-client application:

```
app.before_request_funcs = {
    'orders_client_bp': [before_check_api_server],
    'pizza_client_bp': [before_log_pizza_bp]
}

app.after_request_funcs = {
    'orders_client_bp': [after_check_api_server],
    'pizza_client_bp': [after_log_pizza_bp]
}
```

Both before_request_funcs and after_request_funcs contain the concerned Blueprint names and their corresponding lists of implemented filter method names.

Can we also apply the same exception-handling directives used in the web-based applications of *Chapters 1* and *2*? Let us find out in the following discussion.

Handling exceptions

In RESTful applications, Flask allows the endpoint function to trigger error handlers that return error messages in JSON format. The following snippets are the error handlers of our ch03 application:

```
@app.errorhandler(404)
def not_found(e):
    return jsonify(error=str(e)), 404


@app.errorhandler(400)
def bad_request(e):
    return jsonify(error=str(e)), 400


def server_error(e):
    print(e)
    return jsonify(error=str(e)), 500
app.register_error_handler(500, server_error)
```

Error handlers can also return the JSON response through the jsonify(), make_response(), or Response class. As shown in the given error handlers, the implementation is the same with the web-based error handlers except for the jsonify() method, which serializes the captured error message to the JSON type instead of using render_template().

Custom exception classes must include both the *HTTP Status Code* and error message in the JSON message. The customization must include a to_dict() method that will convert the payload and other external parameters to a dictionary object for the jsonify() to serialize. The following is a custom exception class raised by our INSERT repository transactions and endpoint functions:

```
class DuplicateRecordException(HTTPException):
    status_code = 500

    def __init__(self, message, status_code=None, payload=None):
        super().__init__()
        self.message = message
        if status_code is not None:
            self.status_code = status_code
        self.payload = payload

    def to_dict(self):
        rv = dict(self.payload or ())
        rv['message'] = self.message
        return rv
```

After raising this `DuplicateRecordException`, the following error handler will access its `to_dict()` instance method and convert it to JSON through `jsonify()`. It will also access the `status_code` for the response:

```
@app.errorhandler(DuplicateRecordException)
def insert_record_exception(e):
    return jsonify(e.to_dict()), e.status_code
```

Any database INSERT transaction in our application will throw or raise the given `Database RecordException`, triggering this `insert_record_exception()` handler. But for Python-related exceptions, the following error handler will also render the built-in exception messages in JSON format:

```
@app.errorhandler(Exception)
def handle_built_exception(e):
    if isinstance(e, HTTPException):
        return e
    return jsonify(error=str(e)), 500
```

This `handle_built_exception()` handler will always return a JSON-formatted error message and raise the Werkzeug-specific exceptions for other custom handlers to manage. But for raised Python-specific exceptions, `handle_built_exception()` will directly render the JSON error message.

After completing the needed components in building our RESTful application, it is time to consume these API endpoints using a client application.

Consuming API endpoints

Our `ch03-client` project is a web-based Flask application that utilizes the API endpoints created in the `ch03` application. So far, the easiest way to consume a Flask API endpoint is to use the `requests` extension module. To install the `requests` library, run the following command:

```
pip install requests
```

This `requests` module has a `get()` helper method to send an HTTP GET request to a URL to retrieve some server resources. The following view function from the `ch03-client` project retrieves a list of customers and employees from the `ch03` application and passes them as context data to the `add_order.html` template:

```
@current_app.route('/client/order/add', methods = ['GET', 'POST'])
def add_order():
    if request.method == 'POST':
        order_dict = request.form.to_dict(flat=True)
        order_add_api = "http://localhost:5000/order/add"
        response: requests.Response = requests.post(order_add_api,
```

```
json=order_dict)
    customers_list_api = "http://localhost:5000/customer/list/all"
    employees_list_api = "http://localhost:5000/employee/list/all"
    resp_customers:requests.Response = requests.get(customers_list_
api)
    resp_employees:requests.Response = requests.get(employees_list_
api)
    return render_template('add_order.html', customers=resp_customers.
json(), employees=resp_employees.json())
```

The get() method returns a requests.Response object that contains essential details, such as content, url, status_code, json(), encoding, and other headers from the API's server. Our add_order() calls the json() for each GET response to serialize the result in JSON format.

For the HTTP POST transaction, the request module has a post() method to send an HTTP POST request to http://localhost:5000/order/add API. For a successful POST request handshake, the post() requires the URL of the API service and the record or object as the request body in dictionary format.

Aside from the dictionary type, the post() method can also allow the submission of a list of *tuples*, *bytes*, or *file entity types*. It also has various parameter options such as data, json, or files that can accept the appropriate request body types.

Now, other than get() and post() methods, the requests library has other helper methods that can also send other HTTP requests to the server, such as the put() that calls the PUT API service, delete() that calls DELETE API service, and patch() for the PATCH API service.

Summary

This chapter has proven to us that some components apply to both API-based and web-based applications, but there are specific components that fit better in API transactions than in web-based ones. It provided details on Flask's JSON de-serialization applied to request bodies and serialization of outgoing objects to be part of the API responses. The many options of capturing the request body through request.json, request.data, and request.get_json() and generating responses through its jsonify() or make_response() and Response class with application/json as a mime-type show Flask's flexibility as a framework.

The chapter also showcased Flask's ability to adapt to different third-party JSON providers through sub-classing its JSONProvider class. Moreover, the many options for providing our API endpoints with route filtering mechanisms also show that the platform can manage the application's incoming requests and outgoing responses like any good framework. Regarding error handling mechanisms, the framework can provide error handlers for web-based applications that render templates and those that send JSON responses for RESTful applications.

When consuming the API endpoints, this chapter exhibited that Flask could support typical Python REST client modules, such as `requests`, without any additional workaround.

So, we have seen that Flask can support building web-based and API-based applications even though it is lightweight and a microframework.

The next chapter will discuss simplifying and organizing Flask implementations using popular third-party Flask module extensions.

4

Utilizing Flask Extensions

Flask is popular due to its extensions, which are installable external or third-party modules or plugins that add support and even enhance some built-in features that may seem redundant to create, such as form handling, session handling, authentication procedures, and caching.

Applying Flask extensions to project development can save time and effort compared to re-creating the same features again. Also, these modules can have interdependence with other essential Python and Flask modules without requiring too much configuration, which is convenient for adding new features to the baseline project. Despite the positive factors, however, there are also some side effects of installing extensions for our Flask applications, such as having collisions with some installed modules and version problems with the current Flask version, which results in us having to downgrade some Flask extensions or the Flask version itself. Version collisions, deprecation, and non-support remain the core concerns when utilizing Flask extensions; therefore, it is advisable to read the documentation of every Flask extension before installing each on the platform.

This chapter will showcase the same project components that were created in *Chapters 1* to *3*, including web forms, REST services, backend databases, web sessions, and look-and-feel, but using their respective Flask extension modules. Moreover, this chapter will also show you how to apply caching and integrate mail features into the application.

This chapter will cover the following topics:

- Applying database migration with Flask-Migrate
- Designing the UI using Bootstrap-Flask
- Building Flask forms with Flask-WTF
- Building RESTful services with Flask-RESTful
- Implementing session handling with Flask-Session
- Applying caching using Flask-Caching
- Adding mail features with Flask-Mail

Technical requirements

This chapter will highlight two prototypes for an *Online Complaint Management System* that utilizes different popular Flask 3.0 extensions. These extensions will build the complaint, administration, login, and report modules. The `ch04-web` project will consist of the form-based side, while the `ch04-api` project contains RESTful services to cater to the various complaint details. Both applications will utilize `Blueprints` to organize their directory structure and use *SQLAlchemy* to perform CRUD transactions with their PostgreSQL database. All these projects have been uploaded at `https://github.com/PacktPublishing/Mastering-Flask-Web-Development/tree/main/ch04`.

Applying database migration with Flask-Migrate

The significant third-party Flask module to use when building an application is a module that will manage the data model layer, and that is the **Flask-Migrate** extension. Although it is sometimes appropriate to customize database migration using *Alembic*, *Flask-Migrate* offers easy setup and configuration with less coding and fast results.

> **Important note**
>
> Alembic is a lightweight and fast database migration tool for SQLAlchemy that can be customized to support various database backends.

Database migration is a way of deriving and generating the database schema from the Flask model classes and allowing the changes to be monitored and audited in these schemas throughout the application's lifespan, such as adding and dropping table columns, modifying table constraints, and renaming columns without ruining the current data. All these mechanisms are managed by *Flask-Migrate*.

Now, let's understand how to set up a database backend of our application using *Flask-Migrate* instead of manually creating the table schemas.

Installing Flask-Migrate and Flask-SQLAlchemy

First, since our applications will utilize *SQLAlchemy* as the ORM choice, install `flask-sqlalchemy` through the `pip` command:

```
pip install flask-sqlalchemy
```

Second, enable `SQLAlchemy` by creating the `engine`, `db_session`, and `Base` classes since our prototypes will utilize the *declarative approach* of database connectivity. This setup can be found in the `/model/config.py` module of both applications.

Now, using the `Base` class, create the model classes, which will become the basis of the database migration. The following code snippets show how to implement model classes using SQLAlchemy's `Base` class:

```
class Complaint(Base):
    __tablename__ = 'complaint'
    id = Column(Integer, Sequence('complaint_id_seq', increment=1),
primary_key = True)
    cid = Column(Integer, ForeignKey('complainant.id'), nullable =
False)
    catid = Column(Integer, ForeignKey('category.id'), nullable =
False)
    ctype = Column(Integer, ForeignKey('complaint_type.id'), nullable =
False)

    category = relationship('Category', back_populates="complaints")
    complainants = relationship('Complainant', back_
populates="complaints")
    complaint_type = relationship('ComplaintType', back_
populates="complaints")
    complaint_details = relationship('ComplaintDetails', back_
populates="complaint", uselist=False)
```

The declarative approach uses the `Base` class to create an SQLAlchemy model class that will depict the schema of its corresponding table. For instance, a given `Complaint` class corresponds to the `complaint` table with the `id`, `cid`, `catid`, and `ctype` columns, as defined by the `Column` helper class with the matching column type classes. All column metadata must be correct since *Flask-Migrate* will derive the table schema details from this metadata during migration. All column metadata, including the *primary*, *unique*, and *foreign key constraints*, will be part of this database migration. After migration, the following model classes will generate sub-tables for the `complaint` table:

```
class Category(Base):
    __tablename__ = 'category'
    id = Column(Integer, Sequence('category_id_seq', increment=1),
primary_key = True)
    name = Column(String(45), nullable = False)

    complaints = relationship('Complaint', back_populates="category")
    ... ... ... ... ... ...
class ComplaintType(Base):
    __tablename__ = 'complaint_type'
    id = Column(Integer, Sequence('complaint_type_id_seq',
increment=1), primary_key = True)
    name = Column(String(45), nullable = False)
```

```
    complaints = relationship('Complaint', back_populates="complaint_
type")
    ... ... ... ... ... ...
class ComplaintDetails(Base):
    __tablename__ = 'complaint_details'
    id = Column(Integer, Sequence('complaint_details_id_seq',
increment=1), primary_key = True)
    compid = Column(Integer, ForeignKey('complaint.id'), nullable =
False, unique=True)
    statement = Column(String(100), nullable = False)
    status = Column(String(50))
    resolution = Column(String(100))
    date_resolved = Column(Date)
    complaint = relationship('Complaint', back_populates="complaint_
details")
    ... ... ... ... ... ...
```

The `Category`, `ComplaintType`, and `ComplaintDetails` classes all reference the parent `Complaint`, as depicted by their respective `relationship()` parameters.

With SQLAlchemy set up, install the `flask-migrate` extension module:

```
pip install flask-migrate
```

Before running the migration commands from the extension module, create a module file (not `main. py`) to provide the necessary helper classes to run the migration commands locally. The following `manage.py` file of our prototypes will run the module's `install`, `manage`, and `upgrade` CLI commands:

```
from flask_migrate import Migrate
from flask_sqlalchemy import SQLAlchemy
from model.config import Base
from main import app
import toml

app.config.from_file('config-dev.toml', toml.load)
db = SQLAlchemy(app, metadata=Base.metadata)
migrate = Migrate(app, db)
```

If our SQLAlchemy setup uses the usual `flask-sqlalchemy` approach, where the instance of the SQLAlchemy class creates the model classes, instantiating the `Migrate` class just involves passing the app instance and the SQLAlchemy instance. In this approach, SQLAlchemy is still vital to the migration process, but its explicit instantiation will depend on the `Base.metadata` constructor parameter aside from the app instance. The instantiation of the `Migration` class also requires the app instance and the derived SQLAlchemy instance, as shown in the given module script.

Now, if the migration setup is ready and correct, the `migrate` instance provided by `manage.py` can run the `init` CLI command. This execution will generate the Alembic files needed for the migration process.

Setting up the Alembic configuration

Flask-Migrate uses Alembic to establish and manage database migrations. Running the `init` CLI command from the `migrate` instance will generate the Alembic configuration files inside the project directory. The following Python command runs Flask-Migrate's `init` CLI command using our `manage.py` file:

```
python -m flask --app manage.py db init
```

In the preceding command, `db` specifies the SQLAlchemy `db` instance that's passed to the `migrate` instance, while `init` is the CLI command that is part of the `flask_migrate` module. Running the preceding command will create logs that list all the folders and files generated by the `init` command, as depicted in *Figure 4.1*:

```
(ch04-env) C:\Alibata\Training\Source\flask\mastering\ch04-web>python -m flask --app mana
ge.py db init
Creating directory 'C:\\Alibata\\Training\\Source\\flask\\mastering\\ch04-web\\migrations' ... done
Creating directory 'C:\\Alibata\\Training\\Source\\flask\\mastering\\ch04-web\\migrations\\versions' ... done
Generating C:\Alibata\Training\Source\flask\mastering\ch04-web\migrations\alembic.ini ... done
Generating C:\Alibata\Training\Source\flask\mastering\ch04-web\migrations\env.py ... done
Generating C:\Alibata\Training\Source\flask\mastering\ch04-web\migrations\README ... done
Generating C:\Alibata\Training\Source\flask\mastering\ch04-web\migrations\script.py.mako ... done
Please edit configuration/connection/logging settings in 'C:\\Alibata\\Training\\Source\\flask\\mastering\\ch04-web\\
migrations\\alembic.ini' before proceeding.
```

Figure 4.1 – The init CLI command logs

All the Alembic files are inside the `migrations` folder and are auto-generated by the preceding command. The `migrations` folder contains the main Alembic file, `env.py`, which can be tweaked or further configured to support some additional migration requirements. *Figure 4.2* shows the content of the `migrations` folder:

Figure 4.2 – The migrations folder

Aside from env.py, the following files are also included in the migrations folder:

- The alembic.ini file, which contains the default Alembic configuration variables.

- The script.py.mako file, which serves as the template file for the migration files.

There will also be a versions folder that will contain the migration scripts after running the migrate command.

Creating the migrations

After generating the Alembic files, the migrate CLI command will be ready to start the *initial migration*. Running the migrate command for the first time generates all the tables from the ground up based on the SQLAlchemy model classes. The Python command to run the migrate CLI command is as follows:

```
python -m flask --app manage.py db migrate -m "Initial"
```

Figure 4.3 shows the log messages after running this initial migration:

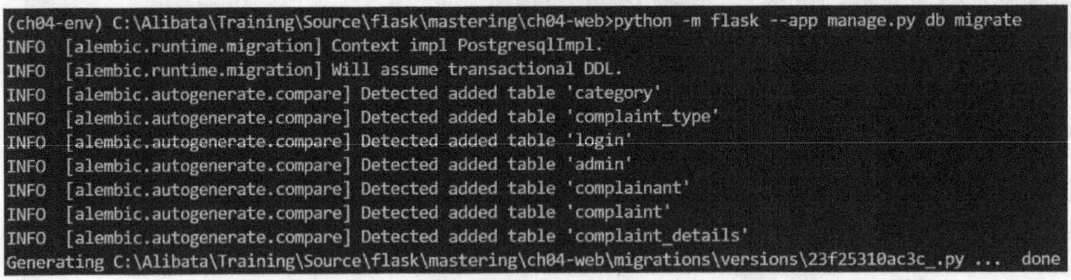

```
(ch04-env) C:\Alibata\Training\Source\flask\mastering\ch04-web>python -m flask --app manage.py db migrate
INFO  [alembic.runtime.migration] Context impl PostgresqlImpl.
INFO  [alembic.runtime.migration] Will assume transactional DDL.
INFO  [alembic.autogenerate.compare] Detected added table 'category'
INFO  [alembic.autogenerate.compare] Detected added table 'complaint_type'
INFO  [alembic.autogenerate.compare] Detected added table 'login'
INFO  [alembic.autogenerate.compare] Detected added table 'admin'
INFO  [alembic.autogenerate.compare] Detected added table 'complainant'
INFO  [alembic.autogenerate.compare] Detected added table 'complaint'
INFO  [alembic.autogenerate.compare] Detected added table 'complaint_details'
Generating C:\Alibata\Training\Source\flask\mastering\ch04-web\migrations\versions\23f25310ac3c_.py ...    done
```

Figure 4.3 – The migrate CLI command logs

A successful initial migration will create an alembic_version table in the database. *Figure 4.4* shows the content of the ocms database after the initial database migration:

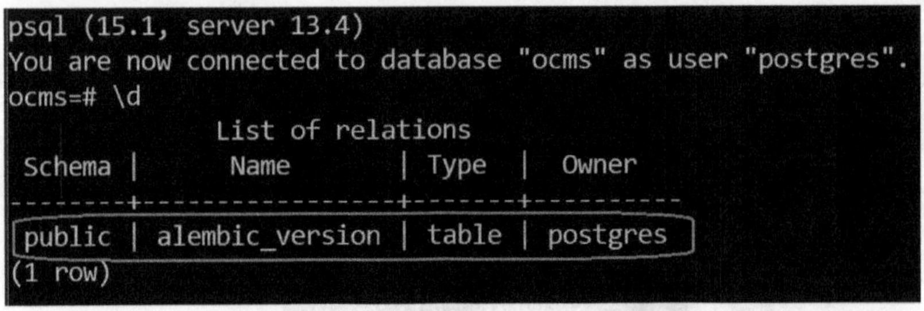

```
psql (15.1, server 13.4)
You are now connected to database "ocms" as user "postgres".
ocms=# \d
                  List of relations
   Schema  |        Name       | Type  |  Owner
 ----------+-------------------+-------+----------
   public  | alembic_version   | table | postgres
 (1 row)
```

Figure 4.4 – The alembic_version table

Every execution of the `migrate` command creates a migration script with a filename similar to its assigned unique *version number*. Flask-Migrate logs these version numbers in the `alembic_version` table and places all migration scripts inside the `migrations` folder under the `/versions` sub-directory. The following is a sample of this migration script:

```python
"""empty message

Revision ID: 9eafa601a7db
Revises:
Create Date: 2023-06-08 06:51:46.327352

"""
from alembic import op
import sqlalchemy as sa

# revision identifiers, used by Alembic.
revision = '9eafa601a7db'
down_revision = None
branch_labels = None
depends_on = None

def upgrade():
    # ### commands auto generated by Alembic - please adjust! ###
    op.create_table('category',
    sa.Column('id', sa.Integer(), nullable=False),
    sa.Column('name', sa.String(length=45), nullable=False),
    sa.PrimaryKeyConstraint('id')
    )
    op.create_table('complaint_type',
    sa.Column('id', sa.Integer(), nullable=False),
    sa.Column('name', sa.String(length=45), nullable=False),
    sa.PrimaryKeyConstraint('id')
    )
    ... ... ... ... ... ... ...
```

These auto-generated migration scripts sometimes need to be validated, edited, and re-coded because they aren't always the exact depiction of the SQLAlchemy model classes. Sometimes, these scripts do not capture the required changes in the table relationships and metadata that's applied to the models.

Now, to implement the final migration script, the `upgrade` CLI command needs to be executed.

Applying the database changes

The complete Python command to run the `upgrade` CLI command is as follows:

```
python -m flask --app manage.py db upgrade
```

Figure 4.6 shows the log messages after running the `upgrade` CLI command:

```
(ch04-env) C:\Alibata\Training\Source\flask\mastering\ch04-web>python -m flask --app manage.py db upgrade
INFO  [alembic.runtime.migration] Context impl PostgresqlImpl.
INFO  [alembic.runtime.migration] Will assume transactional DDL.
INFO  [alembic.runtime.migration] Running upgrade  -> 23f25310ac3c, empty message
```

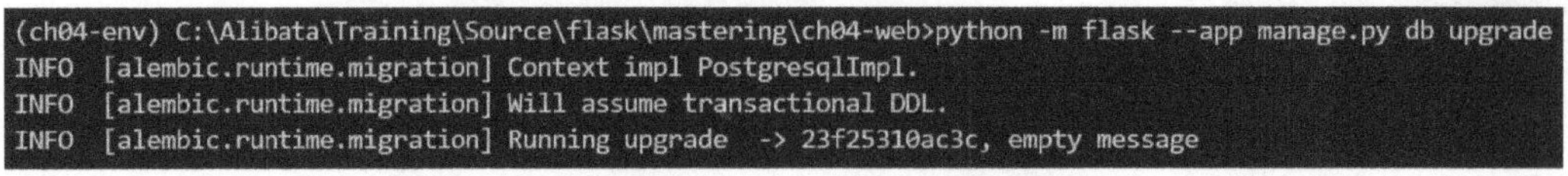

Figure 4.5 – The upgrade CLI command logs

The initial upgrade execution generates all the tables as defined in the initial migration script. Moreover, the succeeding scripts will always modify the schemas depending on the changes that are applied to the model classes. On the other hand, **Flask-Migrate** also has a rollback mechanism that can be implemented by running the `downgrade` CLI command. This command restores the previous version of the database.

Database migration in Flask projects will not be straightforward and seamless without Flask-Migrate. Writing the migration setup and processes from scratch will be time-consuming and rigorous to some extent.

The next extension that can help the development team save time handling Bootstrap's static files and importing them into Jinja2 templates is *Bootstrap-Flask*.

Designing the UI using Bootstrap-Flask

There are several ways to render context data with look-and-feel to the Jinja2 templates without stressing too much about downloading the resources files or referencing the static files from the **content delivery network (CDN)** repository and importing them into the template pages to manage the best UI design for the renditions. One of the most ideal and up-to-date options is **Bootstrap-Flask**, a far different module from the *Flask-Bootstrap* extension module. The latter uses only Bootstrap version 3.0, while *Bootstrap-Flask* can support up to *Bootstrap 5.0*. So, it is recommended to uninstall Flask-Bootstrap and other UI-related modules first before setting up Flask-Bootstrap to avoid unexpected conflicts. Allowing only Bootstrap-Flask to manage the UI designs can provide better results.

But first, let's install *Bootstrap-Flask* by running the following `pip` command:

```
pip install bootstrap-flask
```

Next, we'll set up the Bootstrap module with the desired Bootstrap framework distribution.

Setting up the UI module

For the module to work with the Flask platform, it must be set up in the `main.py` module. The `bootstrap_flask` module has `Bootstrap4` and `Bootstrap5` core classes that must be wired to the Flask instance before we can apply the framework's assets. An application can only use one Bootstrap distribution: our `ch04-web` application utilizes the `Bootstrap4` class to maintain consistency from *Chapter 3*'s Bootstrap preference. The following `main.py` module instantiates `Bootstrap4`, which enables the extension module:

```
from flask import Flask
from flask_bootstrap import Bootstrap4
import toml
from model.config import init_db

init_db()
app = Flask(__name__, template_folder='pages', static_
folder="resources")
app.config.from_file('config-dev.toml', toml.load)
bootstrap = Bootstrap4(app)
```

The `Bootstrap4` class requires the `app` instance as its constructor argument to proceed with the instantiation. After this setup, the Jinja2 templates can now load the necessary built-in resource files and start the web design process.

Applying the Bootstrap files and assets

Bootstrap-Flask has a `bootstrap.load_css()` helper function that loads the CSS resources into the Jinja2 template and a `bootstrap.load_js()` helper function that loads all Bootstrap JavaScript files. The following is the `login.html` template of the `ch04-web` application with the preceding two helper functions:

```
<!DOCTYPE html>
<html lang="en">
  <head>
    <meta charset="utf-8" />
    <meta http-equiv="x-ua-compatible" content="ie=edge" />
    <meta name="viewport" content="width=device-width, initial-
scale=1" />
    <title>Online Complaint Management System</title>
    {{ bootstrap.load_css() }}
  </head>
  <body>

    ... ... ... ... ... ...
    {{ bootstrap.load_js() }}
```

```
    </body>
</html>
```

It is always the standard to call `bootstrap.load_css()` in `<head>`, which is the appropriate markup to call the `<style>` tag. Calling `bootstrap.load_js()` in `<head>` is also feasible, but for many, the custom is to load all the JavaScript files in the last part of the `<body>` content, which is why `bootstrap.load_css()` is present there. On the other hand, if there are custom `styles.css` or JavaScript files for the applications, the module can allow their imports in the Jinja2 templates, so long as there are no conflicts with the Bootstrap resources.

After loading the CSS and JavaScript, we can start designing the pages with the Bootstrap components. The following code shows the content of the given `login.html` page with all the needed *Bootstrap 4* components:

```
<body>
    <section class="position-relative py-4 py-xl-5">
        <div class="container position-relative">
            <div class="row mb-5">
                <div class="col-md-8 col-xl-6 text-center mx-auto">
                    <h2 class="display-3">User Login</h2>
                </div>
            </div>
            <div class="row d-flex justify-content-center">
                <div class="col-md-6 col-xl-4">
                    <div class="card">
                        <div class="card-body text-center d-flex flex-
column align-items-center">
                            <form action="{{ request.path }}" method =
"post">
                                ... ... ... ... ... ...
                            </form>
                        </div>
                    </div>
                </div>
            </div>
        </div>
    </section>
    ... ... ... ... ... ...
</body>
```

Figure 4.6 shows the published version of the given login.html web design:

Figure 4.6 – The login.html page using Bootstrap 4

Aside from the updated Bootstrap support, the Bootstrap-Flask module has macros and built-in configurations that applications can use to create a better UI design.

Utilizing built-in features

The extension module has five built-in *macros* that Jinja2 templates can import to create fewer HTML codes and manageable components. These built-in macros are as follows:

- bootstrap4/form.html: This can render *Flask-WTF* forms or form components and their hidden error messages.
- bootstrap4/nav.html: This can render navigations and breadcrumbs.
- bootstrap4/pagination.html: This can provide paginations to *Flask-SQLAlchemy* data.
- bootstrap4/table.html: This can render table-formatted context data.
- bootstrap4/utils.html: This can provide other utilities, such as rendering flash messages, icons, and resource reference code.

The module also has built-in *configuration variables* to enable and disable some features and customize Bootstrap components. For instance, BOOTSTRAP_SERVE_LOCAL disables the process of loading built-in CSS and JavaScript when set to false and allows us to refer to CDN or local resources in /static instead. In addition, BOOTSTRAP_BTN_SIZE and BOOTSTRAP_BTN_STYLE can customize buttons. The **TOML** or any configuration file is where all these configuration variables are registered and set.

Next, we'll focus on *Flask-WTF*, a module supported by *Bootstrap-Flask*.

Building Flask forms with Flask-WTF

Flask-WTF is a flexible Flask extension module that uses classes to create forms, form fields, validations, and form renditions. It uses the `WTForms` library to enhance form handling in Flask applications. Instead of using HTML markup, Flask-WTF provides the necessary utilities to manage the web forms in a Pythonic way through form models.

Creating the form models

Form models must extend the `FlaskForm` core class to create and render the `<form>` tag. Its attributes correspond to the form fields defined by the following helper classes:

- `StringField`: Defines and creates a text input field that accepts string-typed data.
- `IntegerField`: Defines and creates a text input field that accepts integers.
- `DecimalField`: Defines and creates a text input field that asks for decimal values.
- `DateField`: Defines and creates a text input field that supports `Date` types with the default format of `yyyy-mm-dd`.
- `EmailField`: Defines and creates a text input that uses a regular expression to manage email-formatted values.
- `SelectField`: Defines and creates a combo box.
- `SelectMultipleField`: Defines and creates a combo box with a list of options.
- `TextAreaField`: Defines a text area for multi-line text input.
- `FileField`: Defines and creates a file upload field for uploading files.
- `PasswordField`: Defines a password input field.

The following code shows a form model that utilizes the given helper classes to build the `Complainant` form *widgets* for the `add_complainant()` view:

```python
from flask_wtf import FlaskForm
from wtforms import StringField, IntegerField, SelectField, DateField, EmailField
from wtforms.validators import InputRequired, Length, Regexp, Email

class ComplainantForm(FlaskForm):
    id = SelectField('Choose Login ID: ',
validators=[InputRequired()])
    firstname = StringField('Enter firstname:',
validators=[InputRequired(), Length(max=50)])
    middlename = StringField('Enter middlename:',
validators=[InputRequired(), Length(max=50)])
```

```
    lastname = StringField('Enter lastname:',
validators=[InputRequired(), Length(max=50)])
    email = EmailField('Enter email:', validators=[InputRequired(),
Length(max=20), Email()])
    mobile = StringField('Enter mobile:', validators=[InputRequired(),
Length(max=20), Regexp(regex=r"^(\+63)[-]{1}\d{3}
            [-]{1}\d{3}[-]{1}\d{4}$", message="Valid phone number
format is +63-xxx-xxx-xxxx")])
    address = StringField('Enter address:',
validators=[InputRequired(), Length(max=100)])
    zipcode = IntegerField('Enter zip
code:',  validators=[InputRequired()])
    status = SelectField('Enter status:', choices=[('active',
'ACTIVE'),
        ('inactive','INACTIVE'), ('blocked','BLOCKED')],
validators=[InputRequired()])
    date_registered = DateField('Enter date registered', format='%Y-
%m-%d', validators=[InputRequired()])
```

Here, `firstname`, `middlename`, `lastname`, and `address` are input-type text boxes with varying lengths and are required form parameters. For specific input types, `date_registered` is a required form parameter of the `Date` type with a date format of `yyyy-mm-dd`, while `email` is an email-type text box. On the other hand, the `status` and `id` form parameters are combo boxes, but the difference is the absence of options in `id`. The `status` form parameter has its `choices` options already defined in the form, while in the `id` form parameter, the view function will populate these fields at runtime. The following is a snippet of the `add_complainant()` view that manages the `id` parameter's options:

```
@complainant_bp.route('/complainant/add', methods=['GET', 'POST'])
def add_complainant():
    form:ComplainantForm = ComplainantForm()
    login_repo = LoginRepository(db_session)
    users = login_repo.select_all()
    form.id.choices = [(f"{u.id}", f"{u.username}") for u in users]
    if request.method == 'GET':
        return render_template('complainant_add.html', form=form), 200
    else:
        if form.validate_on_submit():
            details = dict()
            details["id"] = int(form.id.data)
            details["firstname"] = form.firstname.data
            details["lastname"]  = form.lastname.data
            ... ... ... ... ... ...
            complainant:Complainant = Complainant(**details)
            complainant_repo:ComplainantRepository =
ComplainantRepository(db_session)
```

```
        result = complainant_repo.insert(complainant)
    if result:
        records = complainant_repo.select_all()
        return render_template( 'complainant_list_all.
html', records=records), 200
    else:
        return render_template('complainant_add.
html', form=form), 500
else:
    return render_template('complainant_add.html', form=form),
500
```

The preceding view accesses the `choices` parameter of the `id` form parameter to assign it with a list of `(id, username)` `Tuple`, with `username` as the label and `id` as its value.

On the other hand, the HTTP POST transaction of the `add_complainant ()` view will verify form validation errors after submission through the `form` parameter's `validate_on_submit ()`. If there is none, the view function will extract all form data from the `form` object, insert the complaint details into the database, and render a list of all complainants. Otherwise, it will return the form page with the submitted `ComplainantForm` instance with form data values. Now that we've implemented the form models and the view functions that manage them, we can focus on how to map these models to their respective `<form>` tags using Jinja2 templates.

Rendering the forms

Before returning the WTF form model to the Jinja2 template, the view function must access and instantiate the FlaskForm sub-class, and even populate some fields with values in preparation for the `<form>` mapping. Assigning the model form with the appropriate values can avoid Jinja2 errors during `<form>` loading.

Now, the form rendition happens only when there is an *HTTP GET* request for form loading or when an HTTP POST encounters validation errors during submission, requiring the form page that's showing the current values and the error status to be reloaded. The type of HTTP request determines what values to assign to the form model's fields before rendering the form. Thus, in the given `add_complainant ()` view, checking if `request.method` is a *GET* request means verifying when to render the `complainant_add.html` form template with the `ComplainantForm` instance with base or initialized values. Otherwise, it will be a rendition of a form page with the current form values and validation errors.

The following `complainant_add.html` page maps the `ComplainantForm` fields, with base or current values, to the `<form>` tag:

```
<form action = "{{ request.path }}" method = "post">
        {{ form.csrf_token }}
  <div class="mb-3">{{ form.id(size=1, class="form-control") }}</div>
  <div class="mb-3"> {{ form.firstname(size=50,
placeholder='Firstname', class="form-control") }}
  </div>
        {% if form.firstname.errors %}
            <ul>
                {% for error in form.username.errors %}
                    <li>{{ error }}</li>
                {% endfor %}
            </ul>
        {% endif %}
    ... ... ... ... ... ...
  <div class="mb-3"> {{ form.mobile(size=50, placeholder='+63-XXX-XXX-
XXXX', class="form-control") }}
  </div>
    ... ... ... ... ... ...
  <div class="mb-3"> {{ form.email(size=50,placeholder='xxxxxxx@xxxx.
xxx', class="form-control") }}
  </div>
    ... ... ... ... ... ...
    <div class="mb-3"> {{ form.zipcode(size=50, placeholder='Zip Code',
class="form-control") }}
    </div>
    ... ... ... ... ... ...
</form>
```

Binding the individual model field to `<form>` requires calling the property of the field – for example, `context_name.field()`. So, to render the `firstname` form field of `ComplainantForm`, for instance, the Jinja2 template must call `form.firstname()` inside the `{{ }}` statement. The method call can also include its `kwargs` or *keyword arguments* of widget properties, such as `size`, `placeholder`, and `class`, if there is a change in the default widget settings during the rendition. As shown in the template, the *Flask-WTF* widgets support the Bootstrap components provided by the *Bootstrap-Flask* module extension. Adding custom CSS styles is also feasible with widgets, so long the CSS properties are set in the widget's `kwargs`.

Now, let's explore if Flask can secure form transactions from **cross-site request forgery** (CSRF) problems like in Django's forms.

Applying CSRF

Flask-WTF has built-in CSRF support through its `csrf_token` generation. To enable CSRF through Flask-WTF, instantiate `CSRFProtect` from the `flask_wtf` module in `main.py`, as shown in the following snippet:

```python
from flask_wtf import CSRFProtect


app = Flask(__name__, template_folder='pages', static_
folder="resources")
app.config.from_file('config-dev.toml', toml.load)
bootstrap = Bootstrap4(app)
csrf = CSRFProtect(app)
```

Calling the generated CSRF token (`csrf_token`) using the WTF form context inside the `{{}}` statement enables the token generation per-user access, as shown in the given `complainant_add.html` template. Flask-WTF generates a unique token for every rendition of the form fields. Note that CRSF protection is only possible with Flask-WTF if the `SECRET_KEY` configuration variable is part of the configuration file and has the appropriate hash value.

CRSF protection occurs in every form submission that involves the form model instance. Now, let's discuss the general flow of form submission with the Flask-WTF module.

Submitting the form

After clicking the submit button, the *HTTP POST* request transaction of `add_complainant()` retrieves the form values after the validation, as shown in the preceding snippet. Flask-WTF sends the form data to the view function through the *HTTP POST* request method, requiring the view function to have validation for the incoming `POST` requests. If `request.method` is `POST`, the view must perform another evaluation on the extension's `validate_on_submit()` to check for violation of some form constraints. If the results for all these evaluations are `True`, the view function can access all the form data in `form_object.<field_name>.data`. Otherwise, the view function will redirect the users to the form page with the corresponding error message(s) and *HTTP Status Code 500*.

But what comprises Flask-WTF's form validation framework, or what criteria are the basis of the `validate_on_submit()` result after form submission?

Validating form fields

Flask-WTF has a list of useful built-in validator classes that support core validation rules such as `InputRequired()`, which imposes the HTML-required constraint. Some constraints are specific to widgets, such as the `Length()` validator, which can restrict the input length of the `StringField` values, and `RegExp()`, which can impose regular expressions for mobile and telephone data formats. Moreover, some validators require dependency modules to be installed, such as `Email()`, which needs an email-validator external library. All these built-in validators are ready to be imported from *Flask-WTF*'s `wtforms.validators` module. The `validators` parameter of every *FlaskForm* attribute can accept any callables from validator classes to impose constraint rules.

Violations are added to the field's errors list that can trigger `validate_on_submit()` to return `False`. The form template must render all such error messages per field during redirection after an *HTTP Status Code 500* error.

The module can also support custom validation for some constraints that are not typical. There are many ways to implement custom validations, and one is through the *in-line validator approach* exhibited in the following snippet:

```
class ComplainantForm(FlaskForm):
    … … … … … …
    zipcode = IntegerField('Enter zip
code:',  validators=[InputRequired()])
    … … … … … …

    def validate_zipcode(self, zipcode):
        if not len(str(zipcode.data)) == 4:
            raise ValueError('zip code must be 4 digits')
```

An *in-line validator* is a custom `FlaskForm` function with the method name prefixed with `validate_`, followed by the form field name it validates, that takes that form field as a parameter. The given `validate_zipcode()` checks whether the form field's `zipcode` checks if the input is a four-number value. Otherwise, it throws an exception class. Another approach is to implement validators as *typical FlaskForm functions*, but the validator function needs to be injected explicitly into the `validators` parameter of the field it validates.

Lastly, *a closure-like or callable approach* to validator implementation is also possible. Here `disallow_invalid_dates()` is a closure-type validator that does not allow date input before the given `date_after`:

```
from wtforms.validators ValidationError

class ComplainantForm(FlaskForm):
    def disallow_invalid_dates(date_after):
        message = 'Must be after %s.' % (date_after)
```

```
    def _disallow_invalid_dates(form, field):
        base_date = datetime.strptime(date_after, '%Y-%m-%d').date()
        if field.data < base_date:
            raise ValidationError(message)

    return _disallow_invalid_dates
    … … … … … …
    … … … … … …
    date_registered = DateField('Enter date registered',
format='%Y-%m-%d', validators=[InputRequired(), disallow_invalid_
dates('2000-01-01')])
```

In this approach, a validator function serves as a factory of a closure or local function that implements the validation rule. In the case of disallow_invalid_dates(), the closure is _disallow_invalid_dates(form, field), which raises ValidationError when the field.data or date_registered form value is before the specified boundary date's date_after provided by the validator function. To apply validators, you can call them just like a typical method – that is, with parameter values in the validators parameter of the field class.

Another extension module that is popular nowadays and supports the recent Flask framework is the *Flask-RESTful* module. We'll take a look at it in the next section.

Building RESTful services with Flask-RESTful

The flask-RESTful module uses the *class-based view strategy* of Flask to build RESTful services. It provides a Resource class to create custom resources to build from the ground up HTTP-based services instead of endpoint-based routes.

This chapter specifies another application, ch04-api, that implements RESTful endpoints for managing user complaints and related details. Here's one of the resource-based implementations of our application's API endpoints:

```
from flask_restful import Resource

class ListComplaintRestAPI(Resource):
    def get(self):
        repo = ComplaintRepository(db_session)
        records = repo.select_all()
        complaint_rec = [rec.to_json() for rec in records]
        return make_response(jsonify(complaint_rec), 201)
```

Here, `flask_restful` provides the `Resource` class that creates resources for views. In this case, `ListComplaintRestAPI` sub-classes the `Resource` class to override its `get()` instance method, which will retrieve all complaints from the database through an HTTP *GET* request. On the other hand, `AddComplaintRestAPI` implements the *INSERT* complaint transaction through an HTTP *POST* request:

```
class AddComplaintRestAPI(Resource):
    def post(self):
        complaint_json = request.get_json()
        repo = ComplaintRepository(db_session)
        complaint = Complaint(**complaint_json)
        result = repo.insert(complaint)
        if result:
            content = jsonify(complaint_json)
            return make_response(content, 201)
        else:
            content = jsonify(message="insert complaint record
encountered a problem")
            return make_response(content, 500)
```

The `Resource` class has a `post()` method that needs to be overridden to create *POST* transactions. The following `UpdateComplainantRestAPI`, `DeleteComplaintRestAPI`, and `UpdateComplaintRestAPI` resources implement HTTP *PATCH*, *DELETE*, and *PUT*, respectively:

```
class UpdateComplainantRestAPI(Resource):
    def patch(self, id):
        complaint_json = request.get_json()
        repo = ComplaintRepository(db_session)
        result = repo.update(id, complaint_json)
        if result:
            content = jsonify(complaint_json)
            return make_response(content, 201)
        else:
            content = jsonify(message="update complainant ID
encountered a problem")
            return make_response(content, 500)
```

The `Resource` class' `patch()` method implements the HTTP *PATCH* request when overridden. Like HTTP *GET*, `patch()` can also accept path variables or request parameters by declaring local parameters to the override. The `id` parameter in the `patch()` method of `UpdateComplaintRestAPI` is a path variable for a complainant ID. This is required to retrieve the complainant's profile:

```
class DeleteComplaintRestAPI(Resource):
    def delete(self, id):
        repo = ComplaintRepository(db_session)
```

```
        result = repo.delete(id)
        if result:
            content = jsonify(message=f'complaint {id} deleted')
            return make_response(content, 201)
        else:
            content = jsonify(message="delete complaint record
encountered a problem")
            return make_response(content, 500)
```

The `delete()` override of `DeleteComplaintRestAPI` also has an `id` parameter that's used to delete the complaint:

```
class UpdateComplaintRestAPI(Resource):
    def put(self):
        complaint_json = request.get_json()
        repo = ComplaintRepository(db_session)
        result = repo.update(complaint_json['id'], complaint_json)
        if result:
            content = jsonify(complaint_json)
            return make_response(content, 201)
        else:
            content = jsonify(message="update complaint record
encountered a problem")
            return make_response(content, 500)
```

To utilize all the preceding resources, the Flask-RESTful extension module has an `Api` class that must be instantiated with the `Flask` or `Blueprint` instance as its constructor argument. Since the `ch04-api` project uses blueprints, the following `__init__.py` file of the complaint blueprint module highlights how to create the `Api` instance and map all these resources with their respective URL patterns:

```
from flask import Blueprint
from flask_restful import Api


complaint_bp = Blueprint('complaint_bp', __name__)
from modules.complaint.api.complaint import AddComplaintRestAPI,
ListComplaintRestAPI, UpdateComplainantRestAPI,
UpdateComplaintRestAPI, DeleteComplaintRestAPI

… … … … … …
api = Api(complaint_bp)
api.add_resource(AddComplaintRestAPI, '/complaint/add', endpoint='add_
complaint')
api.add_resource(ListComplaintRestAPI, '/complaint/list/all',
endpoint='list_all_complaint')
api.add_resource(UpdateComplainantRestAPI, '/complaint/update/
complainant/<int:id>', endpoint='update_complainant')
```

```
api.add_resource(UpdateComplaintRestAPI, '/complaint/update',
endpoint='update_complaint')
api.add_resource(DeleteComplaintRestAPI, '/complaint/delete/<int:id>',
endpoint='delete_complaint')
```

The Api class has an add_resource() method that maps every resource to its *URL pattern* and *endpoint* or *view function name*. This script shows how all the complaint module's resource classes are injected into the platform as full-fledged API endpoints. Conflicts on endpoint names and URLs within and outside the blueprint modules will cause compile-time errors, so all details must be unique to each resource.

The next module extension, *Flask-Session*, provides Flask with a better session-handling solution than its built-in implementation.

Implementing session handling with Flask-Session

The **Flask-Session** module, like Flask's built-in session, is easy to configure and use, except the module extension does not store session data in the web browser.

Before you can configure this module, you must install it using the pip command:

```
pip install flask-session
```

Then, import the Session class into the main.py module to instantiate and integrate the extension module into the Flask platform. The following main.py snippet shows the configuration of *Flask-Session*:

```
from flask_session import Session

app = Flask(__name__)
app.config.from_file('config-dev.toml', toml.load)
sess = Session()
sess.init_app(app)
```

The Session instance is only used for configuration and not for session handling per se. Flask's session proxy object should always directly access the session data for storage and retrieval.

Afterward, set some Flask-Session configuration variables, such as SESSION_FILE_THRESHOLD, which sets the maximum number of data the session stores before deletion, and SESSION_TYPE, which determines the kind of data storage for the session data. The following are some SESSION_TYPE options:

- null (default): This utilizes NullSessionInterface, which triggers an Exception error.
- redis: This utilizes RedisSessionInterface to use *Redis* as a data store.
- memcached: This utilizes MemcachedSessionInterface to use *memcached*.

- `filesystem`: This utilizes `FileSystemSessionInterface` to use the *filesystem* as the datastore.

- `mongodb`: This utilizes `MongoDBSessionInterface` to use the MongoDB database.

- `sqlalchemy`: This uses `SqlAlchemySessionInterface` to apply the SQLAlchemy ORM for a relational database as session storage.

The module can also recognize Flask session config variables such as `SESSION_LIFETIME`.

The following configuration variables are registered in the `config-dev.toml` file for both applications:

```
SESSION_LIFETIME = true
SESSION_TYPE = "filesystem"
SESSION_FILE_THRESHOLD = 600
SESSION_PERMANENT = true
```

Lastly, start the Flask server to load all the configurations and check the module's integration. The module will establish database connectivity to the specified data storage at server startup. In our case, the *Flask-Session* module will create a `flask_session` directory inside the project directory when the application starts.

Figure 4.7 shows the `flask_session` folder and its content:

Figure 4.7 – The session files inside the flask_session folder

With everything set up, utilize Flask's `session` to handle session data. This can be seen in `login_db_auth()`, which stores `username` as a session attribute for other views' reach:

```
from flask import session

@login_bp.route('/login/auth', methods=['GET', 'POST'])
def login_db_auth():
    authForm:LoginAuthForm = LoginAuthForm()
    ... ... ... ... ... ...
    if authForm.validate_on_submit():
        repo = LoginRepository(db_session)
        username = authForm.username.data
    ... ... ... ... ... ...
```

```
        if user == None:
            return render_template('login.html', form=authForm) , 500
        elif not user.password == password:
            return render_template('login.html', form=authForm) , 500
        else:
            session['username'] = request.form['username']
            return redirect('/ch04/login/add')
    else:
      return render_template('login.html', form=authForm) , 500
```

Similar to *Flask-Session*, another extension module that can help build a better enterprise-grade Flask application is the *Flask-Caching* module.

Applying caching using Flask-Caching

Flask-Caching provides enhanced caching support for views, data rendered by Jinja2 templates, and the custom functions, such as repository or service functions, of any Flask application. It also supports built-in caching of Flask and allows you to customize a cache through the BaseCache class from its flask_caching.backends.base module.

Before we can configure Flask-Caching, we must install the flask-caching module via the pip command:

```
pip install flask-caching
```

Then, we must register some of its configuration variables in the configuration file, such as CACHE_TYPE, which sets the cache type suited for the application, and CACHE_DEFAULT_TIMEOUT, which sets the caching timeout. The following are the applications' caching configuration variables declared in their respective config-dev.toml files:

```
CACHE_TYPE = "FileSystemCache"
CACHE_DEFAULT_TIMEOUT = 300
CACHE_DIR = "./cache_dir/"
CACHE_THRESHOLD = 800
```

Here, CACHE_DIR sets the cache folder for the filesystem cache type, while CACHE_THRESHOLD sets the maximum number of cached items before it starts deleting some.

Afterward, to avoid cyclic collisions, create a separate module file, such as `main_cache.py`, to instantiate the `Cache` class from the `flask_caching` module. Access to the `cache` instance must be done from `main_cache.py`, not `main.py`, even though the final setup of the extension module occurs in `main.py`. The following snippet integrates the *Flask-Caching* module into the Flask platform:

```
from main_cache import cache

app = Flask(__name__, template_folder='pages', static_
folder="resources")
app.config.from_file('config-dev.toml', toml.load)
cache.init_app(app)
```

Finally, run the Flask server to check for compile-time errors and if the module successfully created the cache. In our setup, a directory called `cache_dir` must be created inside the main folder, as shown in *Figure 4.8*:

Figure 4.8 – The cache files in cache_dir

If the setup is successful, components can now access `main_cache.py` for the cache instance. It has a `cached()` decorator that can provide caching for various functions. First, it can cache views, usually with an HTTP *GET* request to retrieve bulk records from the database. The following view function from the `complainant.py` module of the `ch04-web` application caches all its results to `cache_dir` for optimal performance:

```
from main_cache import cache

@complainant_bp.route('/complainant/list/all', methods=['GET'])
@cache.cached(timeout=50, key_prefix="all_complaints")
def list_all_complainant():
    repo:ComplainantRepository = ComplainantRepository(db_session)
    records = repo.select_all()
    return render_template('complainant_list_all.html',
records=records), 200
```

The exact area to decorate the `cached()` decorator is between the function definition and the route decorator of the view function. The decorator needs `key_prefix` to generate `cache_key`. If not specified, *Flask-Caching* will use the default `request.path` as the `cache_key` value. Note that `cache_key` is the key that's used to access the cached value of the function and is solely for the module to access. The given `list_all_complainant()` caches the rendered list of complaints with `prefix_key` set to `all_complaints`.

Moreover, the endpoint functions of the resource-based API created by Flask-RESTful can also cache their returned values through the `@cached()` decorator. The following code shows `ListComplaintDetailsRestAPI` from the `ch04-api` application, which caches the list of `ComplaintDetails` records into `cache_dir`:

```
class ListComplaintDetailsRestAPI(Resource):
    @cache.cached(timeout=50)
    def get(self):
        repo = ComplaintDetailsRepository(db_session)
        records = repo.select_all()
        compdetails_rec = [rec.to_json() for rec in records]
        return make_response(jsonify(compdetails_rec), 201)
```

As shown in the preceding code snippet, the decorator is placed above the overridden method of the `Resource` class. This rule is also valid with the other class-based views.

The module can also cache repository and service functions that retrieve large amounts of data during user access. The following code shows a `select_all()` function that retrieves data from the `login` table:

```
from main_cache import cache

class LoginRepository:
    def __init__(self, sess:Session):
        self.sess = sess
    ... ... ... ... ... ...
    @cache.cached(timeout=50, key_prefix='all_login')
    def select_all(self) -> List[Any]:
        users = self.sess.query(Login).all()
        return users
    ... ... ... ... ... ...
```

Moreover, the module also supports the *memoization* process to store values, similar to caching, but for custom functions that are frequently accessed. The `cache` instance has a `memoize()` decorator that manages these functions to improve performance. The following code shows the `@memoize` decorated method of `ComplaintRepository`:

```
from main_cache import cache


class ComplaintRepository:
    def __init__(self, sess:Session):
        self.sess = sess


    @cache.memoize(timeout=50)
    def select_all(self) -> List[Any]:
        complaint = self.sess.query(Complaint).all()
        return complaint
```

The given `select_all()` method will cache all the queried records for 50 seconds to improve its data retrieval performance. To clear the caches after server startup, always call `cache.clear()` in the `main.py` module after blueprint registration.

To be able to send complaints through emails, let's showcase a popular extension module called *Flask-Mail*.

Adding mail features using Flask-Mail

Flask-Mail is an extension module that handles sending emails to an email server without too much configuration.

First, install the `flask-mail` module using the `pip` command:

```
pip install flask-mail
```

Then, create a separate module script, such as `mail_config.py`, to instantiate the `Mail` class. This approach solves the cyclic collisions that occur when views or endpoint functions access the `mail` instance for the utility methods.

Despite the separate module, the `main.py` module still needs to access the `mail` instance to integrate the module into the Flask platform. The following `main.py` snippet shows how to set up the *Flask-Mail* module with Flask:

```
from mail_config import mail


app = Flask(__name__, template_folder='pages', static_folder="resources")
app.config.from_file('config-dev.toml', toml.load)
mail.init_app(app)
```

Afterward, the setup requires some configuration variables to be set in the config file. The following configuration variables are the most essential settings for our applications in this chapter:

```
MAIL_SERVER ="smtp.gmail.com"
MAIL_PORT = 465
MAIL_USERNAME = "your_email@gmail.com"
MAIL_PASSWORD = "xxxxxxxxxxxxxxxxx"
MAIL_USE_TLS = false
MAIL_USE_SSL = true
```

These details pertain to a *Gmail* account, but you can replace them with *Yahoo!* account details. Preferably, MAIL_USE_SSL must be set to true. Note that MAIL_PASSWORD is a token that's generated from the *app password* of the email account and not the actual one to establish a secure connection to the mail server. After server startup, Flask-Mail is ready to use if there are no compiler errors.

The following email_complaint() view function uses the mail instance to send a complaint through the *Flask-WTF* and *Flask-Mail* extension modules:

```
from flask_mail import Message
from mail_config import mail


@complaint_bp.route("/complaint/email")
def email_complaint():
    form:EmailComplaintForm = EmailComplaintForm()
    if request.method == 'GET':
        return render_template('email_form.html', form=form), 200
    if form.validate_on_submit():
        try:
            recipients = [rec for rec in str(form.to.data).split(';')]
            msg = Message(form.subject, sender = 'your_email@gmail.
com', recipients = recipients)
            msg.body = form.message.data
            mail.send(msg)
            form:EmailComplaintForm = EmailComplaintForm()
            return render_template('email_.html', form=form,
message='Email sent.'), 200
        except:
            return render_template('email_.html', form=form), 500
```

Here, EmailComplaintForm provides details for the Message() attributes, except for the sender, which is the email address that's assigned to the MAIL_USERNAME configuration variable. The mail instance provides the send() utility method to send the message to the recipient(s).

Summary

This chapter explained how to utilize Flask extension modules when building applications. Most extension modules will let us focus on the requirements rather than the complexities of configurations and setups, such as the *Flask-WTF* and *Bootstrap-Flask* modules. Some will shorten development time instead of programming the snippets repeatedly or handling the details all over again, such as *Flask-Migrate* on database migrations and *Flask-Mail* for sending messages to email servers. Some modules can enhance the built-in features of the Flask framework and provide better configuration options, such as *Flask_Caching* and *Flask-Session*. Finally, a few will organize the concepts and the implementations of the components, such as *Flask-RESTful*.

Although problems may arise with version conflicts and outdated modules, the many advantages the extension modules provide supersede all of their drawbacks. But above all, the software requirements must always come first when applying these extension modules. You should always be able to choose the appropriate options among these modules that will best address the needed requirements, not because they can provide shortcuts.

The next chapter will be about the Flask framework and asynchronous programming.

Part 2:
Building Advanced
Flask 3.x Applications

In this part, you will learn how to extend your Flask skills to build enterprise applications with enhanced performance using asynchronous views and API endpoints. You will learn to use the asynchronous SQLAlchemy to build asynchronous repository transactions and NoSQL databases to manage non-structured or semi-structured big data. Overall, this part will lead you to implement features that utilize asynchronous background tasks, upload XLSX and CSV files to generate charts, graphs, and tabular data, generate PDF documents, use WebSockets and **Server-Sent Events** (**SSE**), implement non-BPMN and BPMN workflows, and secure the application from web attacks.

This part includes the following chapters:

- *Chapter 5, Building Asynchronous Transactions*
- *Chapter 6, Developing Computational and Scientific Applications*
- *Chapter 7, Using Non-Relational Data Storage*
- *Chapter 8, Building Workflows with Flask*
- *Chapter 9, Securing Flask Applications*

5

Building Asynchronous
Transactions

After rigorous discussion on the core components and advanced features of the Flask 3.0 framework, this chapter will explore Flask's capability to manage requests and responses asynchronously and its ability to execute asynchronous services and repository transactions.

Flask was originally a standard Python framework that ran on the **Web Server Gateway Interface (WSGI)**-based platform popular in managing blocking processes. But Flask 3.0 supports the creation and execution of non-blocking view and API functions. It can run transactions using some `asyncio` utilities and build asynchronous repository transactions with SQLAlchemy 2.x.

This chapter will also explore other avenues that help provide Flask applications with the fastest performance using asynchronous mechanisms, such as Celery tasks, task queues, WebSocket, and server push. The chapter will also introduce **Quart**, the asynchronous Flask-based platform that can build and run all components asynchronously compared to the original Flask framework.

Here are the topics that this chapter will highlight:

- Creating asynchronous Flask components
- Building an asynchronous SQLAlchemy repository layer
- Implementing async transactions with `asyncio`
- Utilizing asynchronous signal notifications
- Constructing background tasks with Celery and Redis
- Building WebSockets with asynchronous transactions
- Implementing asynchronous **Server-Sent Events (SSE)**
- Applying reactive programming with RxPy
- Choosing Quart over Flask 2.x

Technical requirements

This chapter will highlight an *Online Voting System* prototype with some asynchronous tasks and background processes to manage the high bandwidth of candidates' applications and election-related submissions from different areas and to cater to the simultaneous retrieval of vote tallies from various parties. The system is composed of three separate projects, namely `ch05-api`, which has the API endpoints, `ch05-web`, which implements the SSE, WebSocket, and template-based results, and `ch05-quart`, which provides another platform for the app using the Quart framework. All these projects use the **application factory design** pattern and are available at `https://github.com/PacktPublishing/Mastering-Flask-Web-Development/tree/main/ch05`.

Creating asynchronous Flask components

Flask 2.3 and up to the current version support running asynchronous API endpoint and web-based view functions over its WSGI-based platform. However, to fully use this feature, install the `flask[async]` module using the following `pip` command:

```
pip install flask[async]
```

After installing the `flask[async]` module, implementing synchronous views using the `async/await` design pattern can now be feasible.

Implementing asynchronous views and endpoints

Like Django or FastAPI, creating asynchronous views and endpoints in the Flask framework involves applying the `async/await` keywords. The following web view from `ch05-web` renders a welcome greeting message to the users with the description of our *Online Voting* application:

```
@current_app.route('/ch05/web/index')
async def welcome():
    return render_template('index.html'), 200
```

Another asynchronous view function from the other application, `ch05-api`, is showcased in the following API endpoint that adds new login credentials to the **database (DB)**:

```
@current_app.post('/ch05/login/add')
async def add_login():
    async with db_session() as sess:
            repo = LoginRepository(sess)
            login_json = request.get_json()
            login = Login(**login_json)
            result = await repo.insert(login)
            if result:
                content = jsonify(login_json)
```

```
                    return make_response(content, 201)
            else:
                    raise DuplicateRecordException("add login credential
 has failed")
```

Both the given view and API functions use `async` routes to manage their respective request and response objects. Defining these routes with `async` creates coroutines that Flask 2.x can surprisingly run using the `run()` utility of the `asyncio` module. But how does the Flask framework manage the coroutine executions despite the pitfalls of its WSGI platform?

Flask spawns a **worker thread** to run these views and endpoints. With `async`, the framework creates a *sub-thread* from the worker thread to create an **event loop** that will execute the coroutine using the `asyncio` utilities. Despite the asynchronous processes, there are still limitations on how far `async` can push through since the environment is still within the WSGI, a synchronous platform. However, for not-so-complex non-blocking transactions, the `flask[async]` framework is enough to improve software quality and performance.

However, async components are not limited to view and API functions and also have Flask event handlers.

Implementing the async before_request and after_request handlers

Aside from views and endpoints, the Flask 3.0 framework allows the implementation of asynchronous `before_request` and `after_request` handlers, just like the following `ch05-api` handlers that log every API request transaction:

```
@app.before_request
async def init_request():
    app.logger.info('executing ' + request.endpoint + ' starts')


@app.after_request
async def return_response(response):
    app.logger.info('executing ' + request.endpoint + ' stops')
    return response
```

These event handlers still use the `app` instance created in `main.py` to create the log files using the built-in `logging` module.

On the other hand, the `flask[async]` module can allow the creation of asynchronous error handlers.

Creating asynchronous error handlers

Flask 2.x can decorate coroutines with `@errorhandler` to manage raised exceptions and HTTP status codes. The following are the asynchronous error handlers of the `ch05-api` project placed in `main.py`:

```python
@app.errorhandler(404)
async def not_found(e):
    return jsonify(error=str(e)), 404


@app.errorhandler(400)
async def bad_request(e):
    return jsonify(error=str(e)), 400


@app.errorhandler(DuplicateRecordException)
async def insert_record_exception(e):
    return jsonify(e.to_dict()), e.code


async def server_error(e):
    return jsonify(error=str(e)), 500


app.register_error_handler(500, server_error)
```

Now, all these asynchronous Flask components can also await other asynchronous operations such as the repository layer of the application. Indeed, the asynchronous Flask environment is open to integration with asynchronous third-party extension modules such as the async SQLAlchemy 2.x.

Building an asynchronous SQLAlchemy repository layer

The updated `flask-sqlalchemy` extension module supports SQLAlchemy 2.x that provides API utilities, which use the `asyncio` environment with `greenlet` as the main library, allowing propagation of the `await` keyword in the APIs' internal processes. Our `ch05-web` and `ch05-api` projects have the async transactions that call these awaited SQLAlchemy **Create-Read-Update-Delete (CRUD)** operations using a new DB configuration in our projects' `/models/config.py` file that utilizes an `asyncpg` driver to build a session for non-blocking repository transactions.

Setting up the DB connectivity

To start with the configuration, install the `asyncpg` DB driver or dialect that the `asyncio`-driven SQLAlchemy module requires using the `pip` command:

```
pip install asyncpg
```

Also, include the greenlet library in the installation if it is not yet part of the virtual environment:

```
pip install greenlet
```

The setup also requires a **connection string** or **DB URL** for the DB connectivity that includes the installed `asyncpg` protocol, user credentials, the DB server host address, port, and the schema name. Our projects use the connection string, `postgresql+asyncpg://postgres:admin2255@` `localhost:5433/ovs`, to generate the `AsyncEngine` instance of the connection using `create_async_engine()` of the SQLAlchemy framework, the asynchronous version of its `create_engine()` utility. Aside from the DB URL, the method requires its `future` parameter to be set to `True`, `pool_pre_pring` to be set to `True`, and `poolclass` to be set to a connection pooling strategy such as `NullPool`. `poolclass` manages the threads the SQLAlchemy will utilize during its CRUD operations, and setting it to `NullPool` will restrict one Python thread to run only one event loop for one CRUD operation. `pool_pre_ping`, on the other hand, helps connection pooling for the pessimistic approach of handling disconnection. If it determines the DB connection as non-usable or invalid, that connection and its previous ones will be immediately recycled before executing a new operation. Most importantly, the `future` parameter must be set to `True` to enable asynchronous features of SQLAlchemy 2.x, or else the asynchronous SQLAlchemy setup will not work.

After its successful creation, the `sessionmaker` callable will need the `AsyncEngine` instance to instantiate the session that every CRUD operation requires. However, this time, the session will be of the `AsyncSession` type, and the `async_scoped_session` callable will help derive the object with its provided `scopefunc` parameter to manage lightweight thread-local session-based operations in the repository layer. Every repository class will require this `AsyncSession` instance to implement every necessary DB transaction of the asynchronous Flask platform.

Now, there is nothing new with the `declarative_base()` method, for it will still provide the needed helper classes to generate the model classes for the repository layer, like in the standard SQLAlchemy setup. The following is the complete module script of the specified SQLAlchemy setup:

```
from sqlalchemy.ext.asyncio import create_async_engine, AsyncSession,
async_scoped_session
from sqlalchemy.orm import declarative_base, sessionmaker
from sqlalchemy.pool import NullPool
from asyncio import current_task


DB_URL = "postgresql+asyncpg:// postgres:admin2255@localhost:5433/ovs"
engine = create_async_engine(DB_URL, future=True, echo=True, pool_pre_
ping=True, poolclass=NullPool)
db_session = async_scoped_session(sessionmaker(engine, expire_on_
commit=False, class_=AsyncSession), scopefunc=current_task)
Base = declarative_base()
def init_db():
    import app.model.db
```

The `echo` parameter of the given `create_async_engine` enables logging for the `AsyncEngine`-related transactions. Now, the `init_db()` method from the preceding configuration exposes the model classes to the different areas of the application. These model classes, built using the `Base` instance, help auto-generate the table schemas of our DB through the *Flask-Migrate* extension module, which still works with `flask[async]` and `flask-sqlalchemy` integration.

Let us now use the derived `async_scoped_session()` to build repository classes.

Building the asynchronous repository layer

The asynchronous repository layer of the application requires the `AsyncSession` and the model classes to be created in the setup. The following is a `VoterRepository` class implementation that provides CRUD transactions for managing the `Voter` records:

```
from sqlalchemy import update, delete, insert
from sqlalchemy.future import select
from sqlalchemy.orm import Session
from app.model.db import Voter
from datetime import datetime


class VoterRepository:

    def __init__(self, sess:Session):
        self.sess:Session = sess
```

Like in the standard SQLAlchemy repository, all executions are session managed, so the `Session` object is always part of the constructor parameters of the repository class, like in the preceding `VoterRepository`.

Every operation under the `AsyncSession` scope requires an `await` process to finish its execution, which means every repository transaction must be *coroutines*. Every repository transaction requires an event loop to pursue its execution because of the `async/await` design pattern delegated by `AsyncSession`.

The best-fit approach to applying the asynchronous *INSERT* operation is to utilize the `insert()` method from SQLAlchemy utilities. The `insert()` method will establish the *INSERT* command, which `AsyncSession` will *execute*, *commit*, or *roll back* asynchronously. The following is `VoterRepository`'s INSERT transaction:

```
    async def insert_voter(self, voter: Voter) -> bool:
        try:
            sql = insert(Voter).values(mid=voter.mid, precinct=voter.
precinct, voter_id=voter.voter_id, last_vote_date=datetime.
strptime( voter.last_vote_date, '%Y-%m-%d').date())
            await self.sess.execute(sql)
```

```
        await self.sess.commit()
        await self.sess.close()
        return True
    except Exception as e:
        print(e)
    return False
```

As depicted in the preceding snippet, the transaction awaits the `execute()`, `commit()`, and `close()` methods to finish their respective tasks, which is a clear indicator that a repository operation needs to be a coroutine before executing these `AsyncSession` member methods. The same applies to the following UPDATE transaction of the repository:

```
    async def update_voter(self, id:int, details:Dict[str, Any]) ->
bool:
        try:
            sql = update(Voter).where(Voter.id == id).values(**details)
            await self.sess.execute(sql)
            await self.sess.commit()
            await self.sess.close()
            return True
        except Exception as e:
            print(e)
        return False
```

The preceding `update_voter()` also uses the same asynchronous approach as `insert_voter()` using the `AsyncSession` methods. `update_voter()` also needs an event loop from Flask to run successfully as an asynchronous task:

```
    async def delete_voter(self, id:int) -> bool:
        try:
            sql = delete(Voter).where(Voter.id == id)
            await self.sess.execute(sql)
            await self.sess.commit()
            await self.sess.close()
            return True
        except Exception as e:
            print(e)
        return False
```

For the query transactions, the following are the repository's coroutines that implement its SELECT operations:

```python
async def select_all_voter(self):
    sql = select(Voter)
    q = await self.sess.execute(sql)
    records = q.scalars().all()
    await self.sess.close()
    return records
```

Both select_all_voter() and select_voter() use the select() method from the sqlalchemy or sqlalchemy.future module. With the same objective as the insert(), update(), and delete() utilities, the select() method establishes a *SELECT* command object, which requires the asynchronous execute() utility for its execution. Thus, both query implementations are also coroutines:

```python
async def select_voter(self, id:int):
    sql = select(Voter).where(Voter.id == id)
    q = await self.sess.execute(sql)
    record = q.scalars().all()
    await self.sess.close()
    return record
```

In SQLAlchemy, the *INSERT*, *UPDATE*, and *DELETE* transactions technically utilize the model attributes that refer to the primary keys of the models' corresponding DB tables, such as id. Conventionally, SQLAlchemy recommends updating and removing retrieved records based on their **primary keys**. However, there are special cases in *UPDATE* and *DELETE* operations when record searches are based on non-primary keys or arbitrary values, like in the following update_precinct() and delete_voter_by_precinct() of the repository:

```python
async def update_precinct(self, old_prec:str,   new_prec:str) -> bool:
        try:
            sql = update(Voter).where(Voter.precinct == old_prec).
values(precint=new_prec)
            sql.execution_options(synchronize_session= "fetch")
            await self.sess.execute(sql)
            await self.sess.commit()
            await self.sess.close()
            return True
        except Exception as e:
            print(e)
        return False
```

`update_precinct()` searches a `Voter` record with an existing `old_prec` (old precinct) and replaces it with `new_prec` (new precinct). There is no `id` primary key used to search the records for updating. The same scenario is also depicted in `delete_voter_by_precinct()`, which uses the `precinct` non-primary key value for record removal. Both transactions do not conform with the ideal **object-relational mapper** persistence:

```python
async def delete_voter_by_precinct(self, precint:str) -> bool:
    try:
        sql = delete(Voter).where(Voter.precinct == precint)
        sql.execution_options(synchronize_session= "fetch")
        await self.sess.execute(sql)
        await self.sess.commit()
        await self.sess.close()
        return True
    except Exception as e:
        print(e)
    return False
```

In this regard, it is mandatory to perform `execution_options()` to apply the necessary synchronization strategy, preferably the `fetch` strategy, before executing the *UPDATE* and *DELETE* operations that do not conform with the ORM persistence. This mechanism provides the session with the resolution to manage the changes reflected by these two operations. For instance, the `fetch` strategy will let the session retrieve the primary keys of those records retrieved through the arbitrary values and will eventually update the in-memory objects or records affected by the operations and merge them into the actual table records. This setup is essential for the asynchronous SQLAlchemy operations.

After building the repository layer, let us call these CRUD transactions in our view or API functions.

Utilizing the asynchronous DB transactions

To call the repository transactions, the asynchronous view and endpoint functions require an asynchronous context manager to create and manage `AsyncSession` for the repository class. The following is an `add_login()` API function that adds a new `Login` credential to the DB:

```python
from app.model.db import Login
from app.repository.login import LoginRepository
from app.model.config import db_session

@current_app.post('/ch05/login/add')
async def add_login():
    async with db_session() as sess:
        async with sess.begin():
            repo = LoginRepository(sess)
            login_json = request.get_json()
```

```
login = Login(**login_json)
result = await repo.insert_login(login)
if result:
    content = jsonify(login_json)
    return make_response(content, 201)
else:
    abort(500)
```

The view function uses the `async with` context manager to localize the session for the coroutine or task execution. It opens the session for that specific task that will run the `insert_login()` transaction of `LoginRepository`. Then, eventually, the session will be closed by the repository or the context manager itself.

Now, let us focus on another way of running asynchronous transactions using the `asyncio` library.

Implementing async transactions with asyncio

The `asyncio` module is an easy-to-use library for implementing asynchronous tasks. Compared to the `threading` module, the `asyncio` utilities use an event loop to execute each task, which is lightweight and easier to control. Threading uses one whole thread to run one specific operation, while `asyncio` utilizes only a single event loop to run all registered tasks concurrently. Thus, constructing an event loop is more resource friendly than running multiple threads to build concurrent transactions.

`asyncio` is seamlessly compatible with `flask[async]`, and the clear proof is the following API function that adds a new voter to the DB using the task created by the `create_task()` method:

```
from app.model.db import Member
from app.repository.member import MemberRepository
from app.model.config import db_session
from asyncio import create_task, ensure_future, InvalidStateError
from app.exceptions.db import DuplicateRecordException


@current_app.post("/ch05/member/add")
async def add_member():
    async with db_session() as sess:
        async with sess.begin():
            repo = MemberRepository(sess)
            member_json = request.get_json()
            member = Member(**member_json)
            try:
                insert_task = create_task(repo.insert(member))
                await insert_task
                result = insert_task.result()
                if result:
```

```
                    content = jsonify(member_json)
                    return make_response(content, 201)
                else:
                    raise DuplicateRecordException("insert member
record failed")
            except InvalidStateError:
                abort(500)
```

The `create_task()` method requires a coroutine to create a task and schedule its execution in an event loop. So, coroutines are not tasks at all, but they are the core inputs for generating these tasks. Running the scheduled task requires the `await` keyword. After its execution, the task returns a `Future` object that requires the task's `result()` built-in method to retrieve its actual returned value. The given API transaction creates an *INSERT* task from the `insert_login()` coroutine and retrieves a `bool` result after execution.

Now, `create_task()` automatically utilizes Flask's internal event loop in running its tasks. However, for complex cases such as executing scheduled tasks, `get_event_loop()` or `get_running_loop()` are more applicable to utilize than `create_task()` due to their flexible settings. `get_event_loop()` gets the current running event loop, while `get_running_loop()` uses the running event in the current system's thread.

Another way of creating tasks from the coroutine is through `asyncio`'s `ensure_future()`. The following API uses this utility to spawn a task that lists all user accounts:

```
@current_app.get("/ch05/member/list/all")
async def list_all_member():
    async with db_session() as sess:
        async with sess.begin():
            repo = MemberRepository(sess)
            list_member_task = ensure_future(repo.select_all_member())
            await list_member_task
            records = list_member_task.result()
            member_rec = [rec.to_json() for rec in records]
            return make_response(member_rec, 201)
```

The only difference between `create_task()` and `ensure_future()` is that the former strictly requires coroutines, while the latter can accept coroutines, `Future`, or any awaitable objects. `ensure_future()` also invokes `create_task()` to wrap a `coroutine()` argument or directly return a `Future` result from a `Future` parameter object.

On the other hand, `flask[async]` supports creating and running multiple tasks concurrently using `asyncio`. Its `gather()` method has two parameters:

- The first parameter is the sequence of coroutines, `Future`, or any awaitable objects.
- The second parameter is `return_exceptions`, which is set to `False` by default.

The following is an endpoint function that inserts multiple profiles of candidates using concurrent tasks:

```
@current_app.post('/ch05/candidates/party')
async def add_list_candidates():
    candidates = request.get_json()
    count_rec_added = 0
    results = await gather( *[insert_candidate_task(data) for data in
candidates])
    for success in results:
        if success:
            count_rec_added = count_rec_added  + 1
    return jsonify(message=f'there are {count_rec_added} newly added
candidates'), 201
```

The given API expects a list of candidate profile details from `request`. A service named `insert_candidate_task()` will create a task that will convert the dictionary of objects to a `Candidate` instance and add the model instance to the DB through the `insert_candidate()` transaction of `CandidateRepository`. The following code showcases the complete implementation of this service task:

```
from asyncio import create_task
... ... ... ... ... ...
async def insert_candidate_task(data):
    async with db_session() as sess:
        async with sess.begin():
            repo = CandidateRepository(sess)
            insert_task = create_task(repo.insert_
candidate(    Candidate(**data)))
            await insert_task
            result = insert_task.result()
            return result
```

Since our SQLAlchemy connection pooling is `NullPool`, which means connection pooling is disabled, we cannot utilize the same `AsyncSession` for all the `insert_candidate()` transactions. Otherwise, `gather()` will throw `RuntimeError` object. Thus, each `insert_candidate_task()` will open a new localized session for every `insert_candidate()` task execution. To add connection pooling, replace `NullPool` with `QueuePool`, `AsyncAdaptedQueuePool`, or `SingletonThreadPool`.

Now, the `await` keyword will concurrently run the sequence of tasks registered in `gather()` and propagate all results in the resulting `tuple` of `Future` once these tasks have finished their execution successfully. The order of these `Future` objects is the same as the sequence of the awaitable objects provided in `gather()`. If a task has encountered failure or exception, it will not throw any exception and pre-empt the other task execution because `return_exceptions` of `gather()` is `False`. Instead, the failed task will join as a typical awaitable object in the resulting `tuple`.

By the way, the given `add_list_candidates()` API function will return the number of successful INSERT tasks that persisted in the candidate profiles.

The next section will discuss how to de-couple Flask components using the event-driven behavior of Flask **signals**.

Utilizing asynchronous signal notifications

Flask has a built-in lightweight event-driven mechanism called signals that can establish a loosely coupled software architecture using subscription-based event handling. It can trigger single or multiple transactions depending on the purpose. The `blinker` module provides the building blocks for Flask signal utilities, so install `blinker` using the `pip` command if it is not yet in the virtual environment.

Flask has built-in signals and listens to many Flask events and callbacks such as `render_template()`, `before_request()`, and `after_request()`. These signals, such as `request_started`, `request_finished`, `message_flashed`, and `template_rendered`, are found in the `flask` module. For instance, once a component connects to `template_rendered`, it will run its callback method after `render_template()` finishes posting a Jinja template. However, our target is to create custom *asynchronous signals*.

To create custom signals, import the `Namespace` class from the `flask.signals` module and instantiate it. Use its instance to define and instantiate specific custom signals, each having a unique name. The following is a snippet from our applications that creates an event signal for election date verification and another for retrieving all the election details:

```
from flask.signals import Namespace

election_ns = Namespace()
check_election = election_ns.signal('check_election')
list_elections = election_ns.signal('list_elections')
```

Each named signal must have an assigned function or event, either asynchronous or standard, that will serve as their implementation. `check_election_event`, for instance, has the following asynchronous method that uses `ElectionRepository` to verify an election date:

```
@check_election.connect
async def check_election_event(app, election_date):
    async with db_session() as sess:
        async with sess.begin():
            repo = ElectionRepository(sess)
            records = await repo.select_all_election()
            election_rec = [rec.to_json() for rec in records if rec.
election_date == datetime.strptime(election_date, '%Y-%m-%d').date()]
            if len(election_rec) > 0:
```

```
            return True
        return False
```

Meanwhile, our `list_all_election()` API endpoint has the following `list_elections_event()` that returns a list of records in JSON format:

```
@list_elections.connect
async def list_elections_event(app):
    async with db_session() as sess:
        async with sess.begin():
            repo = ElectionRepository(sess)
            records = await repo.select_all_election()
            election_rec = [rec.to_json() for rec in records]
            return election_rec
```

Event or signal functions must accept a *sender* or *listener* as the first local parameter argument, followed by the other custom `args` objects essential to the event transaction. If the event mechanism is part of the class scope, the value of the function must be `self` or the class instance itself. Otherwise, if the signal is for a global event handling, its first argument must be the Flask `app` instance.

A signal has a `connect()` function or decorator that registers an event or function as its implementation. These events will execute once a caller emits the signals. Flask components can emit signals by invoking the signal's `send()` or `send_async()` utility with the event function arguments. The following `verify_election()` endpoint checks from the DB through the `check_election` signal if an election happens on a particular date:

```
@current_app.post('/ch05/election/verify')
async def verify_election():
    election_json = request.get_json()
    election_date = election_json['election_date']
    result_tuple = await check_election.send_async(current_app,
election_date=election_date)
    isApproved = result_tuple[0][1]
    if isApproved:
        return jsonify(message=f'election for {election_date} is
approved'), 201
    else:
        return jsonify(message=f'election for {election_date} is
disabled'), 201
```

If the event function is a standard Python function, send the notification for its execution through the signal's `send()` method. However, if it is an asynchronous method, like in our case, use `send_async()` to create and run the task for the coroutine with `await` to extract its `Future` value.

Generally, signals can employ the de-coupling of components in a scalable application to reduce dependencies and improve modularity and maintainability. This can also help build applications to have a distributed architecture design. However, as the requirements become complicated and the subscribers of the signals become numerous, the notifications can slow down the performance of the whole application. So, it is a good design if the caller and the event function can lessen the dependencies on each other's parameters, returned values, and conditions. The subscribers must have an independent scope as to the event functions. Also, it is a good programming approach to create an event function that is flexible and not too narrow in its objectives so that many components can subscribe to it.

After exploring how Flask supports event handling using its signals, let us now learn how to create background processes using its platform.

Constructing background tasks with Celery and Redis

It is impossible to create background processes or transactions in Flask using its `flask [async]` platform. The event loop that runs tasks for the asynchronous view or endpoint will not allow the spawning of another event loop that will cater to background tasks because it cannot wait for the background processes to finish once the view or endpoint finishes its processing. However, with some third-party components, such as task queues, background processing is feasible for the Flask platform.

One of the solutions is to use Celery, which is an asynchronous task queue that can run processes outside the context of the application. So, while the event loop is running the view or endpoint coroutines, they can entrust to Celery the management of the background transactions.

Setting up the Celery task queue

There are a few considerations when writing the background processes with Celery, and the first is to install the `celery` extension module using the `pip` command:

```
pip install celery
```

Then, we designate some local workers in the WSGI server to run tasks with the background jobs in the Celery queue, but in our application, our Flask server will only use a single worker to run all the processes.

Let us now install the Redis server, which will serve as the message broker to the Celery.

Installing the Redis DB

After designating the worker, Celery requires a message broker for its workers to communicate with the client application about running the background jobs. Our applications use the Redis DB as the broker. So, install Redis in Windows using the **Windows Subsystem for Linux (WSL2)** shell or by downloading the Windows installer at `https://github.com/microsoftarchive/redis/releases`.

The next step is to add the necessary Celery configuration variables, including `CELERY_BROKER_URL`, to the `app` instance.

Setting up the Celery client configuration

Since our projects use TOML files for setting the configuration environment variables, Celery will fetch all its configuration details from these files as TOML variables. The following is a snapshot of the `config_dev.toml` file that contains Celery setup variables:

```
CELERY_BROKER_URL = "redis://127.0.0.1:6379/0"
CELERY_RESULT_BACKEND = "redis://127.0.0.1:6379/0"

[CELERY]
celery_store_errors_even_if_ignored = true
task_create_missing_queues = true
task_store_errors_even_if_ignored = true
task_ignore_result = false
broker_connection_retry_on_startup = true
celery_task_serializer = "pickle"
celery_result_serializer = "pickle"
celery_event_serializer = "json"
celery_accept_content = ["pickle", "application/json", "application/x-
python-serialize"]
celery_result_accept_content = ["pickle", "application/json",
"application/x-python-serialize"]
```

The two most important variables needed by the Celery client module are `CELERY_BROKER_URL` and `CELERY_RESULT_BACKEND`, which provide the address, port, and DB name of the Redis broker and backend server, respectively. Redis has DBs 0 to 15, but our application utilizes only DB 0 for default purposes. Since the `CELERY_RESULT_BACKEND` is not that important in this setup, setting `CELERY_RESULT_BACKEND` as the defined broker URL or removing it from the configuration is acceptable.

Then, create the `CELERY` TOML dictionary to contain the details needed by the Celery instance in managing the background task executions. First, `celery_store_errors_even_if_ignored` and `task_store_errors_even_if_ignored` must be `True` to enable audit trail features for logging errors during Celery execution. `broker_connection_retry_on_startup` should be `True` in case Redis is still in shutdown mode. On the other hand, `task_ignore_result` must be `False` since some of our coroutine jobs will be returning some values to the caller. Moreover, `task_create_missing_queues` is set to `True` in case there are undefined task queues that the application can utilize during traffic. By the way, the default task queue's name is `celery`.

Other details are about the mime-type of resources that tasks can accept for their coroutines (`celery_accept_content`) and the returned values that these background processes can return to the invoker (`celery_result_accept_content`). The task serializers are also part of the details because they are the mechanisms that convert the task's incoming arguments and returning values to be in their acceptable state and valid mime-type types.

Now, let us focus on building the Celery client modules of our projects, starting with the instantiation of the Celery instance.

Creating the Client instance

Since all projects in this chapter use the application factory approach, the setup recognizing the application as a Celery client happens in `app/__init__.py`. However, the exact `Celery` class instantiation occurs in another module, `celery_config.py`, to avoid circular import errors. The following snippet shows the instantiation of the `Celery` class in `celery_config.py`:

```
from celery import Celery, Task
from flask import Flask

def celery_init_app(app: Flask) -> Celery:
    class FlaskTask(Task):
        def __call__(self, *args: object, **kwargs: object) -> object:
            with app.app_context():
                return self.run(*args, **kwargs)
    celery_app = Celery(app.name, task_cls=FlaskTask, broker=app.
config["CELERY_BROKER_URL"], backend=app.config["CELERY_RESULT_
BACKEND"])

    celery_app.config_from_object(app.config["CELERY"])
    celery_app.set_default()
    return celery_app
```

From the preceding snippet, the instantiation of the `Celery` class strictly requires the Celery application name, `CELERY_BROKER_URL`, and the worker task. The first parameter, the Celery application name, can have any prescribed name or just the Flask app's name since the Celery client module will run background jobs (`FlaskTask`) in the app's thread.

After instantiating the Celery, the Celery instance, `celery_app`, needs to load the CELERY TOML dictionary from the Flask app to configure the task queue and its message broker. Lastly, `celery_app` must invoke `set_default()` to seal the configuration. Now, `app/__init__.py` will import the `celery_init_app()` factory to eventually pursue the creation of the Celery client out of the Flask application.

Let us now build the Celery client module with custom tasks.

Implementing the Celery tasks

To avoid circular import problems, it is not advisable to import `celery_app` and use it to decorate functions with the `task()` decorator. The `shared_task()` decorator from the `celery` module is enough proxy to define functions as Celery tasks. Here is a Celery task that adds a new vote to a candidate:

```python
from celery import shared_task
from asyncio import run


@shared_task
def add_vote_task_wrapper(details):
    async def add_vote_task(details):
        try:
            async with db_session() as sess:
                async with sess.begin():
                    repo = VoteRepository(sess)
                    details_dict = loads(details)
                    print(details_dict)
                    election = Vote(**details_dict)
                    result = await repo.insert(election)
                    if result:
                        return str(True)
                    else:
                        return str(False)
        except Exception as e:
            print(e)
            return str(False)
    return run(add_vote_task(details))
```

Now, a Celery task, such as the one created by `add_vote_task_wrapper()`, must not be a coroutine. A Celery task is a class generated by any callable decorated by `@shared_task`, which means it cannot propagate the `await` keyword outwards with the `async` function call. However, it can enclose an asynchronous local method to handle all the operations asynchronously, such as `add_vote_task()`, which wraps and executes the INSERT transactions for new vote details. The Celery task can apply the `asyncio`'s `run()` utility method to run its async local function.

Since our Celery app does not ignore the result, our task returns a Boolean value converted into a string, a safe object type that a task can return to the caller. Although it is feasible to use pickling, through the `pickle` module, to pass an argument to or transport return values from Celery tasks to the callers, it might open vulnerabilities that can pose security risks to the application, such as accidentally exposing confidential information stored in the pickled object or unpickling/de-serializing malicious objects.

Another approach to manage the Celery task's input arguments and returned values, especially if they are collection types, is through the `loads()` and `dumps()` utilities of the `json` module. This `loads()` function deserializes a JSON string into a Python object while `dumps()` serializes Python objects (e.g., dictionaries, lists, etc.) into a JSON formatted string. However, sometimes, using `dumps()` to convert these objects to strings is not certain. There are data in the string payload that can cause serialization error, because Celery does not support their default format, such as `time`, `date`, and `datetime`. In this scenario, the `dumps()` method needs a custom serializer to convert these temporal data types to their equivalent *ISO 8601* formats. The following Celery task has the same problem, thus the presence of a custom `json_date_serializer()`:

```python
@shared_task
def list_all_votes_task_wrapper():
    async def list_all_votes_task():
      async with db_session() as sess:
        async with sess.begin():
            repo = VoteRepository(sess)
            records = await repo.select_all_vote()
            vote_rec = [rec.to_json() for rec in records]
            return dumps(vote_rec, default=json_date_serializer)
    return run(list_all_votes_task())


def json_date_serializer(obj):
    if isinstance(obj, time):
        return obj.isoformat()
    raise TypeError ("Type %s not …" % type(obj))
```

Among the many ways to implement a date serializer, `json_date_serializer()` uses the `time`'s `isoformat()` method to convert the time object to an *ISO 8601* or *HH:MM:SS:ssssss* formatted string value so that the task can return the list of vote records without conflicts on the `date` types.

Running the Celery worker server

After creating the Celery tasks, the next step is to run the built-in Celery server through the following command to check whether the server can recognize them:

```
celery -A main.celery_app worker --loglevel=info -P solo
```

main in the command is the `main.py` module, and `celery_app` is the Celery instance found in the `main.py` module. The `loglevel` option creates a console logger for the server, and the `P` option indicates the *concurrency pool*, which is `solo` in the given command. *Figure 5.1* shows the screen details after the server started.

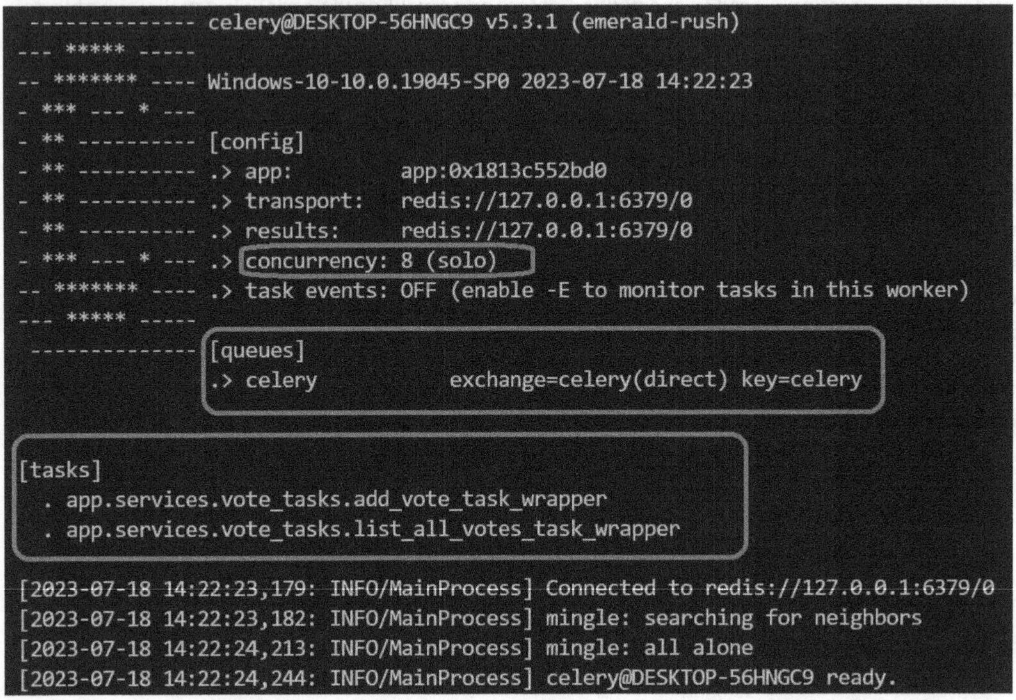

```
-------------- celery@DESKTOP-56HNGC9 v5.3.1 (emerald-rush)
--- ***** -----
-- ******* ---- Windows-10-10.0.19045-SP0 2023-07-18 14:22:23
- *** --- * ---
- ** ---------- [config]
- ** ---------- .> app:         app:0x1813c552bd0
- ** ---------- .> transport:   redis://127.0.0.1:6379/0
- ** ---------- .> results:     redis://127.0.0.1:6379/0
- *** --- * --- .> concurrency: 8 (solo)
-- ******* ---- .> task events: OFF (enable -E to monitor tasks in this worker)
--- ***** -----
-------------- [queues]
               .> celery            exchange=celery(direct) key=celery


[tasks]
  . app.services.vote_tasks.add_vote_task_wrapper
  . app.services.vote_tasks.list_all_votes_task_wrapper

[2023-07-18 14:22:23,179: INFO/MainProcess] Connected to redis://127.0.0.1:6379/0
[2023-07-18 14:22:23,182: INFO/MainProcess] mingle: searching for neighbors
[2023-07-18 14:22:24,213: INFO/MainProcess] mingle: all alone
[2023-07-18 14:22:24,244: INFO/MainProcess] celery@DESKTOP-56HNGC9 ready.
```

Figure 5.1 – Server details after Celery server startup

Celery server fetched the `add_vote_task_wrapper()` and `list_all_votes_task_wrapper()` tasks, as indicated in *Figure 5.1*. Thus, Flask views and endpoints can now use these tasks to cast and view the votes from users. Aside from the list of ready-to-use tasks, the server logs also show details of the default task queue, `celery`. Also, it indicates the concurrency pool type, which is `solo`, and has a concurrency worker limit of 8. Among the `prefork`, `eventlet`, `gevent`, and `solo` concurrency options, our applications use `solo` and `eventlet`. However, to use `eventlet`, install the `eventlet` module using the `pip` command:

```
pip install eventlet
```

Our application uses the solo Celery execution pool because it runs within the worker process, which makes a task's performance fast. This pool is fit for running resource-intensive tasks. Other better options are eventlet and gevent, which spawn greenlets, sometimes called green threads, cooperative threads, or coroutines. Most Input/Output-bound tasks run better with eventlet or gevent because they generate more threads and emulate a multi-threading environment for efficiency.

Once the Celery server loads and recognizes the tasks with a worker managing the message queues, Flask view and endpoint functions can invoke the tasks now using Celery utility methods.

Utilizing the Celery tasks

Once the Celery worker server runs with the list of tasks, Flask's async views and endpoints can now access and run these tasks like signals. These tasks will execute only when the caller invokes their built-in delay() or apply_async() methods. The following endpoint function runs add_vote_task_wrapper() to cast a vote for a user:

```
@current_app.post('/ch05/vote/add')
async def add_vote():
    vote_json = request.get_json()
    vote_str = dumps(vote_json)
    task = add_vote_task_wrapper.apply_async(args=[vote_str])
    result = task.get()
    return jsonify(message=result), 201
```

The given add_vote() endpoint retrieves the request JSON data and converts it to a string before passing it as an argument to add_vote_task_wrapper(). Without using the await keyword, the Celery task has apply_async(), which the invoker can use to trigger its execution with the argument. apply_async() returns an AsyncResult object with a get() method that returns the returned value, if any. It also has a traceback variable that retrieves an exception stack trace when the execution raises an exception.

From creating asynchronous background tasks, let us move on to WebSocket implementation with asynchronous transactions.

Building WebSockets with asynchronous transactions

WebSocket is a well-known bi-directional communication between a server and browser-based clients. Many popular frameworks such as Spring, JSF, Jakarta EE, Django, FastAPI, Angular, and React support this technology, and Flask is one of them. However, this chapter will focus on implementing WebSocket and its client applications using the asynchronous paradigm.

Creating the client-side application

Our WebSocket implementation with the client-side application is in the ch05-web project. Calling /ch05/votecount/add from the vote_count.py view module will give us the following HTML form in *Figure 5.2*, which handles the data entry for the final vote tally per precinct or election district:

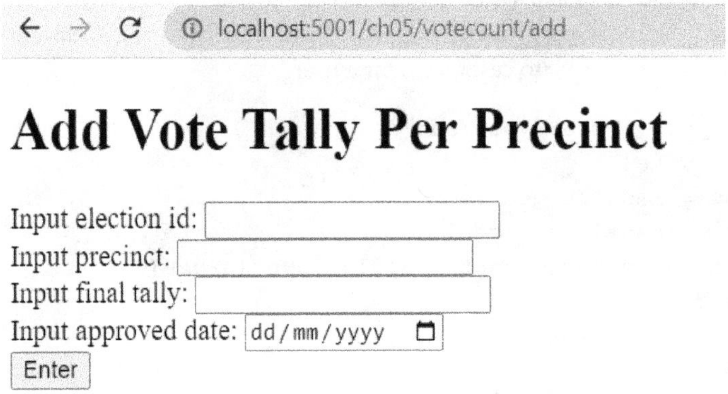

Figure 5.2 – Client-side application for adding final vote counts

Our WebSocket captures election data from officers and then updates DB records in real time. It retrieves a string message from the server as a response. The HTML form and the **JavaScript (JS)** implementation of WebSocket are in pages/vote_count_add.html of ch05-web. The following snippet is the JS code that communicates with our server-side WebSocket:

```
<script>
    const add_log = (message) => {
        document.getElementById('add_log').innerHTML +=
`<span>${message}</span><br>`;
    };


    const socket = new WebSocket('ws://' + location.host + '/ch05/
vote/save/ws');
    socket.addEventListener('message', msg => {
        add_log('server: ' + msg.data);
    });
```

```javascript
        document.getElementById('vote_form').onsubmit = data => {
            data.preventDefault();
            const election_id = document.getElementById('election_
id');
            const precinct = document.getElementById('precinct');
            const final_tally = document.getElementById('final_
tally');
            const approved_date = document.getElementById('approved_
date');

            var vote_count = new Object();
            vote_count.election_id = election_id.value;
            vote_count.precinct   = precinct.value;
            vote_count.final_tally = final_tally.value;
            vote_count.approved_date = approved_date.value;
            var vote_count_json = JSON.stringify(vote_count);

            add_log('client: ' + vote_count_json);
            socket.send(vote_count_json);
            election_id.value = '';
            precinct.value = '';
            final_tally.value = '';
            approved_date.value = '';
        };
    </script>
```

The preceding JS script will connect to the Flask server through `ws://localhost:5001/ch05/vote/save/ws` by instantiating the `WebSocket` API. When the connection is ready, the client can ask for vote details from the client through the form components. Submitting the data will create a JSON object out of the form data before sending the JSON formatted details to the server through the `WebSocket` connection.

On the other hand, to capture the message from the server, the client must create a listener to the message emitter by calling the WebSocket's `addEventListener()`, which will watch and retrieve any JSON message from the Flask server. The custom `add_log()` function will render the message to the front end using the `<span>` tag.

Next, let us focus on the WebSocket implementation per se using the `flask-sock` module.

Creating server-side transactions

There are many ways to implement a server-side message emitter, such as WebSocket, in Flask, and many Flask extensions can provide support for it, such as flask-socketio, flask-sockets, and flask-sock. This chapter will use the flask-sock module to create WebSocket routes because it can implement WebSocket communication with minimal configuration and setup. So, to start, install the flask-sock extension using the pip command:

```
pip install flask-sock
```

Then, integrate the extension to Flask by instantiating the Sock class with the app instance as its required argument. The following app/__init__.py snippet shows the flask-sock setup:

```
from flask_sock import Sock
sock = Sock()

def create_app(config_file):
    app = Flask(__name__, template_folder='../app/pages', static_
folder="../app/resources")
    app.config.from_file(config_file, toml.load)
    init_db()
    sock.init_app(app)
```

After that simple configuration, import the sock instance in /api/ votecount_websocket. py module to define the WebSocket routes. ws://localhost:5001/ch05/vote/save/ ws, which was invoked by the preceding JS code, has the following route implementation:

```
from app import sock

@sock.route('/ch05/vote/save/ws')
def add_vote_count_server(ws):
    async def add_vote_count():
        while True:
            vote_count_json = ws.receive()
            vote_count_dict = loads(vote_count_json)
            async with db_session() as sess:
                repo = VoteCountRepository(sess)
                vote_count = VoteCount(**vote_count_dict)
                result = await repo.insert(vote_count)
                if result:
                    ws.send("data added")
                else:
                    ws.send("data not added")
    run(add_vote_count())
```

The `Sock` instance has a `route()` decorator that defines WebSocket implementation. WebSocket route function or handler is always non-asynchronous with a required parameter that accepts an injected WebSocket object from `Sock`. This `ws` object has a `send()` method that emits data to the client application, a `receive()` utility that accepts messages from the client, and `close()` to employ forced disconnection of the two-way communication when runtime exceptions or server-related problems occur.

The WebSocket handler usually holds an *open loop process* where it can receive a message first through `receive()` and then emit its message using `send()` continuously, depending on the purpose of the messaging.

In the case of `add_vote_count_server()`, which needs to await asynchronous `VoteCountRepository`'s INSERT transaction, an `async` local method similar to the Celery task must be present inside the WebSocket route function. This local method will encase the asynchronous operations, and the `asyncio`'s `run()` will execute it inside the route function.

Now, to witness the exchange of messages, *Figure 5.3* shows a snapshot of the communication between our JS client and the `add_vote_count_server()` handler at runtime:

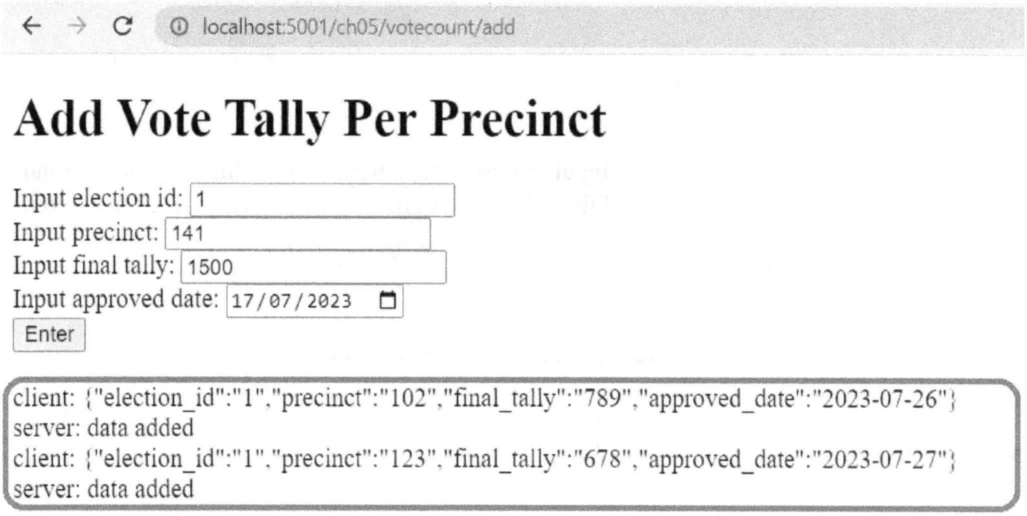

Figure 5.3 – A message exchange between a JS client and flask-sock WebSocket

Aside from web-based clients, WebSocket can also propagate or send data to API clients.

Creating a Flask API client application

Another way to connect through WebSocket emitters is through Flask components, not JS codes. Sometimes, the client applications are not web components composed of HTML, CSS, and frontend JS frameworks that support WebSocket communication. For instance, in our ch05-api project, a POST API function, bulk_check_vote_count(), asks for a list of candidates to count the votes they have during the election. The input to the API is a JSON string, such as the following sample data:

```
[
    {
        "election_id": 1,
        "cand_id": "PHL-101"
    },
    {
        "election_id": 1,
        "cand_id": "PHL-111"
    },
    {
        "election_id": 1,
        "cand_id": "PHL-005"
    }
]
```

Then, the API function converts this JSON input to a list of dictionaries containing the candidate and election IDs. Here is the implementation of this API function that serves as a client to a WebSocket:

```
from simple_websocket import Client
from json import dumps


@current_app.post("/ch05/check/vote/counts/client")
def bulk_check_vote_count():
    ws = Client('ws://127.0.0.1:5000/ch05/check/vote/counts/
ws', headers={"Access-Control-Allow-Origin": "*"})
    candidates = request.get_json()
    for candidate in candidates:
            try:
                    print(f'client sent: {candidate}')
                    ws.send(dumps(candidate))
                    vote_count = ws.receive()
                    print(f'client recieved: {vote_count}')
            except Exception as e:
                    print(e)
    return jsonify(message="done client transaction"), 201
```

Since the most compatible WebSocket client extension for `flask-sock` is `simple-websocket`, install this module using the `pip` command:

```
pip install simple-websocket
```

Instantiate the `Client` class from the `simple-websocket` module to connect to the `flask-sock` WebSocket emitter with `Access-Control-Allow-Origin` to allow cross-origin access. Then, the API will send the dictionary-converted-to-string details to the emitter using the `Client`'s `send()` method.

On the other hand, the WebSocket route that will receive the election details from the `bulk_check_vote_count()` client API has the following implementation:

```python
@sock.route("/ch05/check/vote/counts/ws")
def bulk_check_vote_count_ws(websocket):
    async def vote_count():
        While True:
            try:
                candidate = websocket.receive()
                candidate_map = loads(candidate)
                print(f'server received: {candidate_map}')
                async with db_session() as sess:
                    async with sess.begin():
                        repo = VoteRepository(sess)
                        count = await repo.count_votes_by_candidate( candidate_
map["cand_id"], int(candidate_map["election_id"]))
                        vote_count_data = {"cand_id": candidate_map["cand_
id"], "vote_count": count}
                        websocket.send(dumps(vote_count_data))
                        print(f'server sent: {candidate_map}')
            except  Exception as e:
                print(e)
                break
    run(vote_count())
```

Similar to the preceding implementation, our WebSocket route uses `run()` from `asyncio` to execute asynchronous query transactions from `VoteRepository` and extract the total number of votes for each candidate sent by the API client. The emitter will send a newly formed dictionary containing the candidate's ID and counted votes back to the client API in string format. So, the handshake in this setup is between two Flask components, the WebSocket route and an async Flask API.

There are other client-server interactions that `flask[async]` can build, and one of these is the SSE.

Implementing asynchronous SSE

Like the WebSocket, the SSE is a real-time mechanism for sending messages from the server to client applications. However, unlike the WebSocket, it establishes unidirectional communication between the server and client applications.

There are many ways to build server push solutions in Flask, but our applications prefer using the built-in response's `text/event-stream`.

Implementing the message publisher

SSE is a *server push* solution that requires an input source where it can listen for incoming data or messages in real time and push that data to its client applications. One of the reliable sources that will work with SSE is a **message broker**, which can store messages from various resources. It can also help the SSE generator function to listen for incoming messages before yielding them to the clients.

In this chapter, our `ch05-web` application utilizes Redis as the broker, which our `ch05-api` project used for invoking the Celery background tasks. However, in this scenario, there is a need to create a Redis client application that will implement its publisher-subscribe pattern. So, install the *redis-py* extension by using the `pip` command:

```
pip install redis
```

This extension will provide us with the `Redis` client that will connect to the Redis server once instantiated in the `main` module. The following `main.py` snippet shows the setup of the Redis client application:

```
from app import create_app
from redis import Redis

app = create_app('../config_dev.toml')
redis_conn = Redis(
    db = 0,
    host='127.0.0.1',
    port=6379,
    decode_responses=True
)
```

The Redis callable requires details about the DB (db), port, and host address of the installed Redis server as its parameters for setup. Since Celery tasks can return bytes, the Redis constructor should set its decode_response parameter to True to enable binary message data decoding mechanism and receive decoded strings. The instance, redis_conn, will be the key to the message publisher implementation needed by the SSE. In the complaint module of the application, our input source is a form view function that requests the user its statement and voter's ID before pushing these details to the Redis broker. The following is the view that publishes data to the Redis server:

```python
from main import redis_conn
from json import dumps


@current_app.route('/ch05/election/complaint/form', methods =
['GET','POST'])
async def create_complaint():
    if request.method == "GET":
        return render_template('complaint_form.html')
    else:
        voter_id = request.form['voter_id']
        complaint = request.form['complaint']
        record = {'voter_id': voter_id, 'complaint': complaint}
        redis_conn.publish("complaint_channel", dumps(record))
        return render_template('complaint_form.html')
```

The Redis client instance, redis_conn, has a publish() method that stores a message to Redis under a specific topic or channel, a point where a subscriber will fetch the message from the broker. The name of our Redis channel is complaint_channel.

Building the server push

Our SSE will be the subscriber to complaint_channel. It will create a subscriber object first, through redis_conn's pubsub() method, to connect to Redis and eventually use the broker to listen for any published message from the form view. The following is our SSE implementation using the async Flask route:

```python
@current_app.route('/ch05/elec/comaplaint/stream')
async def elec_complaint_sse():
    def process_complaint_event():
        connection = redis_conn.pubsub()
        connection.subscribe('complaint_channel')
        for message in connection.listen():
            time.sleep(1)
            if message is not None and message['type'] == 'message':
                data = message['data']
```

```
        yield 'data: %s\n\n' % data
    return Response(process_complaint_event(), mimetype="text/event-
stream")
```

`process_complaint_event()` in the given SSE route is the *generator function* that creates the subscriber object (`connection`), connects to Redis by invoking the `subscribe()` method, and builds an open loop transaction that will listen continuously from the broker for currently published messages. The message it retrieves from the `listen()` utility of the subscriber object is a JSON entity containing details about the message type, channel, and the `data` published by the form view publisher. `elec_complaint_sse()` only needs to yield the `data` portion of the message. Now, running the `process_complaint_event()` generator requires the SSE route to return Flask's `Response`, which will execute and render it as a `text/event-stream` type object. *Figure 5.4* shows the form view catering to the voters for their complaints:

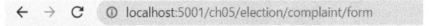

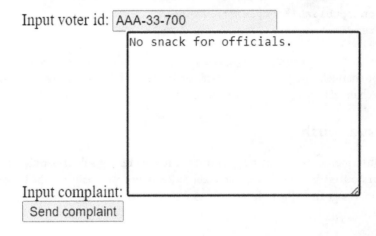

Election Complaint Desk

Input voter id: AAA-33-700

No snack for officials.

Input complaint:
Send complaint

Figure 5.4 – A complaint form view of the published data to Redis

Figure 5.5 provides a snapshot of the SSE client page with the pushed messages from the Redis broker.

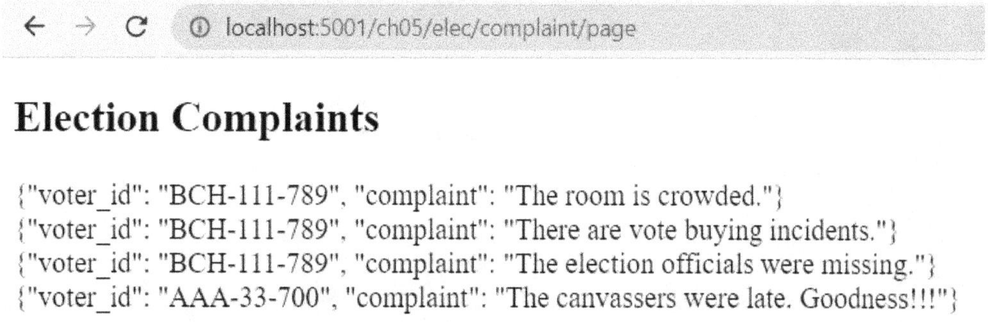

← → C   ⓘ localhost:5001/ch05/elec/complaint/page

Election Complaints

{"voter_id": "BCH-111-789", "complaint": "The room is crowded."}
{"voter_id": "BCH-111-789", "complaint": "There are vote buying incidents."}
{"voter_id": "BCH-111-789", "complaint": "The election officials were missing."}
{"voter_id": "AAA-33-700", "complaint": "The canvassers were late. Goodness!!!"}

Figure 5.5 – An SSE client page rendering pushed data from Redis

Aside from broker messaging, Flask supports other libraries that use publisher-subscriber design patterns in creating its components. The next subject will showcase one of them, the reactivex module.

Applying reactive programming with RxPy

Reactive programming is one of the emerging popular programming paradigms nowadays that focuses on asynchronous data streams and operations that can manage executions, events, repositories, and exception propagations. It utilizes the publisher-subscriber programming approach, which builds asynchronous interactions between software components and transactions.

The library used to apply reactive streams to build services transactions and API functions in this chapter is reactivex, so install the module using the pip command:

```
pip install reactivex
```

The reactivex module has an Observable class that generates data sources for the subscribers to consume. Observer is another API class that pertains to the subscriber entities. reactivex will not be a complete reactive programming library without its *operators*. The following is a vote-counting service implementation that uses the reactivex utilities:

```
from reactivex import Observable, Observer, create
from reactivex.disposable import Disposable
from asyncio import ensure_future

async def extract_precinct_tally(rec_dict):
    del rec_dict['id']
    del rec_dict['election_id']
    del rec_dict['approved_date']
    return str(rec_dict)
```

```python
async def create_tally_data(observer):
    async with db_session() as sess:
        async with sess.begin():
            repo = VoteCountRepository(sess)
            records = await repo.select_all_votecount()
            votecount_rec = [rec.to_json() for rec in records]
            print(votecount_rec)
            for vc in votecount_rec:
                rec_str = await extract_precinct_tally(vc)
                observer.on_next(rec_str)
            observer.on_completed()

def create_observable(loop) -> Observable:
    def on_subscribe(observer: Observer, scheduler):
        task = ensure_future(create_tally_data(observer), loop=loop)
        return Disposable(lambda: task.cancel())
    return create(on_subscribe)
```

All repository transactions involved in the process are asynchronous, which makes the `create_tally_data()` and `extract_precinct_tally()` service operations that utilize these `async` queries also asynchronous. The objective is not to call these `async` services directly from the API layer but to wrap these service transactions in one `Observable` object through `create_observable()` and let the API functions subscribe to it. However, the problem is that `create_observable()` can't be `async` because `reactivex` does not allow `async` to deal with its operators such as `create()`, `from_iterable()`, and `from_list()`.

With that, the `create_observable()` custom function needs a local-scoped subscriber function, `on_subscribe()`, that will invoke `create_task()` or `ensure_future()` with an event loop to create a task for the `create_tally_data()` coroutine and return it as a `Disposable` resource object. A disposable resource link allows for the cleaning up of the resources used by the observable operators during the subscription. Creating the `async` subscriber disposable will help manage the Flask resources.

In connection with this setup, `create_tally_data()` will now emit the vote counts from the repository to the observer or subscriber. The only goal now of `create_observable()` is to return its created `Observable` based on the `on_subscribe()` emissions.

The API transaction needs to run the `create_tally_date()` service and extract all the emitted vote counts by invoking `create_observable()` and subscribing to its returned `Observable` through the `subscribe()` method. The following is the `list_votecount_tally()` endpoint function that creates a subscription to the returned `Observable`:

```python
from app.services.vote_count import create_observable
from asyncio import get_event_loop, Future
```

```
@current_app.get("/ch05/votecount/tally")
async def list_votecount_tally():
    finished = Future()
    loop = get_event_loop()

    def on_completed():
        finished.set_result(0)

    tally = []
    disposable = create_observable(loop).subscribe(
                    on_next = lambda i: tally.append(i),
                    on_error = lambda e: print("Error Occurred: {0}".
format(e)),
                    on_completed = on_completed)
    await finished
    disposable.dispose()
    return jsonify(tally=tally), 201
```

subscribe() has three callback methods that are all active and ready to run anytime when triggered:

- on_next(): This executes when Observer receives emitted data.
- on_error(): This executes when Observable encounters an exception along its operators.
- on_completed(): This runs when Observable completes its task.

Our on_next() callback adds all the emitted data to the tally list.

Now, the execution of the Observable operations will not be possible without the event loop. The API function needs the currently running event loop for the create_tally_data() coroutine execution, and thus its get_event_loop() invocation. The API will return the tally list once it disposes of the task running in Observable.

Even though our framework is asynchronous Flask or the solutions applied to our applications are reactive and asynchronous, Flask will remain a WSGI-based framework, unlike FastAPI. The platform is still not 100% asynchronous friendly. However, if the application requires a 100% Flask environment, replace Flask with one of its variations called the *Quart* framework.

Choosing Quart over Flask 2.x

Quart is a Flask framework in and out but with a platform that runs entirely on asyncio. Many of the core features from Flask are part of the Quart framework, except for the main application class. The framework has its Quart class to set up an application.

Moreover, Quart supports the **HTTP/2** protocol, which allows faster interaction between its components and the server. Therefore, it is best to use servers such as the `hypercorn` server, which supports HTTP/2 request-response transactions.

Since Quart and Flask are almost the same, migration of Flask applications to Quart is seamless and straightforward. `ch05-quart` is a product of migrating our `ch05-web` and `ch05-api` projects into using the Quart platform. The following is the `app/__init__.py` configuration of that project:

```python
from quart import Quart
import toml
from app.model.config import init_db

from app.api.home import home, welcome
from app.api.login import add_login, list_all_login

def create_app(config_file):
    app = Quart(__name__, template_folder='../app/pages', static_
folder="../app/resources")
    app.config.from_file(config_file, toml.load)
    init_db()


    app.add_url_rule('/ch05/home', view_func=home, endpoint='home')
    app.add_url_rule('/ch05/welcome', view_func=welcome,
endpoint='welcome')
    app.add_url_rule('/ch05/login/add', view_func=add_login,
endpoint='add_login')
    app.add_url_rule('/ch05/login/list/all', view_func=list_all_login,
endpoint='list_all_login')

    return app
```

The Quart framework has a `Quart` class to build the application. Its constructor parameters, such as `template_folder` and `static_folder`, are the same as those of Flask. The framework can also recognize TOML configuration files.

On the repository layer, the framework has a `quart-sqlalchemy` extension module that supports asynchronous ORM operations for Quart applications. There is no need to rewrite the model and repository classes during the migration because all the helper classes and utilities are the same as the `flask-sqlalchemy` extension. The same `init_db()` from the project's application factory will set up and load the helper functions, methods, and model classes of the `quart-sqlalchemy` ORM.

Quart also supports blueprint, application factory design, or even the hybrid approach in building the application. However, the current version, *Quart 0.18.4*, does not have an easy way to manage the asynchronous request context so that modules inside the application can access the `current_app` proxy for view or API implementation. That's why, from the given configuration, the views and endpoints can be defined inside `create_app()` using `add_url_rule()`. Decorating them with `route()` in their respective module script using the `app` object or `current_app` raises an exception. Now, the following are the view and endpoint implementations in the Quart platform:

```python
from quart import jsonify, render_template, request, make_response

async def add_login():
    async with db_session() as sess:
            repo = LoginRepository(sess)
            login_json = request.get_json()
            login = Login(**login_json)
            result = await repo.insert(login)
            if result:
                content = jsonify(login_json)
                return await make_response(content, 201)
            else:
                content = jsonify(message="insert complaint details
record encountered a problem")
                return await make_response(content, 500)

async def welcome():
    return await render_template('index.html'), 200
```

Compared to Flask, `render_template()` and `make_response()` here in Quart need the `await` keyword. Another difference is Quart's use of `hypercorn` to run its applications instead of the Werkzeug server. So, install `hypercorn` using the `pip` command:

```
pip install hypercorn
```

Then, run the application with the `hypercorn main:app` command.

So far, in general, Quart has been a promising asynchronous framework. Let's hope that collaborations with the creator, support groups, and enthusiasts can help upgrade and expand this framework in the near future.

Summary

Flask 2.2 is now at par with the other frameworks that support and utilize asynchronous solutions to improve the application's runtime performance. Its view and API functions can now be `async` and runnable on an event loop created by Flask. Asynchronous services and transactions can now be executed and awaited on the Flask platform as tasks created by `create_task()` and `ensure_future()`.

The latest *SQLAlchemy[async]* can easily integrate with the Flask application to provide asynchronous CRUD transactions. Also, creating asynchronous tasks to break down the sequence of blocking transactions in Celery background processes, WebSocket messaging, and Observable operations are now possible with Flask 2.2.

Moreover, designing loosely coupled components, application-scoped cross-cut concern solutions, and some distributed setups is now feasible with Flask 2.2 through the built-in asynchronous signals.

There is even a 100% asynchronous Flask framework called Quart that can build fast-performing request-response transactions.

Although the purpose of asynchronous support in Flask is for performance, there is still a boundary on when it can be part of our applications. Some components or utilities will degrade in their running time when used with `asyncio`. Others, such as CRUD operations, will slow down DB access due to DB specifications that do not comply with the asynchronous setup. So, the effect of asynchronous programming still depends on the requirements of the projects and the resources the application uses.

The next chapter will bring us to the computational world of Flask, which deals with `numpy`, `pandas`, graphs, charts, statistics, file serializations, and other scientific solutions that Flask can provide.

6

Developing Computational and Scientific Applications

Computational scientists always choose easy-to-use, effective, and accurate GUI-based applications for their discovery, analysis, synthesis, data mines, and number crunches to save time and effort in arriving at some conclusions for their studies. Although powerful computational tools are available in the market, such as Maple, Matlab, MathCAD, and Mathematica, scientists still prefer mechanisms that can provide them with further customizations to apply their desired precision, accuracy, and calibration for their mathematical and statistical models. In other words, they still prefer custom-made applications that can fit with their laboratory setup and parameters.

Since the utmost priority is to provide scientists with accurate results given an infinite number of data, it is always a challenge as to what application frameworks to use in building scalable, real-time, and fast modules suited to their needs. The ultimate requirement is to create and run asynchronous transactions for complex numerical algorithms, which asynchronous Flask can provide.

The Flask has asynchronous components that can build complex, fast, and real-time applications for scientists. Because of its flexibility, asynchronous features, and wide-ranging support, this framework has the complete building blocks that can provide scientists with their tailor-fit scientific software.

This chapter will cover the following computational building blocks that flask[async] can provide:

- Uploading **Comma-Separated Values (CSV)** and **Microsoft Excel Spreadsheets (XLSX)** documents for computing
- Implementing symbolic computation with visualization
- Using the `pandas` module for data and graphical analysis
- Creating and rendering LaTeX documents
- Building graphical charts with frontend libraries

- Building real-time data plots using WebSocket and **Server-Sent Events (SSE)**
- Using asynchronous background tasks for resource-intensive computations
- Incorporating Julia packages with Flask

Technical requirements

This chapter will highlight a software prototype for an *Online Housing Pricing Prediction and Analysis* application with features expected to appear in many scientific applications. First, it has simple and formal GUIs that capture user data through forms. Forms that will ask for formulas, variable values, and constants with the capability to provide graphical plots, either in real-time or immediately after computations, are used. Second, it is a web application that can be accessible within teams or organizations. Finally, the application can run highly computational tasks asynchronously with the Flask platform.

The test data used in this chapter are from `https://www.kaggle.com/datasets/yasserh/housing-prices-dataset` and `https://data.world/finance/international-house-price-database`. On the other hand, this project uses the `Blueprint` approach for managing the modules and components. All files are available at `https://github.com/PacktPublishing/Mastering-Flask-Web-Development/tree/main/ch06`.

Uploading CSV and XLSX documents for computing

The application will deal with XLSX and CSV files that contain numerical data affecting worldwide house prices, such as the periodic actual and nominal **House Price Index (HPI)** of each country and the nominal and actual **Personal Disposable Income (PDI)** of the customers. Also, some documents will show how factors such as the house area, furnishing status, the main road preference, and the number of bedrooms and bathrooms can affect the housing prices in a country. Our application will upload these documents to the server for data analysis.

Flask has built-in support for a single- or multiple-file-uploading process through an HTML `<form>` with `enctype` of `multipart/form-data`. It stores all uploaded files in the `request.files` dictionary as `FileStorage` instances. `FileStorage` is a thin wrapper class from the `werkzeug` module used by Flask to represent an incoming file. The following is an HTML script that uploads an XLSX document for data analysis using the `pandas` module:

```
<!DOCTYPE html>
<html lang="en">
    ... ... ... ... ... ...
    <body>
      <h1>Data Analysis ... Actual House Price Index (HPI)</h1>
      <form action="{{request.path}}" method="POST" enctype="multipart/
form-data">
```

```
        Upload XLSX file:
        <input type="file" name="data_file"/><br/>
        <input type="submit" value="Upload File"/>
    </form>
  </body><br/>
  {%if df_table == None %}
        <p>No analysis.</p>
  {% else %}
        {{ table | safe}}
  {% endif %}
</html>
```

The following snippet shows the `view` function implementation that renders the given page and accepts the incoming XLSX document:

```
from modules.upload import upload_bp
from flask import render_template, request, current_app
from werkzeug.utils import secure_filename
from werkzeug.datastructures import FileStorage
import os
from pandas import read_excel
from exceptions.custom import (NoneFilenameException,
InvalidTypeException, MissingFileException, FileSavingException)


@upload_bp.route('/upload/xlsx/analysis', methods = ["GET", "POST"])
async def show_analysis():
    if request.method == 'GET':
        df_tbl = None
    else:
        uploaded_file:FileStorage = request.files['data_file']
        filename = secure_filename(uploaded_file.filename)
        if filename == '':
            raise NoneFilenameException()
        file_ext = os.path.splitext(filename)[1]
        if file_ext not in current_app.config['UPLOAD_FILE_TYPES']:
            raise InvalidTypeException()
        if  uploaded_file.filename == '' or uploaded_file == None:
            raise MissingFileException()
        try:
            df_xlsx = read_excel(uploaded_file, sheet_name=2,
skiprows=[1])
            df_tbl = df_xlsx.loc[: , 'Australia':'US'].describe().
to_html()
        except:
```

```
          raise FileSavingException()
    return render_template("file_upload_pandas_xlsx.html", table=df_
tbl), 200
```

Like any form parameter, the view function accesses the file object from `request.files` through the name of the form field. The file object, wrapped in a `FileStorage` wrapper, provides the following attributes:

- `filename`: This provides the raw filename of the file object.

- `stream`: This provides the input stream object that emits Input/Output methods such as `read()`, `write()`, `readline()`, `writelines()`, and `seek()`.

- `headers`: This contains the file's header information.

- `content-length`: This pertains to the content-length header of the file.

- `content-type`: This pertains to the content-type header of the file.

It also contains the following methods that can manage the file at runtime:

- `save(destination)`: This places the file in a destination.

- `close()`: This closes the file, if necessary.

Before accessing the file for reading, writing, transformation, or saving, the view function must apply validation and restriction to the file object received. Here are the following areas of concern where to impose red flags:

- The existence of the actual uploaded file

- A sanitized filename

- The accepted valid extension of the file

- The accepted file size

The given `show_analysis()` view function raises the following custom exception classes when it encounters a problem on the preceding red flags:

- `NoneFilenameException`: This is raised when there is no filename in the request.

- `InvalidTypeException`: This is raised when the sanitized filename gives an empty value.

- `InvalidTypeException`: This is raised when the uploaded file has an extension not supported by the application.

Also, part of the concern is to sanitize the filename of the multipart object before utilizing it for any file transactions. The immediate use of the raw `filename` attribute of the `FileStorage` instance can expose the application to several vulnerabilities because `filename` can have malware-related symbols, some special characters that are suspicious, and characters denoting the file path, such as `../../`, which can cause trouble with the `save()` method. To perform filename sanitation, use the `secure_filename()` utility method of the `werkzeug.utils` module. On the other hand, some of our application's view functions save their uploaded files inside our project's folder, but storing them outside the project directory is still the best practice.

Lastly, always enclose the entire file transactions of the view function with the `try-except` clause and raise the necessary exception classes to log all the underlying problems that will arise at runtime. Now, let us discuss the process after the file uploading with the `pandas` module.

Using the pandas module for data and graphical analysis

The `pandas` module is a popular Python library for data analysis because of its easy-to-apply utility functions and a high-performance tabular data structure called **DataFrame**. However, for the module to work, it needs the numpy module, a low-level library that supports multi-dimensional array objects called `ndarray` and its mathematical operations, and `matplotlib`, a library for visualizations. So, install these two modules first:

```
pip install numpy matplotlib
```

Then, install the `pandas` module:

```
pip install pandas
```

Since our data will be coming from XLSX sheets, install the `openpyxl` dependency module of pandas that deals with reading and writing XLSX documents:

```
pip install openpyxl
```

After installing all the dependency modules, we can start creating the `DataFrame` object.

Utilizing the DataFrame

To read an XLSX document, the `pandas` module has a `read_excel()` method with parameters such as `usecols`, which indicates the columns or range of columns to include, `skiprows`, which selects the rows to skip starting from the column row, and `sheet_name`, which chooses the sheet to read starting from sheet 0. The following from the previous `show_analysis()` view depicts the data retrieval from sheet 2 of the workbook, excluding row 1:

```
df_xlsx = read_excel(uploaded_file, sheet_name=2, skiprows=[1])
```

This result will be similar to the following snapshot from a sample `uploaded_file`:

	A	B	C	D	E	F	G	H	I	J	K	L	M	N
1	Period	Australia	Belgium	Canada	Switzerlan	Germany	Denmark	Spain	Finland	France	UK	Ireland	Italy	Japan
2							Row 1							
3	1975:Q1	38.82	48.86	64.27	96.34	106.39	57.15	36.39	56.47	43.02	34.29	23.05	78.64	100.71
4	1975:Q2	38.29	49.98	64.19	92.34	106.17	57.70	36.48	55.67	43.64	33.12	23.10	75.82	99.24
5	1975:Q3	37.79	50.36	65.01	90.65	106.71	58.91	36.48	55.76	44.31	32.54	23.35	72.92	97.56
6	1975:Q4	37.08	51.93	66.01	89.78	107.07	57.69	36.38	59.07	44.95	31.41	23.90	70.53	95.63
7	1976:Q1	37.24	56.07	67.00	87.56	107.44	58.25	36.19	54.41	45.50	31.33	24.08	67.52	93.61
8	1976:Q2	37.55	56.83	68.34	87.11	108.29	57.36	35.99	53.07	46.22	31.27	23.61	64.46	91.61
20	1979:Q2	36.82	73.77	65.54	91.19	123.73	65.37	37.61	46.22	50.91	35.00	33.62	56.59	92.86
21	1979:Q3	36.59	73.67	67.22	92.83	123.25	65.25	37.30	46.67	51.31	34.97	31.57	57.46	94.24
22	1979:Q4	38.48	73.62	69.75	94.78	123.50	63.52	36.78	47.09	52.24	35.90	31.45	58.19	95.47
23	1980:Q1	39.79	73.84	71.32	97.44	120.58	60.98	35.27	47.22	53.90	36.21	30.95	58.31	96.47
24	1980:Q2	41.22	70.70	72.50	96.84	120.33	57.25	34.71	47.90	54.41	36.09	29.33	60.68	97.45
25	1980:Q3	42.26	68.05	73.77	98.11	121.06	57.26	34.11	48.47	54.73	36.37	29.16	61.80	99.15
26	1980:Q4	42.81	65.18	78.05	97.05	121.86	55.64	33.58	47.86	54.62	35.49	30.39	64.32	100.99
27	1981:Q1	42.38	62.92	83.24	99.59	123.74	53.51	33.27	48.66	54.25	35.15	30.52	74.33	102.33
28	1981:Q2	44.72	61.71	87.47	102.37	123.73	49.60	32.81	48.64	53.77	34.98	30.26	75.55	104.28

⟨ ⟩   Note Sheet 1 | **RHPI** | PDI   RPDI   +

Figure 6.1 – A sample XLSX document containing HPI and PDI data

Figure 6.2 shows a sample `DataFrame` object extracted from an uploaded housing price data set to the `show_analysis()` view function.

	Australia	Belgium	Canada	Switzerland	...	Norway	New Zealand	Sweden	US
0	38.821347	48.864057	64.268656	96.343908	...	56.821609	58.046392	78.722203	55.292562
1	38.287386	49.981693	64.192468	92.336383	...	56.067774	54.896072	79.334548	55.096526
2	37.790320	50.357245	65.006322	90.646865	...	55.567010	53.897218	79.997585	54.624906
3	37.080878	51.931297	66.008740	89.782800	...	55.888116	52.975961	80.682327	54.817585
4	37.243674	56.072096	66.998016	87.556748	...	55.030432	51.327408	81.190213	54.939034
..	...	...	...	...	...	...	...	...	...
141	128.514673	130.488803	136.185021	111.143041	...	125.550934	105.794915	128.657412	86.176343
142	127.443271	130.847991	135.146587	113.167156	...	126.562211	104.444595	128.459749	87.305204
143	126.618848	132.091485	136.399449	114.513336	...	128.931509	102.343491	129.323343	85.550026
144	124.572793	131.201419	143.730300	115.222134	...	131.711763	101.932328	130.563752	82.383445
145	122.356005	130.981873	146.080432	116.225373	...	131.785175	101.594836	128.778981	80.658945

Figure 6.2 – A sample DataFrame from an uploaded file

A `DataFrame` object has easy-to-use properties that can extract a portion of the table, such as `shape`, `size`, `axes`, `at`, `columns`, `indexes`, `ndim`, `iloc`, and `loc`. If the goal is to extract only the columns from Australia to the US, the `loc` property should indicate the range of columns that `DataFrame` object will sift its analysis from, as shown in the following snippet:

```
df_tbl = df_xlsx.loc[: , 'Australia':'US'].describe().to_html()
```

The `loc` property accesses the data values using selected column labels or ranges, while its `iloc` counterpart uses column indices to slice the `DataFrame` instance, like the `df_tbl`. Both properties emit mathematical methods, such as `count()`, `mean()`, `sum()`, `mode()`, `std()`, and `var()`. However, the given view function utilizes the `describe()` method to extract the columnar data from `Australia` to the `US` columns on the actual HPI values quarterly from 1975 to the current year. Here is the actual output of our view when a valid XLSX document on housing datasets is uploaded:

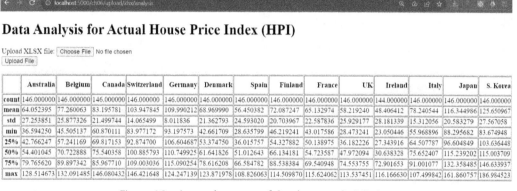

Data Analysis for Actual House Price Index (HPI)

Upload XLSX file: Choose File No file chosen
Upload File

	Australia	Belgium	Canada	Switzerland	Germany	Denmark	Spain	Finland	France	UK	Ireland	Italy	Japan	S. Korea
count	146.000000	146.000000	146.000000	146.000000	146.000000	146.000000	146.000000	146.000000	146.000000	146.000000	144.000000	146.000000	146.000000	146.000000
mean	64.052395	77.260063	83.195781	103.947845	109.990212	68.969990	56.450382	72.087247	65.132974	58.219240	48.406412	78.240544	116.344986	125.650967
std	27.253851	25.877326	21.499744	14.065499	8.011836	21.362793	24.593020	20.703967	22.587836	25.929177	28.181339	15.312056	20.583279	27.567058
min	36.594250	45.505137	60.870111	83.977172	93.197573	42.661709	28.635799	46.219241	43.017586	28.473241	23.050446	55.968896	88.295682	83.674948
25%	42.766247	57.241169	69.817153	92.874700	106.604687	53.374750	36.015757	54.327882	50.138975	36.182226	27.343916	64.507787	96.604849	103.636448
50%	54.401045	70.722888	75.540358	100.885793	110.749925	61.641826	51.012643	66.134181	54.723587	47.972094	30.638328	75.652407	115.239202	115.003709
75%	79.765620	89.897342	85.967710	109.003036	115.090254	78.616208	66.584782	88.538384	69.540948	74.553755	72.901653	91.001077	132.358485	146.633937
max	128.514673	132.091485	146.080432	146.421648	124.247139	123.871978	108.826063	114.509870	115.624062	113.537451	116.166630	107.499842	161.860757	186.984523

Figure 6.3 – A sample output of the show_analysis() view

When rendering data values using Flask, the `DataFrame` object has three utility methods that can provide format-ready results. Here are the three methods:

- `to_html()`: This generates an HTML table format with the datasets.

- `to_latex()`: This creates a LaTeX-formatted result with the data ready for PDF transformation.

- `to_markdown()`: This generates a Markdown-ready template with the data values.

In the case of `show_analysis()`, it uses `to_html()` to render all the captured datasets as an HTML table through `to_html()`. However, the rendition will only work with the `safe` Jinja2 filter because Jinja2 will not automatically HTML-escape all the characters provided by `to_html()` for security purposes. *Figure 6.3* shows a raw result of rendering tabular values from the `DataFrame` instance containing the values using its `to_html()` method.

Rendering graphs and charts using matplotlib

It is easy to plot data when contained in a `DataFrame` object's two-dimension data structure. The `matplotlib` has built-in support for rendering the tabular values as a *line*, *bar*, *pie*, or other graph or chart type. Since our application is a web app, our view functions must render these visuals as images, unlike in a REST application, which returns JSON resources for frontend frameworks.

Now, the first step is to create a `Figure` object. A `Figure` object serves as a canvas of a plot or subplots depending on the visualization approach. It is a plain blank object created by the `figure()` method of the `matplotlib` module or the `Figure` helper class of the `matplotlib.figure` module. It has the following essential properties that need configuration before finalizing the plot:

- `figsize`: This measures the x-axis and y-axis of the canvas' dimensions.

- `dpi`: This gauges the dot pixel per inch for the plot.

- `linewidth`: This measures the borderline of the canvas.

- `edgecolor`: This applies the color of the canvas' borderline.

- `facecolor`: This applies the indicated color to the border area between the canvas borderline and the axes plot borderline.

The following view implementation uploads a file, creates a `DataFrame` object from the uploaded XLSX document, and renders a line graph from the tabular values:

```python
from pandas import read_excel
from numpy import arange
from matplotlib.figure import Figure
from io import BytesIO
import base64


@upload_bp.route("/upload/xlsx/rhpi/plot/belgium", methods =
['GET', 'POST'])
async def upload_xlsx_hpi_belgium_plot():
    if request.method == 'GET':
        data = None
    else:
        ... ... ... ... ...
        try:
            df_rhpi = read_excel(uploaded_file, sheet_name=2,
usecols='C', skiprows=[1])
            array_rhpi = df_rhpi.to_numpy().flatten()
            array_hpi_index = arange(0, array_rhpi.size )
            fig = Figure(figsize=(6, 6), dpi=72, edgecolor='r',
linewidth=2, facecolor='y')

            axis = fig.subplots()
            axis.plot(array_hpi_index, array_rhpi)
            axis.set_xlabel('Quarterly Duration')
            axis.set_ylabel('House Price Index')
            axis.set_title("Belgium's HPI versus RHPI")
            ... ... ... ... ... ...
            output = BytesIO()
```

```
        fig.savefig(output, format="png")
        data = base64.b64encode(output.getbuffer())
.decode("ascii")
    except:
        raise FileSavingException()
    return render_template("file_upload_xlsx_form.html", data=data),
200
```

From the preceding implementation, the `Figure` canvas is now 6 inches x 6 inches in dimension, as managed by its `figsize` parameter. By default, a `Figure` canvas is 6.4 and 4.8 inches. Also, the borderline has an added 2units in thickness, with an `edgecolor` value of 'r', a single character shorthand for color red, and a `facecolor` value of 'y' character notation, which means color yellow. *Figure 6.4* shows the outcome of the given details of the canvas:

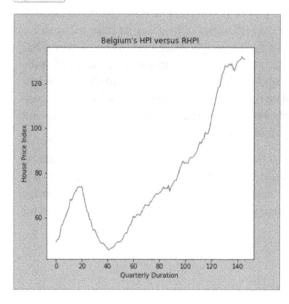

Figure 6.4 – A line graph with a customized Figure instance

The next step is to draw up the data values from the `DataFrame` object using `Axes` or the plot of the `Figure`. `Axes`, not the x-axis and y-axis, is the area on the `Figure` canvas where the visualization will happen. There are two ways to create an `Axes` instance:

- Using the `subplots()` method of the `Figure`.
- Using the `subplots()` method of the `matplotlib` module.

Since there is already an existing `Figure` instance, the former is the appropriate approach to create the plotting area. The latter returns a tuple containing a new `Figure` instance, with `Axes` all in with one method call.

Now, an `Axes` instance has almost all the necessary utilities for setting up any `Figure` component, such as `plot()`, `axis()`, `bar()`, `pie()`, and `tick_params()`. In the given `upload_xlsx_hpi_belgium_plot()`, the goal is to create a Line2D graph of the actual HPI values of Belgium by using the `plot()` method. The extracted DataFrame tabular data focuses only on the `Belgium` column (column C), as indicated by the `usecols` parameter of the `read_excel()` statement:

```
df_rhpi = read_excel(uploaded_file, sheet_name=2, usecols='C',
skiprows=[1])
```

Thus, `plot()`'s x-values or `scalex` will have `ndarray` from 0 to the maximum number of captured HPI values, and its y-values or `scaley` will have the HPI values of Belgium. Its color parameter is set to `#fc0366` to change the default blue color of the line graph. Aside from `plot()`, `Axes` has `set_title()` to add a header title for the image, `set_xlabel()` to add the description of the x-values, `set_ylabel()` for the y-values description, `set_facecolor()` to change the font color of the text, and `tick_params()` to update the color of the x and y tick values. `Axes` also has properties such as `xaxis` and `yaxis` to apply a new color to the x- and y-axis descriptions and spines to adjust the `linewidth` and `edgecolor` of the plot.

After finalizing the plot details, create a `BytesIO` buffer object to contain the `Figure` instance. Saving the `Figure` in `BytesIO` is necessary for decoding the plot as an inline image. The view must pass the base64-encoded image to its Jinja2 template for rendition. Rendering an inline image through the `<url>` tag is a fast way of displaying images. *Figure 6.5* shows the updated line graph for a sample actual HPI dataset for Belgium.

Data Analysis for Belgium's Actual House Price Index (HPI)

Upload XLSX file: [Choose File] No file chosen
[Upload File]

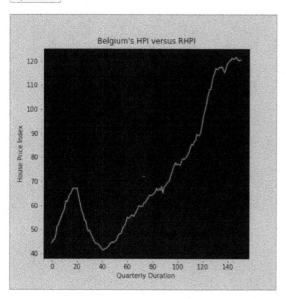

Figure 6.5 – A final line graph for a sample actual HPI data set for Belgium

How about if we have multiple graphs in one `Axes` plot?

Rendering multiple line graphs

Depending on the goal of the visualization, the `pandas` module with `matplotlib` can handle complex graphical renditions of `DataFrame` object's data values. The following view function creates two line graphs that can compare Belgium's actual and nominal HPI values based on a sample dataset:

```
@upload_bp.route("/upload/xlsx/rhpi/hpi/plot/belgium", methods =
['GET', 'POST'])
async def upload_xlsx_belgium_hpi_rhpi_plot():
    if request.method == 'GET':
        data = None
    else:
        … … … … … …
        try:
            df_hpi = read_excel(uploaded_file, sheet_name=1,
```

```
usecols='C', skiprows=[1])
        df_rhpi = read_excel(uploaded_file, sheet_name=2,
usecols='C', skiprows=[1])
        array_hpi = df_hpi.to_numpy().flatten()
        array_hpi_index = arange(0, df_rhpi.size )
        array_rhpi = df_rhpi.to_numpy().flatten()
        array_rhpi_index = arange(0, df_rhpi.size )


        fig = Figure(figsize=(7, 7), dpi=72, edgecolor='#140dde',
linewidth=2, facecolor='#b7b6d4')
        axes = fig.subplots()


        lbl1, = axes.plot(array_hpi_index ,array_hpi,
color="#32a8a2")
        lbl2, = axes.plot(array_rhpi_index ,array_rhpi,
color="#bf8a26")
        axes.set_xlabel('Quarterly Duration')
        axes.set_ylabel('House Price Index')
        axes.legend([lbl1, lbl2], ["HPI", "RHPI"])
        axes.set_title("Belgium's HPI versus RHPI")
        ... ... ... ... ... ...
    except:
        raise FileSavingException()
    return render_template("file_upload_xlsx_sheets_form.html",
data=data), 200
```

Compared to the previous upload_xlsx_hpi_belgium_plot() view, upload_xlsx_belgium_hpi_rhpi_plot() utilizes two sheets from the workbook of an uploaded file, namely sheet[1] for the nominal HPI and sheet[2] for the actual HPI values of Belgium. It derives separate DataFrame object's tabular values from each worksheet and plots a Line2D graph to compare the trend between the two datasets. Similar to the previous vector transformation in this chapter, this view still uses numpy to flatten the extracted vertical vector from the DataFrame's to_numpy() utility method. By the way, the view function only uses one Axes plot for both graphs.

Moreover, the view also showcases the inclusion of a **legend** into the plot. For complex cases such as in multiple graphs, adding legends to the output is helpful to the analysts. There are many ways to add legends into the plot with Axes, but this view captures the Line2D objects from the plot() method calls and maps each plot with a string label using the Axes' legend() method. *Figure 6.6* shows the result of running upload_xlsx_belgium_hpi_rhpi_plot() with an uploaded XLSX document.

Comparative Analysis between Belgium's Actual and Nominal HPI

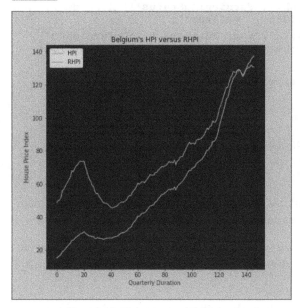

Figure 6.6 – Two line graphs in one Axes plot

Up next, we will see how to plot a pie chart with Flask.

Rendering a pie chart from a CSV file

The pandas module can also read data from CSV files through its read_csv() method. Unlike in read_excel(), the pandas module does not need any dependency to read valid CSV files. The following view uses read_csv() to create a DataFrame of values for plotting a pie chart:

```
from pandas import read_csv
@upload_bp.route("/upload/csv/pie", methods = ['GET', 'POST'])
async def upload_csv_pie():
    if request.method == 'GET':
        data = None
    else:
        ... ... ... ... ... ...
        try:
```

```
              df_csv = read_csv(uploaded_file)
              matplotlib.use('agg')
              fig = plt.figure()
              axes = fig.add_subplot(1, 1, 1)
              explode = (0.1, 0, 0)
              axes.pie(df_csv.groupby(['FurnishingStatus']) ['Price'].
     count(), colors=['#bfe089', '#ebd05b', '#e67eab'],
                   labels =["Furnished","Semi-Furnished", "Unfurnished"],
     autopct ='% 1.1f %%', shadow = True, startangle = 90, explode=explode)
              axes.axis('equal')
              axes.legend(loc='lower right',fontsize=7, bbox_to_anchor =
     (0.75, -01.0) )

              … … … … … …

          except:
              raise FileSavingException()
       return render_template("file_upload_csv_pie_form.html",
     data=data), 200
```

The pandas module can also read data from CSV files through its read_csv() method. Unlike in read_excel(), the pandas module does not need any dependency to read valid CSV files.

On the other hand, the Axes' pie() method has several parameters to consider before reaching the appropriate pie diagram for the data values. Here are some of the parameters used by the upload_csv_pie() view function:

- explode: This provides a list of fraction digits that indicate spaces around the wedges that will make them stand out.

- colors: This provides a list of matplotlib's built-in named colors or hexadecimal formatted color code set to each of the widgets.

- labels: This provides a list of string values assigned to each widget.

- autopct: This provides a string-formatted percentage value of each widget.

- shadow: This allows adding a shadow around the pie chart.

- startangle: This provides an angle of rotation for the pie chart to start with its first wedge.

The goal of the given `upload_csv_pie()` is to generate a pie chart based on the number of projected house prices (`Price`) per furnishing status (`FurnishingStatus`), namely the `Furnished`, `Semi-furnished`, and `Fully-furnished` houses. The `groupby()` method of the `df_csv` DataFrame extracts the needed data values for the `pie()` method. Now, running this view function will render the following chart:

Housing Price Status based on Furnishing Status

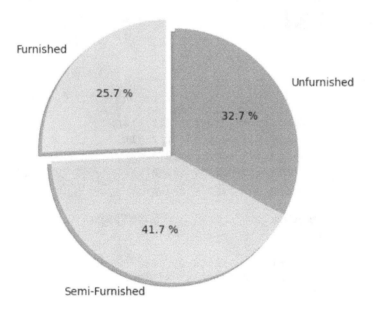

Figure 6.7 – Pie chart on Furnishing Status preference

If saving the pie chart figure produces the following warning message, `UserWarning: Starting a Matplotlib GUI outside of the main thread will likely fail.`, add `matplotlib.use('agg')` anywhere before creating the `Figure` instance to enable the non-interactive backend mode for writing files outside the main thread.

How about if we have multiple `Axes` plots in one `Figure`?

Rendering multiple Axes plots

A Figure can contain more than one plot of different graphs and charts. Scientific applications mostly have GUIs that render several charts of varying data calibration, transformation, and analytics. The following view function uploads an XLSX document and creates four plots on a `Figure` to create different graphs of the DataFrame data values extracted from the document:

```
@upload_bp.route("/upload/xlsx/multi/subplot", methods = ['GET',
'POST'])
async def upload_xlsx_multi_subplots():
    if request.method == 'GET':
        data = None
    else:
        ... ... ... ... ... ...
        try:
            df_xlsx = read_excel(uploaded_file, sheet_name=2,
skiprows=[1])
            fig = plt.figure(figsize=(12, 12))
            axes1 = fig.add_subplot(2, 2, 1)
            axes2 = fig.add_subplot(2, 2, 2)
            axes3 = fig.add_subplot(2, 2, 3)
            axes4 = fig.add_subplot(2, 2, 4)
```

The first plot, `axes1`, creates two line graphs of the actual HPI values of Australia and Belgium for all the quarterly periods, as indicated in the following code block:

```
            axes1.plot(df_xlsx.index.values, df_xlsx['Australia'],
'green', df_xlsx.index.values, df_xlsx['Belgium'], 'red',)
            axes1.set_xlabel('Quarterly Duration')
            axes1.set_ylabel('House Price Index')
            axes1.set_title('RHPI between Australia .........')
```

The second plot, `axes2`, generates a bar chart depicting the mean HPI values of all countries in the tabular values, as shown in the following code block:

```
index = arange(df_xlsx.loc[: , 'Australia':'US'].shape[1])
axes2.bar(index, df_xlsx.loc[: , 'Australia':'US'].
mean(), color=(0.1, 0.1, 0.1, 0.1), edgecolor='blue')
axes2.set_xlabel('Country ID')
axes2.set_ylabel('Mean HPI')
axes2.set_xticks(index)
axes2.set_title('Mean RHPI among countries')
```

The third plot, `axes3`, plots all HPI values of each country in the tabular values from 1975 to the current year, creating multiple line graphs:

```
axes3.plot(df_xlsx.loc[: , 'Australia':'US'])
axes3.set_xlabel('Quarterly Duration')
axes3.set_ylabel('House Price Index')
axes3.set_title('RHPI trend among countries')
```

The last plot, `axes4`, builds a grouped bar chart showing the HPI values of Japan, South Korea, and New Zealand quarterly in 1975:

```
width = 0.3
axes4.bar(df_xlsx.loc[0:3, 'Japan'].index.values-width,
df_xlsx.loc[0:3, 'Japan'], width=width, color='#d9182b', label="JP")
axes4.bar(df_xlsx.loc[0:3, 'S. Korea'].index.values, df_
xlsx.loc[0:3, 'S. Korea'], width=width, color='#f09ec1', label="SK")
axes4.bar(df_xlsx.loc[0:3, 'New Zealand'].index.
values+width, df_xlsx.loc[0:3, 'New Zealand'], width=width,
color='#000', label="NZ")
axes4.set_xlabel('Quarterly Duration')
... ... ... ... ... ...
axes4.legend()
```

The given `axes4` setup uses the `plot()` label parameter to assign codes for each bar plot needed by its `legend()` method in forming the diagram's legends. Running the view function will give us the following multiple graphs:

Multiple Plots In One Figure

Upload XLSX file: [Choose File] No file chosen
[Upload File]

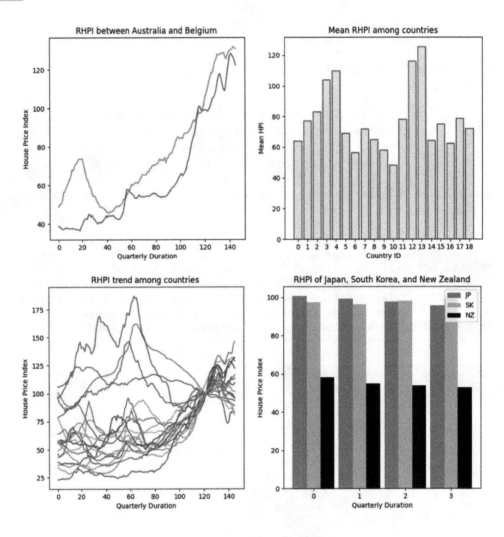

Figure 6.8 – A Figure with multiple plots

Flask's asynchronous components can also support more advanced, informative, and complex mathematical and statistical graphs plotted on a `Figure` with the `seaborn` module. Also, it can create regression plots using various regression techniques using the `statsmodels` module. The next topic will highlight the solving of nonlinear and linear equations with the `sympy` module.

Implementing symbolic computation with visualization

Symbolic computation is an algorithmic approach to model-based analysis and solving solutions for variables or symbols in a formula, mathematical expression, program code, or numerical model representable in a computer. It mainly involves representing numbers in a closed form or exact representation. Many scientific applications have GUI forms that allow users to input these mathematical or statistical models with the experimental values of their variables. Others provide plotting of these equations or formulas using the `matplotlib` and `numpy` modules.

For Flask to recognize symbolic expressions and formulas in a string expression, install the `sympy` module using the `pip` command:

```
pip install sympy
```

Then, install the `mpmath` module, a prerequisite of the `sympy` module:

```
pip install mpmath
```

After these installations, we can start problem solving.

Solving linear equations

Let us begin with the following asynchronous route implementation that asks for any linear equation with x and y variables only:

```
from modules.equations import eqn_bp
from flask import render_template, request
from sympy import sympify
import gladiator as gl


@eqn_bp.route('/eqn/simple/bivar', methods = ['GET', 'POST'])
async def solve_multivariate_linear():
    if request.method == 'GET':
        soln = None
    else:
        field_validations = (
            ('lineqn', gl.required, gl.type_(str),    gl.regex_('[+\-
]?(([0-9]+\.[0-9]+)|([0-9]+\.?)|(\.?[0-9]+))[+\-/*][xy]([+\-/*]((([0-
9]+\.[0-9]+)|([0-9]+\.?)|(\.?[0-9]+))[+\-/*][xy])*([+\-/*]((([0-9]+\.
[0-9]+)|([0-9]+\.?)|(\.?[0-9]+)))*')),
```

```
                ('xvar', gl.required, gl.type_(str), gl.regex_('[0-9]+')),
                ('yvar', gl.required, gl.type_(str), gl.regex_('[0-9]+'))
        )
        form_data = request.form.to_dict()
        result = gl.validate(field_validations, form_data )
        if bool(result):
            xval = float(form_data['xvar'])
            yval = float(form_data['yvar'])
            eqn = sympify(form_data['lineqn'], {'x': xval, 'y': yval})
            soln = eqn.evalf()
        else:
            soln = None
    return render_template('simple_linear_mv_form.html', soln=soln),
200
```

Assuming that `xvar` and `yvar` are valid form parameter values convertible to `float` and `lineqn` is a valid two-variate string expression with x and y variables, the `sympify()` method of the `sympy` module can convert `lineqn` to a symbolic formula with `xvar` and `yvar` values assigned to the x and y symbols and compute the solution. To extract the exact value of the sympification, the resulting symbolic formula has a method such as `evalf()` that returns a floating-point value of the solution. Now, the `sympify()` method uses the risky `eval()` function, so the mathematical expression, such as `lineqn`, requires sanitation by popular validation tools such as `gladiator` before performing sympification. *Figure 6.9* shows a sample execution of `solve_multivariate_linear()` with a sample linear equation and the corresponding values for its x and y:

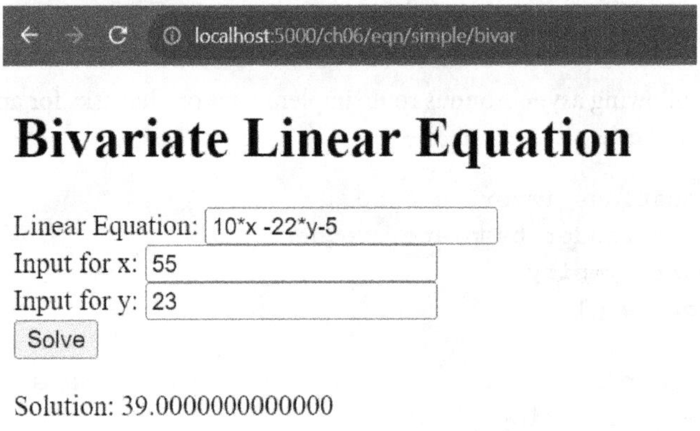

Bivariate Linear Equation

Linear Equation: `10*x -22*y-5`
Input for x: `55`
Input for y: `23`
`Solve`

Solution: 39.0000000000000

Figure 6.9 – Solving a linear equation with x and y variables

Now, not all real-world problems are solvable using linear models. Some require non-linear models to derive their solutions.

Solving non-linear formulas

Flask `async` and `sympy` can also implement a view function for solving non-linear equations. The `sympify()` method can recognize Python mathematical functions such as `exp(x)`, `log(x)`, `sqrt(x)`, `cos(x)`, `sin(x)`, and `pow(x)`. Thus, creating mathematical expressions with the inclusion of these Python functions is feasible with `sympy`. *Figure 6.10* shows a view function that computes a solution of a univariate non-linear equation with one variable.

Figure 6.10 – Solving a non-linear equation with Python functions

The strength of the `sympy` module is to extract the parameter values of an equation or equations based on a given result or solution.

Finding solutions for a linear system

The `sympy` module has a `solve()` method that can solve systems of linear or polynomial equations. The following implementation can find a solution for a system of two polynomial equations:

```
from modules.equations import eqn_bp
from flask import render_template, request
from sympy import symbols, sympify, solve

@eqn_bp.route('/eqn/eqnsystem/solve', methods = ['GET', 'POST'])
async def solve_multiple_eqns():
    if request.method == 'GET':
        soln = None
    else:
        field_validations = (
                ('polyeqn1', gl.required, gl.type_(str)),
                ('polyeqn2', gl.required, gl.type_(str))
            )
```

```
        form_data = request.form.to_dict()
        result = gl.validate(field_validations, form_data )
        if bool(result):
            x, y = symbols('x y')
            eqn1 = sympify(form_data['polyeqn1'])
            eqn2 = sympify(form_data['polyeqn2'])
            soln = solve((eqn1, eqn2),(x, y))
        else:
            soln = None
    return  render_template('complex_multiple_eqns_form.
html',    soln=soln), 200y
```

After the retrieval from `request.form` and a successful validation using `gladiator`, the `polyeqn1` and `polyeqn2` string expressions must undergo sympification through the `sympify()` method to derive their symbolic equations or `sympy` expressions. The function variables, x and y, of these mathematical expressions must have their corresponding `Symbol`-type variables utilizing the `symbols()` function of `sympy`, a vital mechanism for creating `Symbol` variables out of string variables. The `solve()` method requires a tuple of these symbolic equations in its first parameter and a tuple of `Symbols` in its second parameter to find the solutions of the linear system. If the linear equations are not parallel to each other, the `solve()` method will return a feasible solution in a dictionary format with `sympy` variables as keys.

If we execute `solve_multiple_eqns()` with a simple linear system, such as passing the `5*x-3*y-9` equation to `polyeqn1` and the `15*x+3*y+12` equation to `polyeqn2`, `solve()` will provide us with numerical results, as shown in *Figure 6.11*.

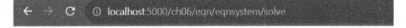

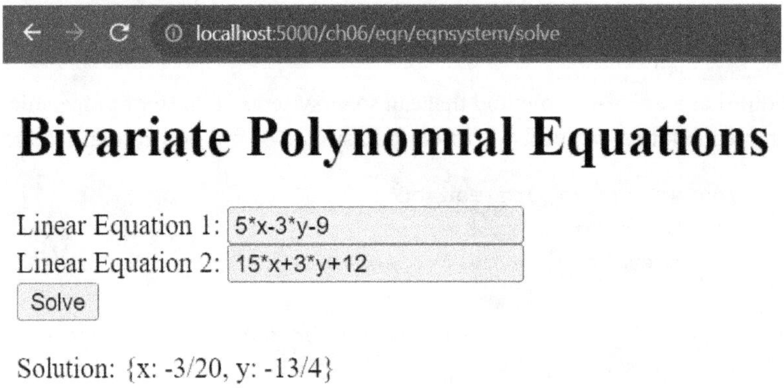

Bivariate Polynomial Equations

Linear Equation 1: `5*x-3*y-9`
Linear Equation 2: `15*x+3*y+12`
Solve

Solution: {x: -3/20, y: -13/4}

Figure 6.11 – Solving simple linear equations

However, if we have polynomials or non-linear equations such as passing the `x**2-10*y+10` quadratic formula to `polyeqn1` and the `10*x+5*y-3` linear expression to `polyeqn2`, the resulting non-linear solutions will be rational values with square roots, as shown in *Figure 6.12*.

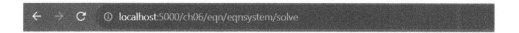

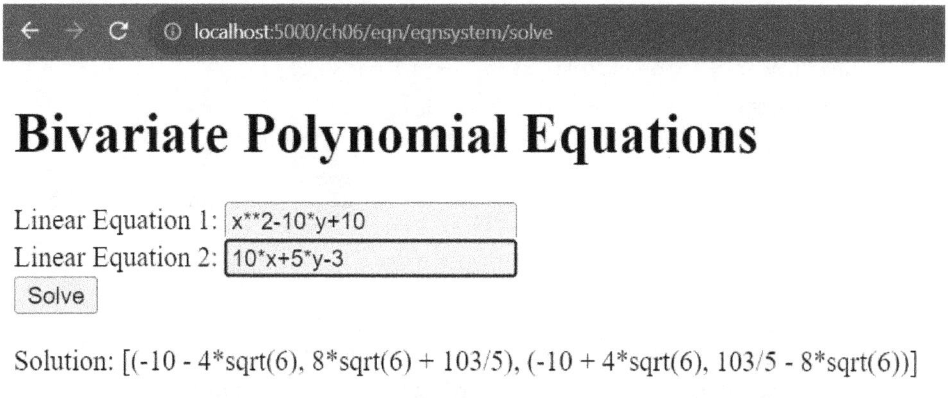

localhost:5000/ch06/eqn/eqnsystem/solve
Bivariate Polynomial Equations
Linear Equation 1: x**2-10*y+10
Linear Equation 2: 10*x+5*y-3
Solve
Solution: [(-10 - 4*sqrt(6), 8*sqrt(6) + 103/5), (-10 + 4*sqrt(6), 103/5 - 8*sqrt(6))]

Figure 6.12 – Solving polynomial system of equations

There are many possible symbolic computations, formulas, and algorithms that Flask can implement with `sympy`. Sometimes, the `scipy` module can help `sympy` solve other mathematical algorithms that are very tedious and complicated, such as approximation problems.

The `sympy` module is also capable of providing graphical analysis through plots.

Plotting mathematical expressions

When it comes to visualization, `sympy` is capable of rendering graphs and charts created by its built-in `matplotlib` library. The following view function accepts two equations from the user and creates a graphical plot for the equations within the specified range of values for x:

```
from sympy import symbols, sympify
from sympy.plotting import plot
import matplotlib

import base64
from io import BytesIO
from PIL import Image

@eqn_bp.route('/eqn/multi/plot', methods = ['GET', 'POST'])
async def plot_two_equations():
    if request.method == 'GET':
        data = None
    else:
        … … … … … …
        form_data = request.form.to_dict()
        result = gl.validate(field_validations, form_data )
```

```
        eqn1_upper = float(form_data['eqn1_maxval'])
        eqn1_lower = float(form_data['eqn1_minval'])
        eqn2_upper = float(form_data['eqn2_maxval'])
        eqn2_lower = float(form_data['eqn2_minval'])
        data = None
        if bool(result) and (eqn1_lower <= eqn1_upper) and (eqn2_lower
<= eqn2_upper):
            matplotlib.use('agg')
            x = symbols('x')
            eqn1 = sympify(form_data['equation1'])
            eqn2 = sympify(form_data['equation2'])
            graph = plot(eqn1, (x, eqn1_lower, eqn1_upper), line_
color='red', show=False)
            graph.extend(plot(eqn2, (x, eqn2_lower, eqn2_upper), line_
color='blue', show=False))
            filename = "./files/img/multi_plot.png"
            graph.save(filename)

            img = Image.open(filename)
            image_io = BytesIO()
            img.save(image_io, 'PNG')

            data = base64.b64encode(image_io.getbuffer())
.decode("ascii")
    return render_template('plot_two_eqns_form.html', data=data), 200
```

After sanitizing the string equations and deriving the `sympy` formulas, the view can directly create a plot for each formula using the `plot()` method in the `sympy.plotting` module, which is almost similar to that in the `matplotlib` module but within the context of `sympy`. The method returns a `Plot` instance that can combine with another `Plot` using its `extend()` method to create multiple plots in one frame. Running the `plot_two_equations()` view will yield line graphs of both `equation1` and `equation2`, as shown in *Figure 6.13*.

Plotting Univariate Polynomial Equations

Equation 1: `2*x**3+10`
Max value: `100`
Min Value: `0`

Equation 2: `30*x**2+100`
Max value: `100`
Min Value: `0`

`Plot`

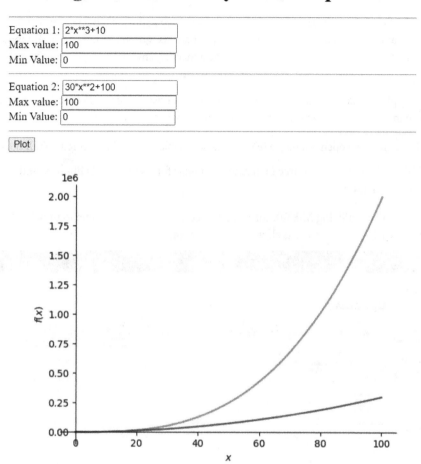

Figure 6.13 – Plotting the two sympy equations

On the other hand, the `Plot` instance has a `save()` method that can store the graphical plot as an image. However, to create an inline image for a Jinja2 rendition, the view needs the `Image` class from **Python's Imaging Library (PIL)** to open and load the saved image before wrapping it in `BytesIO` for `base64` encoding.

Let us examine now how asynchronous Flask can manage those scientific data that need LaTeX serialization or PDF renditions.

Creating and rendering LaTeX documents

LaTex is a high-standard typesetting system used in publishing and packaging technical and scientific papers and literature, especially those documents with charts, graphs, equations, and tabular data. When creating scientific applications, there should be a mechanism for the application to write LaTeX content, save it in a repository, and render it as a response.

But first, our applications will require a LaTeX compiler that assembles and compiles newly created LaTeX documents. Here are two popular tools that offer various LaTeX compilers:

- **TeX Live:** This is an open-source LaTeX tool most suitable for creating secured LaTeX documents.
- **MikTeX:** This is an open-source LaTeX tool popular for its on-the-fly libraries and up-to-date releases.

Our application will be utilizing MikTeX for its LaTeX compilers. Do not forget to update MikTex for the latest plugins using the console, as shown in *Figure 6.14*.

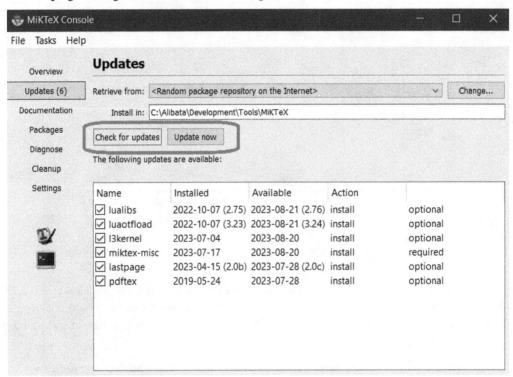

Figure 6.14 – Updating MikTeX using its console

After the MikTeX installation and update, let's create the Flask project by installing the `latex` module.

Rendering LaTeX documents

Asynchronous view functions can create, update, and render LaTeX documents through LaTeX-related modules and `matplotlib` for immediate textual and graphical plots or perform LaTeX to PDF transformation of existing LaTeX documents for rendition. The latter requires the installation of the `latex` module through the `pip` command:

```
pip install latex
```

The `latex` module uses its built-in Jinja libraries to access `latex` files stored in the main project. So, the first step is to create a Jinja environment with all the details that will calibrate the Jinja engine regarding LaTeX file handling. The following snippet shows how to set up the Jinja environment using the `latex.jinja2` module:

```
from jinja2 import FileSystemLoader, Environment
from latex.jinja2 import make_env

environ:Environment = make_env(loader=FileSystemLoader('files'),
enable_async=True,
    block_start_string = '\BLOCK{',
    block_end_string = '}',
    variable_start_string = 'VAR{',
    variable_end_string = '}',
    comment_start_string = '#{',
    comment_end_string = '}',
    line_statement_prefix = '%-',
    line_comment_prefix = '%#',
    trim_blocks = True,
    autoescape = False,)
```

Since ch06-project uses Blueprint to organize the views and the corresponding components, only the rendition module (/modules/rendition) that builds the LaTeX web displays can access this environment configuration. This Jinja environment details, defined in /modules/rendition/__init__.py, declares that the files folder in the project directory will become the root folder for our LaTeX documents. Moreover, it tells Jinja the syntax preferences for some LaTeX commands, such as the BLOCK, VAR, conditional statement, and comment symbols. Instead of having a backslash pipe ("\") in \VAR{}, the setup wants Jinja to recognize the VAR{} statement, an interpolation operator, without the backslash pipe. Violating the given syntax rules will flag an error in Flask. The enable_async property, on the other hand, allows the execution of latex commands in asynchronous view functions, such as the following view implementation that opens a document and updates it for display:

```
from modules.rendition import rendition_bp
from flask import send_from_directory
from jinja2 import FileSystemLoader
from latex.jinja2 import make_env

@rendition_bp.route('/render/hpi/plot/eqns', methods = ['GET',
'POST'])
async def convert_latex():
    tpl = environ.get_template('/latex/hpi_plot.tex')
    outpath=os.path.join('./files/latex','hpi_plot.pdf')
    outfile=open(outpath,'w')
    outfile.write(await tpl.render_async(author='Sherwin John
Tragura', title="Rendering HPI Plot with LaTeX", date=datetime.now().
strftime("%B %d, %Y"), renderTbl=True))
    outfile.close()
    os.system("pdflatex --shell-escape -output-directory=" + './files/
latex' + " " + outpath)
    return send_from_directory('./files/latex', 'hpi_plot.pdf')
```

The given view function uses get_template() of the Environment instance, environ, to create a Jinja2 template of a specific LaTeX document from the /latex sub-directory of the root folder. The template's render_async() function opens the specified LaTeX document for changes, such as passing context values (e.g., author, title, date, and renderTbl) to complete the document.

Afterward, the view function will convert the document into PDF format, which is the necessary approach for this application. os.path.join() will indicate where to save the file. Now, MikTeX offers three compilers to compile and convert the LaTeX document to PDF, namely pdfLaTeX, XeLaTeX, and LuaLaTeX, but our implementation uses pdfLaTeX, which is the default one. os.system() will run the compiler and save the PDF into the specific location. To render the content, Flask has a send_from_directory() method that can display the content of a PDF file saved in the directory. *Figure 6.15* shows the resulting PDF document by running the convert_latex() view function.

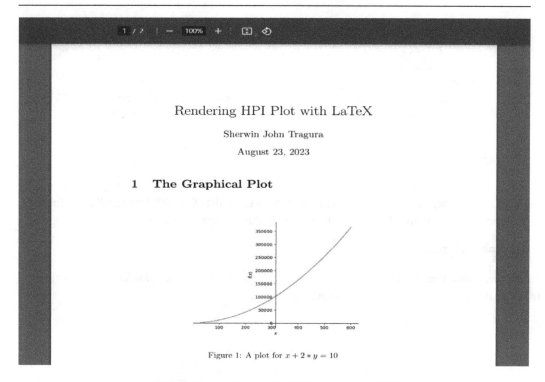

Figure 6.15 – Rendering a LaTeX document as a PDF

Our Flask application does not only render existing LaTeX documents but also creates one before rendering it to the client.

Creating LaTeX documents

So far, the `latex` module with Jinja2 has no LaTeX creation features that Flask can use to build scientific writeups from various data sources. However, other modules, such as `pylatex`, can provide helper classes and methods to serialize LaTeX content at runtime. The following view implementation shows how to generate a LaTeX file with `DataFrame` object's data derived from an uploaded XLSX document:

```
from pylatex import Document, Section, Command, NoEscape, Subsection,
Tabular, Center
from pylatex.utils import italic
from pylatex.basic import NewLine


@rendition_bp.route('/create/hpi/desc/latex', methods = ['GET',
'POST'])
async def create_latex_pdf():
    if request.method == 'GET':
        return render_template("hpi_latex_form.html"), 200
```

```
    else:
        ... ... ... ... ... ...
        ... ... ... ... ... ...
        try:
            df = read_excel(uploaded_file, sheet_name=2, skiprows=[1])
            hpi_data = df.loc[: , 'Australia':'US'].describe().to_
dict()
            hpi_filename = os.path.join('./files/latex','hpi_
analysis')
```

Before anything else, the environment setup must have the MikTeX or TeX Live installation for the LaTeX compilers. Then, install the `pylatex` module through the `pip` command:

```
pip install pylatex
```

To start the transaction, the given `create_latext_pdf()` retrieves an uploaded XLSX document to derive the tabular values for the report generation:

```
            geometry_options = {
                "landscape": True,
                "margin": "0.5in",
                "headheight": "20pt",
                "headsep": "10pt",
                "includeheadfoot": True
            }
            doc = Document(page_numbers=True,
geometry_options=geometry_options, document_
options=['10pt','legalpaper'])
            doc.preamble.append(Command('title', 'Mean HPI per
Country'))
            doc.preamble.append(Command('author', 'Sherwin John C.
Tragura'))
            doc.preamble.append(Command('date', NoEscape(r'\today')))
            doc.append(NoEscape(r'\maketitle'))
```

Then, it sets up a dictionary, `geometry_options`, that consists of the LaTeX document parameters, such as the document orientation (`landscape`), the left, right, top, and bottom margins (`margin`), the vertical height from the bottom part of the header down to the topmost area of first the text (`headsep`), the space from the top margin to the line where to start the header part (`headheight`), and the toggle parameter to include or exclude the document header and footer of the document (`includeheadfoot`). This dictionary is essential to the instantiation of the `pylatex`'s Document `container` class, which will represent the LaTeX document.

Initially, the LaTeX document will be a blank instance with the desired document parameters indicated by its `geometry_option` constructor parameter and the `document_options` list containing other options such as the font size and paper size. Then, to start customizing the document, the `view` function uses the `Command` class to create custom values to the document's title, author, and date without escaping the backslash, thus the use of the `NoEscape` class, and append them to the preamble property of the `Document` instance. This process is similar to calling `\title`, `\author`, and `\ date` commands with custom values interpolated by the `\VAR{ }` command.

Next, the view must append the `\maketitle` command without escaping the backslash to typeset all these added document details. The line following `\maketitle` is always the generation of the body content, in our case, the following section:

```
with doc.create(Section('The Data Analysis')):
        doc.append('Here are the statistical analysis derived
 from the uploaded excel data.')
```

The `pylatex` module classes are equivalent to some LaTeX commands, such as `Axis`, `Math`, `Matrix`, `Center`, `Alignat`, `Alignref`, and `Plot`. The `Command` class is a module class used to run custom or general commands such as `\title`, `\author`, and `\date`. In this `create_latex_pdf()` view, the content generation started with running the `Section` command with a section title, *The Data Analysis*. A section is an organized part of the content that contains combinations of tables, text, plots, and mathematical equations. After that, the view appends a statement in text form. Since there is no backslash to escape, there is no reason to wrap the test with the `NoEscape` class. Then, we create the sub-sections indicated in the following snippet:

```
            with doc.create(Subsection('Statistical analysis
 generated by Pandas')):
                with doc.create(Tabular('| c | c | c | c | c | c |
 c | c | c |')) as table:
                    table.add_hline()
                    table.add_row(("Country", "Count", "Mean",
 "Std Dev", "Min", "25%", "50%", "75%", "Max"))
                    table.add_empty_row()

                    for key, value in hpi_data.items():
                        table.add_hline()
                        table.add_row((key, value['count'],
 value['mean'], value['std'], value['min'], value['25%'], value['50%'],
 value['75%'], value['max']))
                        table.add_empty_row()
                        table.add_hline()
        except:
            raise FileSavingException()
```

After the text, the view appends a `Subsection` command, which will granularize the content of the recently created section. Part of its component is the `Tabular` command that will construct a spreadsheet of HPI values derived from the extracted tabular values. After the assemblage of the LaTeX content, the `create_latex_pdf()` view will now generate the PDF for rendition, as shown in the following snippet:

```
        doc.generate_pdf(hpi_filename, clean_tex=False,
compiler="pdflatex")
        return send_from_directory('./files/latex', 'hpi_analysis.
pdf')
```

The `Document` instance has a `generate_pdf()` method that compiles and generates the LaTeX file, converts the LaTeX file to its PDF form, and saves both files to a specific directory. Once the PDF is available, the view can render the PDF content through Flask's `send_from_directory()` method. *Figure 6.16* displays the generated PDF of the `create_latex_pdf()` view function.

Mean HPI per Country

Sherwin John C. Tragura

August 24, 2023

1   The Data Analysis

Here are the statistical anaysis derived from the uplaoded excel data.

1.1   Statistical analysis generated by Pandas

Country	Count	Mean	Std Dev	Min	25%	50%	75%	Max
Australia	146.0	64.05239489520619	27.253851483569616	36.59425044064457	42.76624749792639	54.40104506445259	79.76562048477005	128.51467289621758
Belgium	146.0	77.26006340976218	25.87732643521116	45.505136574837795	57.24116916860074	70.72288832455669	89.89734197010445	132.0914856874382
Canada	146.0	83.19578136626657	21.499744448231393	60.87011144378709	69.81715255402625	75.54035766251258	85.96771023349093	146.0804315543597
Switzerland	146.0	103.94784470894619	14.065499218846085	83.9771721436297	92.87470021908977	100.8857928989891632	109.00303583249519	146.42164822783676
Germany	146.0	109.9902115073647	8.01183595851763	93.1975726671041	106.60468711065552	110.7499249386849	115.09025425917126	124.24713917162816
Denmark	146.0	68.96999002241503	21.362792573720018	42.661709204184554	53.37475044455957	61.64182616949262	78.6162031274597	123.8719775329469
Spain	146.0	56.45038160318069	24.593019705203208	28.6357991898908	36.01575727735692	51.01264317381302	66.58478182931914	108.82606251509064
Finland	146.0	72.08724746871899	20.703967480731126	46.2192408572517	54.32788177856137	66.13418071830287	88.53838370027415	114.50987030932488
France	146.0	65.13297395013021	22.587836441409614	43.01758556266064	50.138974528388474	54.723587271477115	69.54094753508474	115.62406244427035
UK	146.0	58.219239590607685	25.92917744482758	28.47324138054373	36.18222627030116	47.972093731756054	74.55375528621356	113.53745051966035
Ireland	144.0	48.40641210591301	28.181338611105488	23.05044571203127	27.34391639514842	30.63832762993176	72.90165298683532	116.1666301648695
Italy	146.0	78.2405436701803	15.31205593406923	55.9688963188028	64.50778690006113	75.65240668980306	91.0010769294191	107.4998416858398
Japan	146.0	116.34498642244586	20.58327857503788	88.29568204122862	96.60484942113035	115.23920215340574	132.35848453088147	161.8607570668868
S. Korea	146.0	125.6509672657926	27.567058491806105	83.67494835818991	103.63644806319414	115.0037088103146	146.6339374057635	186.9845226927297
Netherlands	146.0	64.48799564512163	24.88394741033419	35.16040545475672	42.27917074954264	55.3234476307944	93.49806248381164	106.8132330402002
Norway	146.0	74.92090438307272	23.435954714391208	50.17790504723994	55.60923450865594	66.64455378659407	86.2089406424744	131.78517484055075
New Zealand	146.0	62.3648125230	23.41678250339763	35.17714657542279	46.21368270529038	52.00200207150985	65.4849368751785	118.6921177445412
Sweden	146.0	78.8807963971951	21.756357130947876	53.68879774011518	60.00980896489366	75.66524538025425	87.5798040680928	130.56375192012595
US	146.0	72.21444999238926	13.814241323194326	54.6249063375169	63.11931774499625	66.15661398470054	81.86966524795437	105.3828044571705

Figure 6.16 – A PDF generated by the pylatex module

Aside from rendering PDF content, Flask can also utilize popular frontend libraries for displaying graphs and charts. Let us concentrate on how Flask can integrate with these **JavaScript** (**JS**)-based libraries in visualizing datasets.

Building graphical charts with frontend libraries

Most developers prefer rendering graphs and charts using frontend libraries rather than `matplotlib`, which requires complex Python coding to refine presentation and lacks UI-related features such as responsiveness, adaptability, and user interaction. This section will highlight the Chart.js, Bokeh, and `Plotly` libraries, which are all popular libraries with varying strengths and weaknesses as external tools for visualization.

Let's begin with Chart.js.

Plotting with Chart.js

The most common and popular charting library used in many visualization applications is Chart.js. It is 100% JS, is lightweight, is easy to use, and has a straightforward syntax for designing graphs and charts. The following is the Chart.js implementation that displays the mean HPI values of certain countries:

```
<!DOCTYPE html>
<html lang="en">
<head>
   … … … … … …
   … … … … … …
   <script src='https://cdn.jsdelivr.net/npm/chart.js'></script>
</head>
<body>
    <h1>{{ title }}</h1>
    <form action="{{request.path}}" method="POST" enctype="multipart/
form-data">
       Upload XLSX file:
       <input type="file" name="data_file"/><br/>
       <input type="submit" value="Upload File"/>
   </form><br/>
   <canvas id="linechart" width="300" height="100"></canvas>
</body>
<script>
   var linechart = document.getElementById("linechart");
   Chart.defaults.font.family = "Courier";
   Chart.defaults.font.size = 14;
   Chart.defaults.color = "black";
```

Chart.js is available in three sources:

- **Node.js**: By running npm to install the chart.js module.

- **GitHub**: By downloading the `https://github.com/chartjs/Chart.js/releases/download/v4.4.0/chart.js-4.4.0.tgz` file or the latest release available.

- **Content delivery network (CDN)**: By referencing `https://cdn.jsdelivr.net/npm/chart.js`.

Based on the HTML script, our implementation opted for the CDN source.

After referencing Chart.js, create a `<canvas>` tag with the width and height that fits your plot. Then, create a `Chart()` instance with the node or 2D context of `<canvas>` and some configuration options. Moreover, set new and appropriate values to global default properties such as the font name, font size, and font color:

```
new Chart(linechart,{
    type: 'line',
    options: {
        scales: {
            y: {
                beginAtZero: true,
                title: {
                    display: true,
                    text: 'Mean HPI'
                }
            },
            x: {
                offset: true,
                title: {
                    display: true,
                    text: 'Countries with HPI'
                }
            }
        }
    },
    data: {
        borderWidth: ,
        labels : [
            {% for item in labels %}
                "{{ item }}",
            {% endfor %}
        ],
```

The `data` property provides the x-axis labels, data points, and connecting lines. Its `datasets` sub-property contains the look-and-feel details of the plots with the actual data. Both the `label` and `data` lists are context data supplied by its view function:

```
            datasets: [{
                fill : true,
                barPercentage: 0.5,
                barThickness: 20,
                maxBarThickness: 70,
                borderWidth : 1,
                minBarLength: 5,
                backgroundColor: "rgba(230,112,16,0.88)",
                borderColor : "rgba(38,22,6,0.88)",
                label: 'Mean HPI values',
                data : [
                  {% for item in values %}
                    "{{ item }}",
                    {% endfor %}
                ]
            }]
        }
    });
</script>
</html>
```

Now, Chart.js can also build multiple line graphs, varieties of bar graphs, pie charts, and doughnuts, all using the same setup as the given line graph. Running the view function with the given Chart.js script will render a line graph, as indicated in *Figure 6.17*.

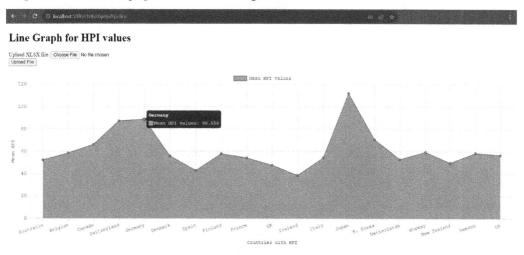

Figure 6.17 – A line graph for HPI values per country

Chart.js supports responsive web design and interactive results, such as the given line graph that provides us with some information during mouse-over on every line dot. Despite its popularity, Chart.js still utilizes HTML canvas, which cannot render efficiently large and complex graphs. Also, it lacks other interactive utilities present in Bokeh and Plotly.

Let us now create graphs using a module friendlier to Python, **Plotly**.

Creating graphs with Plotly

Plotly is also a JS-based library that can render interactive charts and graphs. It is a popular library for various statistical and mathematical projects that require interactive data visualization and 3D graphics effects and can seamlessly plot DataFrame datasets.

To utilize its classes and methods for plotting graphs, install the `plotly` module through the `pip` command:

```
pip install plotly
```

The following view function uses Plotly to create a grouped bar graph about the price and bedroom preferences of buyers categorized according to their furnishing status preference:

```
import json
import plotly
import plotly.express as px


@rendition_bp.route("/plotly/csv/bedprice", methods = ['GET', 'POST'])
async def create_plotly_stacked_bar():
    if request.method == 'GET':
        graphJSON = '{}'
    else:
        ... ... ... ... ... ...
        try:
            df_csv = read_csv(uploaded_file)
            fig = px.bar(df_csv, x='Bedrooms', y='Price',
color='FurnishingStatus', barmode='group')
            graphJSON = json.dumps(fig, cls=plotly.utils.
PlotlyJSONEncoder)
        except:
            raise FileSavingException()
    return render_template('plotly.html', graphJSON=graphJSON,
title="Stacked Bar Graph of Bedroom vs Price per Furnishing Status")
```

Plotly has a `plotly.express` module, which provides several plotting utilities that can set up build graphs with DataFrame as input, similar to `matplotlib`'s methods. In the given `create_plotly_stacked_bar()` view function, the goal is to create a grouped bar chart using the `bar()` method from the `plotly.express` module with the `DataFrame` object's tabular values derived from the uploaded CSV file. The result is a `Figure` in dictionary form containing the details of the desired plot.

After creating the `Figure`, the view function will pass the resulting dictionary to the Jinja2 template for rendition and display using Plotly's JS library. However, JS can only understand the dictionary details if they are in JSON string format. Thus, use the `json.dumps()` method to convert the dictionary `fig` to string.

The following is the Jinja template that will render the graph using the Plotly JS library:

```
<!doctype html>
<html>
    <head>
        <title>Plotly Bar Graph</title>
    </head>
    <body>
        ... ... ... ... ... ...
        {%if graphJSON == '{}' %}
            <p>No plot image.</p>
        {% else %}
            <div id='chart' class='chart'></div>
        {% endif %}
    </body>
    <script src='https://cdn.plot.ly/plotly-latest.js'></script>
    <script type='text/javascript'>
            var graphs = {{ graphJSON | safe }};
            Plotly.plot('chart', graphs, {});
    </script>
</html>
```

The HTML script must reference the latest Plotly library from CDN. Then, a JS script must interpolate the JSON-formatted `Figure` from the view function with a safe filter to spare it from HTML escaping. Also, the JS must apply the `plot()` method of the `Plotly` class library to render the figure through the HTML's `<div>` component. *Figure 6.18* shows the bar graph generated by the `create_plotly_stacked_bar()` view function and displayed by its Jinja template.

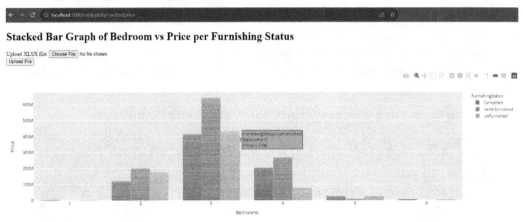

Figure 6.18 – A bar graph created by Plotly

Like Chart.js, the chart provides information regarding a data plot when hovered by the mouse. However, it seems that Chart.js loads faster than Plotly when the data size of the `DataFrame` object's tabular values increases. Also, there is limited support for colors for the background, foreground, and bar shades, so it is hard to construct a more original theme.

The next JS library supports many popular PyData tools and can generate plots directly from `pandas'` `DataFrame`, **Bokeh**.

Visualizing data using Bokeh

Bokeh and Plotly are similar in many ways. They have interactive and 3D graphing features, and both need module installation. However, Bokeh is more Pythonic than Plotly. Because of that, it can transact more with DataFrame objects, especially those with large datasets.

To utilize the library, first install its module using the `pip` command:

```
pip install bokeh
```

Once installed, the module provides a figure class from its `bokeh.plotting` module, which is responsible for setting up the plot configuration. The following view implementation uses Bokeh to create a line graph showing the UK's HPI values through the years:

```python
from bokeh.plotting import figure
from bokeh.embed import components


@rendition_bp.route('/bokeh/hpi/line', methods = ['GET', 'POST'])
def create_bokeh_line():
    if request.method == 'GET':
        script = None
        div = None
    else:
        ... ... ... ... ... ...
        try:
            df = read_excel(uploaded_file, sheet_name=1, skiprows=[1])
            x = df.index.values
            y = df['UK']
            plot = figure(max_width=600, max_height=800,title=None,
toolbar_location="below", background_fill_color="#FFFFCC", x_axis_
label='Period by Quarter ID', y_axis_label='Nominal HPI')
            plot.line(x,y, line_width=4, color="#CC0000")

            script, div = components(plot)
        except:
            raise FileSavingException()
    return render_template('bokeh.html', script=script, div=div,
title="Line Graph of UK's Nominal HPI")
```

After creating the `Figure` instance with the plot details, such as `max_width`, `max_height`, `background_fill_color`, `x_axis_label`, `y_axis_label`, and other related configurations, the view function can now invoke any of its *glyph* or plotting methods, such as `vbar()` for plotting vertical bar graph, `hbar()` for horizontal bar graph, `scatter()` for scatter plots, and `wedge()` for pie charts. The given `create_bokeh_line()` view utilizes the `line()` method to build a line graph with x and y values derived from the tabular values.

After assembling the `Figure` and its plot, call the `components()` function from `bokeh.embed` to wrap the plot instance and extract a tuple of two HTML embeddable components, namely the script that will contain the data of the graph and the `div` component that contains the dashboard embedded in a `<div>` tag. The function must pass these two components to its Jinja template for rendition. The following is the Jinja template that will render the `div` component:

```html
<!DOCTYPE html>
<html lang="en">
    <head>
        <meta charset="utf-8">
        <title>Bokeh HPI</title>
        <script src="https://cdn.bokeh.org/bokeh/release/bokeh-
3.2.2.js"></script>
    </head>
    <body>
        ... ... ... ... ... ...
        {%if div == None and script == None %}
            <p>No plot image.</p>
        {% else %}
        {{ div | safe }}
        {{ script | safe }}
        {% endif %}
    </body>
</html>
```

Be sure to have the latest Bokeh JS library in your HTML script. Since both `div` and `script` are HTML-embeddable components, the template will directly interpolate them with the filter safe. *Figure 6.19* shows the outcome of rendering the `create_bokeh_line()` view function using the datasets:

Line Graph of UK's Nominal HPI

Upload XLSX file: [Choose File] No file chosen
[Upload File]

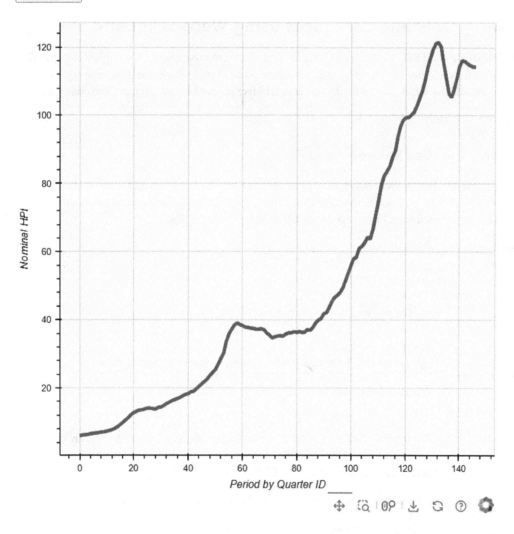

Figure 6.19 – A line graph created by Bokeh

Compared to that of Plotly and Chart.js, the dashboard of Bokeh is so interactive that you can drag the plot in any direction within the canvas. It offers menu options to save, reset, and wheel- or box-zoom the graph. The only problem with Bokeh is its lack of flexibility when going out of the box for more interactive features. But generally, Bokeh has enough utilities and themes to build powerful embeddable graphs.

From the degree of interactivity of the graphs and charts, let us shift our discussions to building real-time visualization approaches with Flask.

Building real-time data plots using WebSocket and SSE

Flask's WebSocket and SSE, discussed in *Chapter 5*, are effective mechanisms for implementing real-time graphical plots. Although other third-party modules can provide Flask with real-time capabilities, these two are still the safest, most flexible, and standard techniques because they are web components.

Let us start with applying WebSocket for real-time charts.

Utilizing the WebSocket

An application can have a WebSocket server that receives data from a form and sends it for plotting to a frontend visualization library. The following `flask-sock` WebSocket server immediately sends all the data it receives from a form page to the Chart.js script for data plotting:

```
@sock.route('/ch06/ws/server/hpi/plot')
def ws_server_plot(ws):
    async def process():
        while True:
            hpi_data_json = ws.receive()
            hpi_data_dict = loads(hpi_data_json)
            json_data = dumps(
                {'period': f"Y{hpi_data_dict['year']} Q{hpi_data_
dict['quarter']}",        'hpi': float(hpi_data_dict['hpi'])})
            ws.send(json_data)
    run(process())
```

The Chart.js script will receive the JSON data as a WebSocket message, scrutinize it, and push it immediately as new labels and dataset values. The following snippet shows the frontend script that manages the WebSocket communication with the `flask-sock` server:

```
const socket = new WebSocket('ws://' + location.host + '/ch06/ws/
server/hpi/plot');
socket.addEventListener('message', msg => {
        const data = JSON.parse(msg.data);
            if (config.data.labels.length === 20) {
                config.data.labels.shift();
```

```
                    config.data.datasets[0].data.shift();
                }
            config.data.labels.push(data.period);
            config.data.datasets[0].data.push(data.hpi);
            lineChart.update();
        });
```

The real-time line chart update occurs at every form submission of the new HPI and date values to the WebSocket server.

Next, let's see how we can use SSE with Redis as the broker storage.

Using SSE

If WebSocket does not fit the requirement, SSE can be a possible solution to real-time data plotting. But first, it requires the installation of the Redis database server and the `redis-py` module and the creation of the `redis-config.py` file for the `Blueprint` approach. The following code shows the configuration of the Redis client instance in our application:

```
from redis import Redis

redis_conn = Redis(
    db = 0,
    host='127.0.0.1',
    port=6379,
    decode_responses=True
)
```

Place this `redis-config.py` file in the project directory with `main.py`.

Now, the role of the Redis server is to create a channel where a view function can push the submitted form data containing the data values. The SSE implementation will subscribe to the message channel, listen to incoming messages, retrieve the recently published message, and yield the JSON data to the frontend plotting library. Our application still uses Chart.js for visualization, and here is a snippet that listens to the event stream for new data plots in JSON format:

```
var source = new EventSource("/ch06/sse/hpi/data/stream");
        source.onmessage = function (event) {
            const data = JSON.parse(event.data);
            if (config.data.labels.length === 20) {
                config.data.labels.shift();
                config.data.datasets[0].data.shift();
            }
            config.data.labels.push(data.period);
            config.data.datasets[0].data.push(data.hpi);
```

```
                lineChart.update();
    };
```

Like the WebSocket approach, the given frontend script will listen to the stream, receive the JSON data, and validate it before pushing it to the current labels and datasets.

Overall, WebSockets and SSE are not limited to web messaging because they can help establish real-time visualization components for many scientific applications to help solve problems that require impromptu analysis.

Let us now focus on how Flask can implement computations that consume more server resources and effort and even create higher contention with other components.

Using asynchronous background tasks for resource-intensive computations

There are implementations of many approximation algorithms and P-complete problems that can create memory-related issues, thread problems, or even memory leaks. To avoid imminent problems when handling solutions for NP-hard problems with indefinite data sets, implement the solutions using asynchronous background tasks.

But first, install the celery client using the pip command:

```
pip install celery
```

Also, install the Redis database server for its broker. Place celery_config.py, which contains celery_init_app(), in the project directory and call the method in the main.py module.

After the setup and installations, create a service package in the Blueprint module folder. ch06-project has the following Celery task in the hpi_formula.py service module found in the internal Blueprint module:

```
@shared_task(ignore_result=False)
def compute_hpi_laspeyre(df_json):
    async def compute_hpi_task(df_json):
        try:
            df_dict = loads(df_json)
            df = DataFrame(df_dict)
            df["p1*q0"] = df["p1"] * df["q0"]
            df["p0*q0"] = df["p0"] * df["q0"]
            print(df)
            numer = df["p1*q0"].sum()
            denom = df["p0*q0"].sum()
            hpi = numer/denom
            return hpi
```

```
        except Exception as e:
            return 0
    return run(compute_hpi_task(df_json))
```

`compute_hpi_laspeyre()` runs an asynchronous task that computes the HPI value using Laspeyre's formula with inputs that include the house price for a particular house preference and the number of customers who bought the house for a specific year. The computation will take longer when given a lot of data, so using an asynchronous Celery task to run the formula when the worst-case scenario happens may improve its execution at runtime.

It is always a good practice to run heavy and resource-intensive computations or processes outside the thread of the view function using asynchronous background tasks. It also employs loose coupling between the request-response transactions and the numerical algorithms, which can help avoid the degradation and starvations of these processes.

Integrating popular numerical and symbolic software into the Flask platform can sometimes save migration time when dealing with existing scientific projects. Let us now explore the capability of Flask to integrate with the Julia language.

Incorporating Julia packages with Flask

Julia is a high-powered compiled programming language that provides mathematical and symbolic libraries. It contains simple syntax for numerical computing and provides better runtime performance for executing its applications.

Although Julia has web frameworks such as Genie, Oxygen, and Bukdu, which can implement Julia-based web applications, it is also possible that Flask applications can run and extract values from Julia functions.

But first, download the latest Julia compiler from `https://julialang.org/downloads/` and install it on your system. Installing an old Julia version into an updated Windows OS will result in a system crash, as indicated in *Figure 6.20*.

```
Please submit a bug report with steps to reproduce this fault, and any error
messages that follow (in their entirety). Thanks.
Exception: EXCEPTION_ACCESS_VIOLATION at 0x915104 -- jl_excstack_state at /cy
gdrive/c/buildbot/worker/package_win64/build/src\rtutils.c:287
in expression starting at none:0
```

Figure 6.20 – System crashes due to Flask running outdated Julia

Let's now take a look at the steps involved in creating and integrating the Julia package into a Flask application.

Creating a custom Julia package

After the installation, go to the project directory of the Flask app through the console and open a Julia shell by running the `julia` command. Then, follow these instructions:

1. Run the command using `Pkg` on the shell.

2. Create a `Julia` package in the Flask app directory by running the following command:

    ```
    Pkg.generate("Ch06JuliaPkg")
    ```

3. Install the `PythonCall` plugin by running the following command:

    ```
    Pkg.add("PythonCall")
    ```

4. Also, install Julia packages such as `DataFrame`, `Pandas`, and `Statistics` for converting and running Python syntax in the Julia environment.

5. Finally, run `Pkg.resolve()` and `Pkg.instantiate()` to finalize the setup.

Next, we'll install the `juliacall` client module and add the Julia-related configuration details to the **TOML** file.

Configuring Julia accessibility in a Flask project

After creating a Julia custom package inside the Flask app, open the app's `config_dev.toml` file and add the following environment variables to integrate Julia into the Flask platform:

*   `PYTHON_JULIAPKG_EXE`: The path to the `julia.exe` file, including the filename (e.g., `C:/Alibata/Development/Language/Julia-1.9.2/bin/julia`).

*   `PYTHON_JULIAPKG_OFFLINE`: Set to `yes` to stop any Julia installation in the background.

*   `PYTHON_JULIAPKG_PROJECT`: The path to the newly created custom Julia package inside the Flask app (e.g., `C:/Alibata/Training/Source/flask/mastering/ch06-web-final/Ch06JuliaPkg/`).

*   `JULIA_PYTHONCALL_EXE`: The path to the Python compiler of the virtual environment, including the filename (e.g., `C:/Alibata/Training/Source/flask/mastering/ch06-web-env/Scripts/python`).

Afterward, install the `juliacall` module through the `pip` command:

```
pip install juliacall
```

After the Flask setup, let us now create the Julia code inside the Julia package.

Implementing Julia functions in the package

After the Python configuration, open `ch06-web-final\Ch06JuliaPkg\src\Ch06JuliaPkg.jl` and create some Julia functions with the imported `PythonCall` package, like in the following snippet:

```
module Ch06JuliaPkg
using PythonCall
const re = PythonCall.pynew() # import re
const np = PythonCall.pynew() # import numpy

function __init__()
    PythonCall.pycopy!(re, pyimport("re"))
    PythonCall.pycopy!(re, pyimport("numpy"))
end

function sum_array(data_list)
    total = 0
    for n in eachindex(data_list)
        total = total + data_list[n]
    end
    return total
end

export sum_array
end # module Ch06JuliaPkg
```

All syntax inside the Julia package must be valid Julia syntax. Thus, the given `sum_array()` is a Julia package. On the other hand, importing Python modules requires the instantiation of `PythonCall` through `pynew()`, and the actual module mapping happens in its `__init__()` initialization method through `pycopy()`.

Creating the Julia service module

To access the functions in the custom Julia package, such as `Ch06JuliaPkg`, create a service module that will activate `Ch06JuliaPkg` and create a Julia module that will execute the Julia commands in Flask in that particular `Blueprint` section. The following is the `\modules\external\services\julia_transactions.py` service module from the external `Blueprint` with the needed `juliacall` executions:

```
import juliacall
from juliacall import Pkg as jlPkg
jlPkg.activate(".\\Ch06JuliaPkg")
jl = juliacall.newmodule("modules.external.services")
jl.seval("using Pkg")
```

```
jl.seval("Pkg.instantiate()")
jl.seval("using Ch06JuliaPkg")
jl.seval("using DataFrames")
jl.seval("using PythonCall")
```

At every startup of the Flask server, the application always activates the Julia package because the application always loads all the services of the blueprints. *Figure 6.21* shows the activation process on the server log of the Flask app:

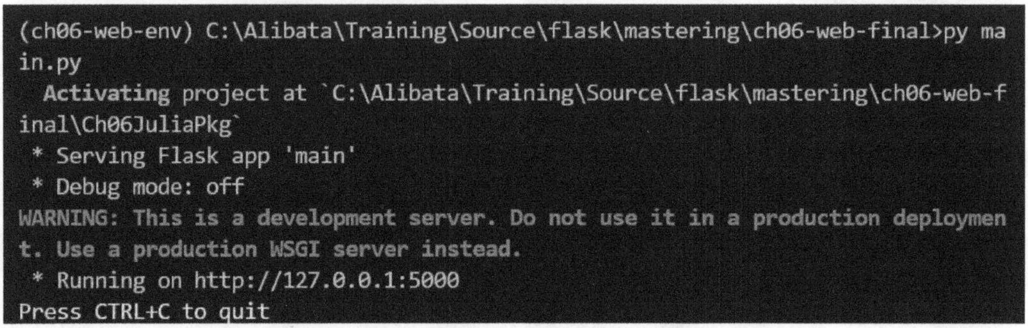

```
(ch06-web-env) C:\Alibata\Training\Source\flask\mastering\ch06-web-final>py ma
in.py
  Activating project at `C:\Alibata\Training\Source\flask\mastering\ch06-web-f
inal\Ch06JuliaPkg`
 * Serving Flask app 'main'
 * Debug mode: off
WARNING: This is a development server. Do not use it in a production deploymen
t. Use a production WSGI server instead.
 * Running on http://127.0.0.1:5000
Press CTRL+C to quit
```

Figure 6.21 – Julia package activation log during server startup

The activation may cause degradation to the startup time of the server, which is a disadvantage for Flask. If this performance glitch worsens, it will be advisable to migrate all implementation to popular Julia web frameworks, such as Oxygen, Genie, and Bukduh, instead of pursuing further the Flask integration.

Now, for the view functions to access Julia functions, add service methods in the `Blueprint` service where the activation happens. In our project, the `modules\external\services\julia_transactions.py` service module implements the following `total_array()` service to expose the `sum_array()` function in `Ch06JuliaPkg`:

```
async def total_array(arrdata):
    result = jl.seval(f"sum_array({arrdata})")
    return result
```

The Julia module or `jl`, using its `seval()` method, is the one that accesses and executes custom or built-in Julia functions in the Flask service. Given that the applications followed all installations and setups given correctly, running `jl.seval()` must not cause any system crash or `HTTP Status 500`. Again, the Python service functions that execute `jl.seval()` must be placed in the service module where the Julia package activation happened.

Summary

Flask 3.0 is the best-fit version of Flask that can build scientific applications because of its asynchronous features and asyncio support. The asynchronous WebSocket, SSE, Celery background tasks, and services, together with the mathematical and computational modules, such as `numpy`, `matplotlib`, `sympy`, `pandas`, `scipy`, and `seaborn`, are the core ingredients in building applications that highlight visualizations, computations, and statistical analysis.

Proven by this chapter, Flask supports LaTeX document generation, updating, and rendition, including its PDF transformation. This feature is crucial in most scientific computing that requires archiving, reporting, and records management.

Flask support on visualization is also clear cut in this chapter, from the real-time data plotting down to the native plots of the `matplotlib` module. Flask can utilize JS-based libraries for data plotting of `DataFrame` object's tabular values seamlessly and in a straightforward manner.

Although not yet stable, the integration of Julia with Flask shows how the interoperability property works with Flask. Using `PythonCall` and `JuliaCall` modules, it is now possible to run existing Julia functions in Flask as long as the setup and configuration are correct.

In conclusion, Flask, particularly the asynchronous version of Flask , is the best option for building web-based scientific applications. The next chapter will discuss how Flask utilizes NoSQL databases and addresses some big data requirements.

7
Using Non-Relational Data Storage

Most applications that use voluminous or big data, and continuously increase every user transaction, do not use relational databases for storage. Applications such as scientific information management systems, sales-related applications, stocks and investment-related software, and location finders are some applications that may utilize data from various types of data structures, such as objects, lists, dictionaries, and bytes. This data can be structured (for example, Excel-formatted medical records and CSV-formatted location data), semi-structured (for example, XML data of sales inventory and emails), and non-structured (for example, images, videos, social media postings, and Word documents). Relational databases do not have the support to manage this data, but **NoSQL** databases do.

NoSQL, which stands for **Not Only SQL**, is a schemaless form of data storage that has no concepts of rows and columns to hold records of information. Like any framework, Flask, when used to build big data applications, can support access to these non-relational databases to manage data for data mining, modeling, analytics, and graphical projections.

The main goal of this chapter is to showcase how to install and configure the different NoSQL databases and how a Flask application can connect to these databases and perform **INSERT**, **UPDATE**, **DELETE**, and **QUERY** transactions.

This chapter will cover the following topics – this will provide you with an initial step toward building big data applications with Flask:

- Managing non-relational data using Apache HBase
- Utilizing the column storage of Apache Cassandra
- Storing search data in Redis
- Handling BSON-based documents with MongoDB
- Managing key-based JSON documents with Couchbase
- Establishing a data relationship with Neo4J

Technical requirements

This chapter highlights a *Tutor Finder* application that accepts students and tutor profiles. The application's main objective is to provide a platform for vying students looking for personal tutors or trainers with different expertise. Aside from profiling, it has a *payment module* for students to pay their tutor's fees based on payment modes, *course modules* for course details, and *search modules* to find the appropriate tutor and student profiles. The application is unique and experimental because it showcases all the NoSQL databases as its backend storage to serve as a specimen for this chapter. On the other hand, the application utilizes the *factory pattern* as its main project structure design. All files are available at `https://github.com/PacktPublishing/Mastering-Flask-Web-Development/tree/main/ch07`.

Managing non-relational data using Apache HBase

One of the most popular NoSQL databases is the **column-oriented database** as it stores its records differently to the row-oriented approach. These column-oriented storage options, which include **HBase**, **Hypertable**, and **Cloudata**, are depicted as sparse and multidimensional sorted maps that hold each record with a unique index key, called the *row key*, a *column key*, which organizes the data, and a *timestamp*. Each form of storage is equivalent to a relational table with records identified by the *row key*, which is of the `byte[]` type. This `byte[]` data is handled by the column families, which are composed of *column qualifiers* or *columns*, each stored in a *cell*. Each column qualifier addresses one *data field* with a *timestamp* that keeps track of the versions of each column field in every update.

Regarding column-oriented databases, this chapter will concentrate solely on integrating Apache HBase into our Flask application. Initially, like any database, we'll design the HBase tables first before integrating them into Flask.

Designing HBase tables

One of the leverages of using relational databases is the availability of many design tools that can assist us with planning and organizing the table schema using different normalization levels. Only a few data modeling tools, such as **DBSchema**, **Moon Modeler**, **Hackolade**, and **draw.io**, have the support to provide design for de-normalized tables of NoSQL databases, such as HBase. In this chapter, we'll use *draw.io* to visualize the `payments` and `bookings` HBase tables using the UML class diagramming approach. *Figure 7.1* shows the UML design for the `payments` and `bookings` tables:

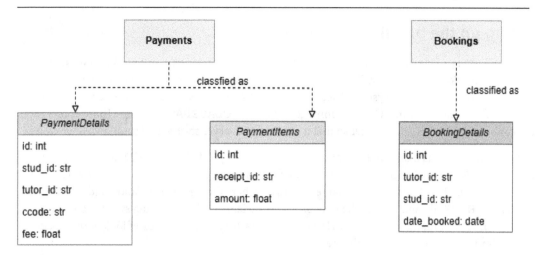

Figure 7.1 – HBase table design for payments and bookings

The payments and bookings contexts signify the two tables of the HBase database. The payments table has two column families, namely PaymentDetails and PaymentItems. The bookings table has one column family, BookingDetails.

Note that PaymentDetails has id, stud_id, tutor_id, ccode, and fee as column qualifiers, while PaymentItems has id, receipt_id, and amount. Furthermore, BookingDetails has id, tutor_id, stud_id, and date_booked columns. A sample record in JSON format will look like this:

```
"payments": {
  "details:id": 1001, "details:stud_id": "STD-001",
  "details:tutor_id": "TUT-001", "details:ccode": "PY-100",
  "details:fee": 5000.00", "items:id": 1001,
  "items:receipt_id": "OR-901", "items:amount": 3000.00"
}
"bookings" : {
  "details:id": 101, "details:tutor_id": TUT-002",
  "details:stud_id": "STD-201",
  "details:date_booked": "2023-10-10"
}
```

The actual name of the column qualifier, when accessed via the Flask app, is a *concatenation* of the column family and the column name itself. An example from the given record is details:stud_id.

Now that we've designed the table structure, let's look at how we can install and configure the Apache HBase and Apache Hadoop platforms. We'll start by using Java.

Setting up the baseline requirements

Apache HBase is a Java-based platform, and all its components depend on **Java Virtual Machine** (**JVM**). So, download the *Java 11* ZIP file from `https://jdk.java.net/java-se-ri/11-MR2` and unzip it to your local filesystem. Then, create a `JAVA_HOME` system environment variable in your Windows, Linux, or macOS environment and update `CLASSPATH` to help HBase access the JDK commands in Java's `/bin` folder so that it can perform server startup and shutdown operations.

Since Apache HBase is a distributed storage that uses **Hadoop File System** (**HDFS**), download the `hadoop-3.3.6/hadoop-3.3.6.tar.gz` file from `https://hadoop.apache.org/releases.html` and unzip it to the same local drive where the HBase installation folder is. Create a `HADOOP_HOME` file and update the `CLASSPATH` variable of the operating system to make the Hadoop `/bin` commands available to HBase during startup. Let's have a brief look at the Apache Hadoop framework to understand this better.

Configuring Apache Hadoop

Apache Hadoop is a Java-based framework that manages scalable big data processing across a distributed cluster setup. It is popular due to its *MapReduce* algorithm, which performs data processing in parallel across cluster(s) of nodes, making the framework's distributed operations fast. Moreover, the framework has **HDFS** a filesystem. This is where it contains the input and output datasets.

The *MapReduce* process starts with these big datasets being stored in HDFS by the Flask application. They're passed through the internal Hadoop servers, which run the `Map()` function to break down this data into tuples of key-value pairs. Then, these groups of key-value data blocks undergo another process called `Reduce()`, which is performed by other Hadoop servers. This exposes these data blocks to various reduce functions, such as summation, averaging, concatenation, compression, ordering, and shuffling, and then saves them as output datasets in HDFS.

Aside from the *MapReduce* distributed data process, HBase needs Hadoop's HDFS because of the high latency batch processing operations it gives to the HBase platform. In return, HBase enables read/write transactions to access the data stored in HDFS and can provide a thrift server so that third-party applications can access the big datasets.

> **Important note**
>
> A *thrift server* is a Hive-compatible interface in HBase that enables multi-language support, allowing applications developed in Python, Java, C#, C++, NodeJS, Go, PHP, and JavaScript to access big data. The term *Hive*, on the other hand, refers to a client application that runs on top of Hadoop and has powerful SQL utilities that are used to implement CRUD operations for large datasets.

Here's the step-by-step procedure to set up a Hadoop platform with a single-node cluster:

1. Go to the installation folder of *Apache Hadoop 3.3.6* and open /etc/hadoop/core-site. xml. Then, set the fs.defaultFS property. This indicates the default location of *NameNode* (master node) in the cluster – in our case, the default filesystem. Its value is a URL address to which *DataNode* (slave node) will send a heartbeat. *NameNode* contains the metadata of the data stored in HDFS, while *DataNode* contains the big datasets. Here is our core-site.xml file:

```
<configuration>
    <property>
        <name>fs.defaultFS</name>
        <value>hdfs://localhost:9000</value>
    </property>
</configuration>
```

2. Inside the same installation folder, create a custom folder called data with two sub-folders called datanode and namenode, as shown in *Figure 7.2*. These folders will eventually contain the configuration files of *DataNode* and *NameNode*, respectively:

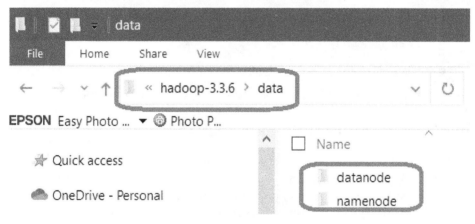

Figure 7.2 – The DataNode and NameNode config folders

3. Next, open /etc/hadoop/hdfs-site.xml and declare the newly created datanode and namenode folders as the final config locations of their respective nodes. Also, set the dfs.replication property to 1 since we only have a single node cluster for our *Tutor Finder* project. Here is our hdf-site.xml file:

```
<configuration>
  <property>
        <name>dfs.replication</name>
        <value>1</value>
    </property>
```

```
    <property>
        <name>dfs.namenode.name.dir</name>
        <value>
    file:///C:/Alibata/Development/Database/hadoop-3.3.6/data/
namenode</value>
    </property>
    <property>
        <name>dfs.datanode.data.dir</name>
        <value>
    file:///C:/Alibata/Development/Database/hadoop-3.3.6/data/
datanode</value>
    </property>
</configuration>
```

4. Since our project will use Hadoop installed in Windows, download the `hadoop-3.3.6-src.tar.gz` file from `https://hadoop.apache.org/releases.html` and compile the Hadoop source files using Maven to generate Hadoop binaries for Windows, such as `winutils.exe`, `hadoop.dll`, and `hdfs.dll`. Drop these files into the `/bin` folder.

5. Format the new active *NameNode*(s) by running the following command at the command line:

```
hdfs namenode -format
```

This command will clean up the *NameNode*(s) if they have existing stored metadata.

Now, we can start setting up a version of Apache HBase that's compatible with Apache Hadoop 3.3.6.

Configuring Zookeeper and Apache HBase

Apache HBase depends on Apache Zookeeper when running its clusters, so the next step is to install and configure a Zookeeper server. **Apache Zookeeper** is a high-performance service that manages distributed and cloud-based applications by providing synchronization and centralized services and maintains details of these applications. Note that this project utilizes Zookeeper bundled with HBase, so you shouldn't install Zookeeper separately unless the setup involves multiple clusters.

Now, download *Apache HBase 2.5.5*, the most compatible HBase distribution, to Apache Hadoop 3.3.6. Unzip it to the folder where Hadoop resides. Then, configure HBase by performing the following steps:

1. First, create an `HBASE_HOME` system environment variable that registers the HBase installation folder.

2. Create two folders inside the installation folder, `hbase` and `zookeeper`. These will serve as the root folders of HBase and the built-in Zookeeper server, respectively.

3. Inside the installation folder, open /conf/hbase-site.xml. Here, set the hbase.
    rootdir property so that it points to the hbase folder and the hbase.zookeeper.
    property.dataDir property so that it points to the zookeeper folder. Now, register
    the hbase.zookeeper.quorum property. This will indicate the Zookeeper server's host.
    Then, set the hbase.cluster.distributed property. This will specify the type of HBase
    server setup. The following is our hbase-site.xml file:

```
<configuration>
  <property>
    <name>hbase.cluster.distributed</name>
    <value>false</value>
  </property>
  <property>
    <name>hbase.tmp.dir</name>
    <value>./tmp</value>
  </property>
  <property>
    <name>hbase.rootdir</name>
    <value>
     file:///C:/Alibata/Development/Database/hbase-2.5.5/hbase</
value>
  </property>
  <property>
    <name>hbase.zookeeper.property.dataDir</name>
    <value>
       /C:/Alibata/Development/Database/hbase-2.5.5/zookeeper</
value>
  </property>
  <property>
     <name>hbase.zookeeper.quorum</name>
    <value>localhost</value>
  </property>
    ... ... ... ... ... ...
</configuration>
```

4. Next, open /bin/hbase.cmd if you're on Windows and search for the java_arguments
    property. Remove %HEAP_SETTINGS% so that the new statement will be as follows:

```
set java_arguments=%HBASE_OPTS% -classpath "%CLASSPATH%" %CLASS%
%hbase-command-arguments%
```

5. Open /conf/hbase-env.cmd and add the following JAVA_HOME and HBASE_* details to the file:

```
set JAVA_HOME=%JAVA_HOME%
set HBASE_CLASSPATH=%HBASE_HOME%\lib\client-facing-thirdparty\*
set HBASE_HEAPSIZE=8000
set HBASE_OPTS="-Djava.net.preferIPv4Stack=true"
set SERVER_GC_OPTS="-verbose:gc" "-Xlog:gc*=info:stdout"
"-XX:+UseG1GC" "-XX:MaxGCPauseMillis=100" "-XX:-ResizePLAB"
%HBASE_GC_OPTS%
set HBASE_USE_GC_LOGFILE=true
set HBASE_JMX_BASE="-Dcom.sun.management.jmxremote.ssl=false"
"-Dcom.sun.management.jmxremote.authenticate=false"
set HBASE_MASTER_OPTS=%HBASE_JMX_BASE% "-Dcom.sun.management.
jmxremote.port=10101"
set HBASE_REGIONSERVER_OPTS=%HBASE_JMX_BASE% "-Dcom.sun.
management.jmxremote.port=10102"
set HBASE_THRIFT_OPTS=%HBASE_JMX_BASE% "-Dcom.sun.management.
jmxremote.port=10103"
set HBASE_ZOOKEEPER_OPTS=%HBASE_JMX_BASE% -Dcom.sun.management.
jmxremote.port=10104"
set HBASE_REGIONSERVERS=%HBASE_HOME%\conf\regionservers
set HBASE_LOG_DIR=%HBASE_HOME%\logs
set HBASE_IDENT_STRING=%USERNAME%
set HBASE_MANAGES_ZK=true
```

Our *Tutor Finding* project uses *Java JDK 11* to run the HBase database server. So, the usual garbage collectors that work with Java 1.8 are now deprecated and invalid. The most suitable GC option for the HBase platform that uses Java JDK 11 to achieve better server performance is *G1GC*.

6. Finally, go to the /bin folder and run the start-hbase command to start the server. *Figure 7.3* shows a snapshot of the HBase logs while at startup:

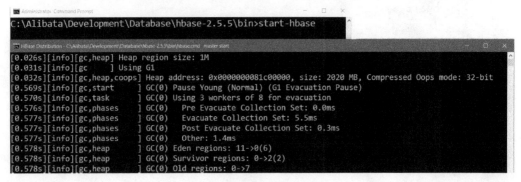

Figure 7.3 – Starting up the HBase server

7. To stop the server, run `stop-hbase`, then `hbase master stop  --shutDownCluster`.

The HBase server log, shown in *Figure 7.4*, shows the Zookeeper server fetching all the Hadoop configuration files to handle all the Hadoop cluster(s) and providing the necessary operational services:

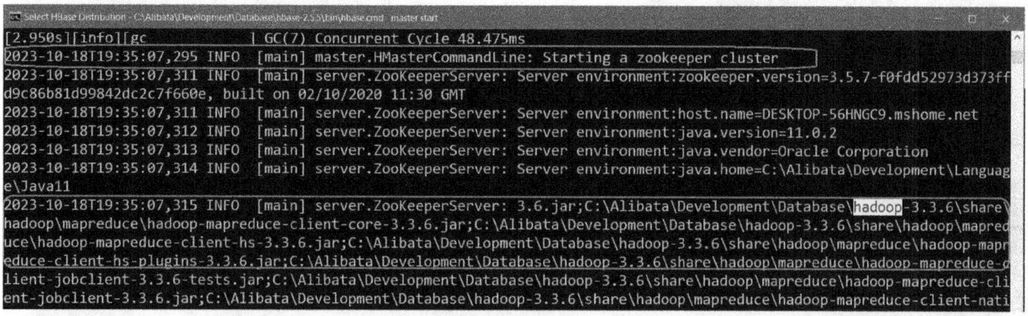

Figure 7.4 – Starting up Zookeeper with the Hadoop cluster(s)

Now that we've made these server configurations, let's run the HBase client so that we can create our `payments` and `bookings` tables.

Setting up the HBase shell

Apache HBase has a built-in interactive shell client created via Java that can communicate with HDFS for big data. The command to spawn the shell is `hbase  shell`. In Apache HBase 2.5.5, running this command will give us the following error message:

```
This file has been superceded by packaging our ruby files into a jar
and using jruby's bootstrapping to invoke them. If you need to source
this file for some reason it is now named 'jar-bootstrap.rb' and is
located in the root of the file hbase-shell.jar and in the source tree
at 'hbase-shell/src/main/ruby'.
```

The reason behind this error is the missing JAR files that the client shell requires from the installation. So, to fix this error, download `jansi-1.18.jar` and `jruby-complete-9.2.13.0.jar` from the Maven repository and place them in the `/lib` directory. Then, go to the `/lib` folder and run the following command to open the client shell:

```
java -cp hbase-shell-2.5.5.jar;client-facing-thirdparty/*;* org.jruby.
JarBootstrapMain
```

Figure 7.5 shows the given command opening the HBase shell:

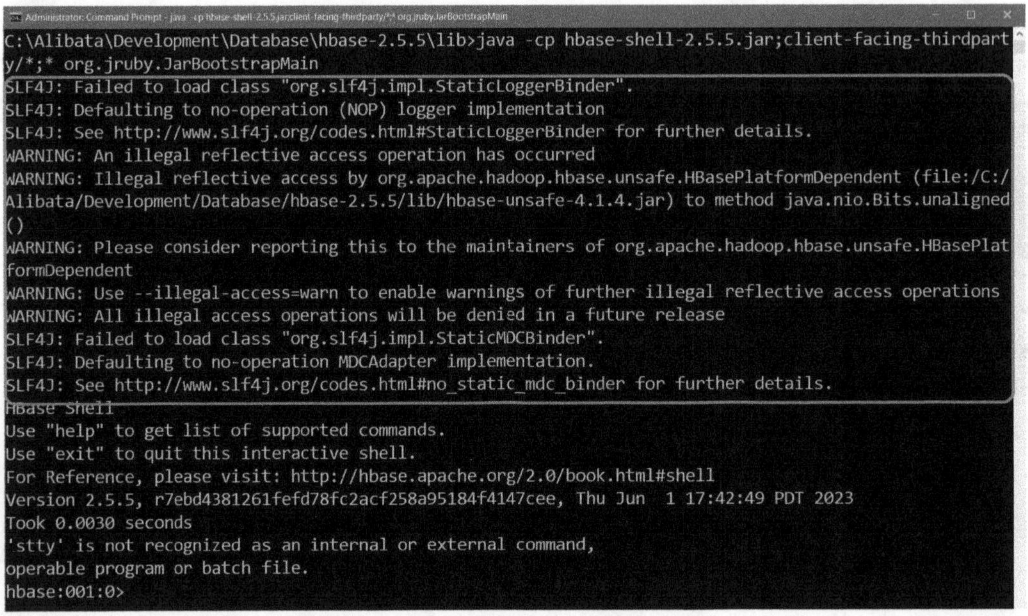

```
C:\Alibata\Development\Database\hbase-2.5.5\lib>java -cp hbase-shell-2.5.5.jar;client-facing-thirdpart
y/*;* org.jruby.JarBootstrapMain
SLF4J: Failed to load class "org.slf4j.impl.StaticLoggerBinder".
SLF4J: Defaulting to no-operation (NOP) logger implementation
SLF4J: See http://www.slf4j.org/codes.html#StaticLoggerBinder for further details.
WARNING: An illegal reflective access operation has occurred
WARNING: Illegal reflective access by org.apache.hadoop.hbase.unsafe.HBasePlatformDependent (file:/C:/
Alibata/Development/Database/hbase-2.5.5/lib/hbase-unsafe-4.1.4.jar) to method java.nio.Bits.unaligned
()
WARNING: Please consider reporting this to the maintainers of org.apache.hadoop.hbase.unsafe.HBasePlat
formDependent
WARNING: Use --illegal-access=warn to enable warnings of further illegal reflective access operations
WARNING: All illegal access operations will be denied in a future release
SLF4J: Failed to load class "org.slf4j.impl.StaticMDCBinder".
SLF4J: Defaulting to no-operation MDCAdapter implementation.
SLF4J: See http://www.slf4j.org/codes.html#no_static_mdc_binder for further details.
HBase Shell
Use "help" to get list of supported commands.
Use "exit" to quit this interactive shell.
For Reference, please visit: http://hbase.apache.org/2.0/book.html#shell
Version 2.5.5, r7ebd4381261fefd78fc2acf258a95184f4147cee, Thu Jun  1 17:42:49 PDT 2023
Took 0.0030 seconds
'stty' is not recognized as an internal or external command,
operable program or batch file.
hbase:001:0>
```

Figure 7.5 – Invoking the HBase shell

The warnings that appear in the logs are due to the collisions of SL4J log libraries from Hadoop's `/common/lib` and HBase's `/lib/client-facing-thirdparty`. Removing redundancies from among these logger libraries can fix these warnings. Now that we've finalized the table designs and set up the HBase environment, we'll build the HBase tables.

Creating the HBase tables

The HBase client application has different commands ready to pursue administrative, table, data manipulation, cluster-related, and general operations for HBase datasets. It can interact with HBase storage based on these commands. *Figure 7.6* shows common general-purpose commands, such as `whoami`, which checks the user information that's been logged in the shell, and `version`, which specifies the version of the running HBase. It also shows the `status` command, which specifies the status of the server and the average load value – that is, the average number of regions per region server across all servers:

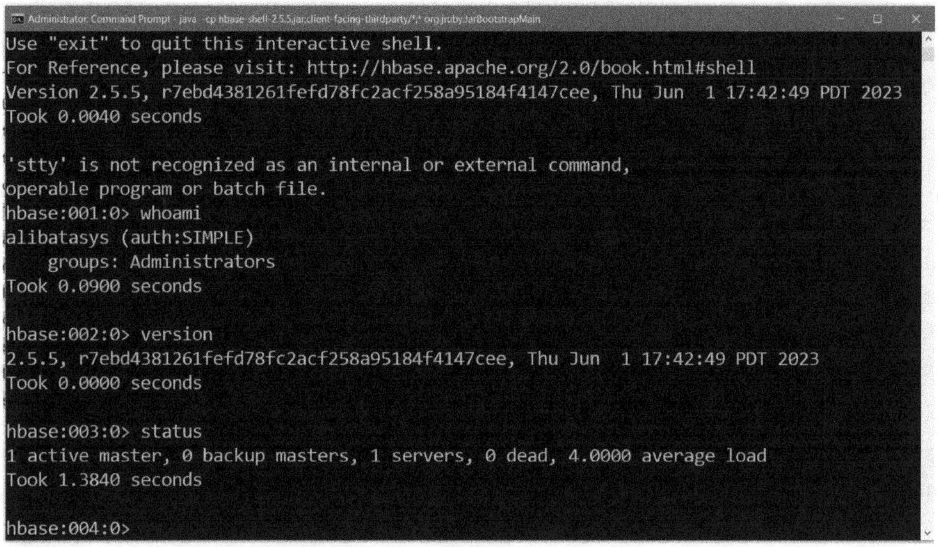

Figure 7.6 – Running general-purpose HBase commands

Most enterprise applications rely on DBA for table design and creation. For HBase database users, the data modelers allow the application's data layer to generate the tables at every server startup. But often, developers build the tables before development using the HBase shell. In our application, for instance, the `payments` and `bookings` tables are generated beforehand using the HBase `create` command. *Figure 7.7* shows how to use the `create` command:

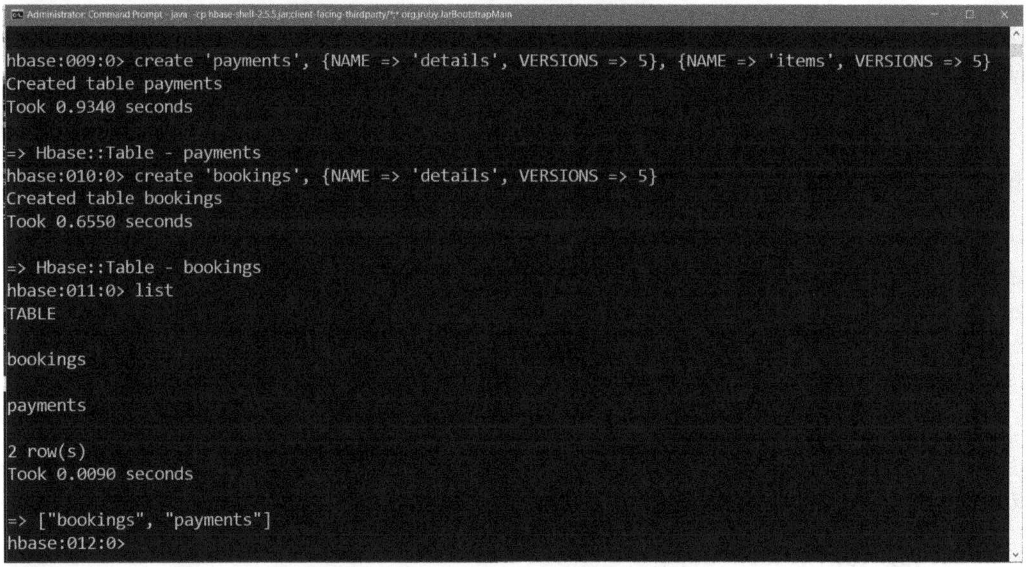

Figure 7.7 – Using the create and list commands

To create an HBase table, use the `create` command with the following parameters:

- The single or double-quoted table name (for example, `'bookings'` or `"payments"`).
- The quoted column family's name(s) or dictionaries containing the column family's attributes, including `NAME` and other properties such as `VERSIONS`, with their values all in quotes.

Figure 7.7 shows the `payments` table being created with the `details` and `items` column families, each with only a maximum of five versions. The `VERSIONS` property sets the maximum allowable number of updates imposed on the column family's columns. So, if the `payments` table has `VERSIONS` set to 5, the maximum number of allowed updates on the values of its column families is at most five times only. The timestamp that's given to each cell storage of the column qualifier traces these updates.

Now, to view all the tables, use the `list` command. There is also the `describe` command, which you can use to check the metadata information of each table (for example, `describe "bookings"`). To drop a table, disable the table first (for example, `disable "bookings"`) before dropping it (for example, via `drop "bookings"`).

After creating the tables in HBase storage, we can integrate our HBase database into our Flask application.

Establishing an HBase connection

Many modern Python libraries that can integrate HBase into Flask are proprietary, such as this CData Python driver (`https://www.cdata.com/drivers/hbase/download/python/`), which can utilize SQLAlchemy to manage HBase storage. But there is one reliable and popular Python driver in the *PyPI* repository that can integrate any Python application into Hbase: the HappyBase library.

The happybase module is a standard Python library that uses the *Python Thrift* library to connect to any HBase database using the *Thrift* service, which is already part of the Apache HBase 2.5.5 platform.

To utilize the `happybase` module, install it using the `pip` command:

```
pip install happybase
```

For *Tutor Finder* to establish a connection to HBase and create multiple threads for reusable connections, the application factory function in `__init__.py` must import `ConnectionPool` from the happybase module and provide it the `host` and `port` values of the Thrift gateway, as well as the number of connections in the pool. The following script shows the application factory function, `create_app()`, that initiates the `happybase` setup:

```
from flask import Flask
import toml
import happybase


def create_app(config_file):
    app = Flask(__name__)
```

```
    app.config.from_file(config_file, toml.load)
    global pool
    pool = happybase.ConnectionPool(size=5, host='localhost',
port=9090)
    with app.app_context():
        import modules.api.hbase.payments
        import modules.api.hbase.bookings
```

The entry point to the HBase platform is the `Connection` class. The `Connection` class creates an open socket to the HBase database through the Thrift service. But `ConnectionPool` provides faster access than the single `Connection` instance, especially if the Flask application is in asynchronous mode. The only requirement is for the application to use a `with` context manager for the connection pool to spawn a `Connection` instance, assign a thread to it, and dispose of the thread when the transaction ends, eventually returning the connection's state to the pool.

Let's use `ConnectionPool` to build the repository layer.

Building the repository layer

The `ConnectionPool` instance from `create_app()` provides the `Connection` instance that implements the CRUD transactions. But it needs a `with` context manager to spawn a `Connection` instance or reuse a connection state from the pool so that the thread can run the CRUD transactions using the `happybase` utility methods. The following script shows the repository class that uses the `ConnectionPool` instance to implement the CRUD transactions for the `payments` table:

```
from typing import Dict, List, Any
from happybase import Table


class PaymentRepository:
    def __init__(self, pool):
        self.pool = pool


    def upsert_details(self, rowkey, tutor_id, stud_id, ccode, fee) ->
bool:
        record = {'details:id' : str(rowkey), 'details:tutor_id':
tutor_id, 'details:stud_id': stud_id, 'details:course_code': ccode,
'details:total_package': str(fee)}
        try:
            with self.pool.connection() as conn:
                tbl:Table = conn.table("payments")
                tbl.put(row=str(rowkey).encode('utf-8'), data=record)
            return True
        except Exception as e:
            print(e)
        return False
```

The `PaymentRepository` class requires a `ConnectionPool` instance (`pool`) as a constructor argument for its instantiation. The `pool` object has a `connection()` method that returns an HBase connection that provides the `happybase` utility methods for CRUD transactions. With the help of a thread, the connection object has a `table()` utility that accesses the HBase table and returns a `Table` object that provides several methods to execute database transactions, such as `put()`.

The `put()` method performs both *INSERT* and *UPDATE* transactions. It requires `rowkey` as its primary parameter for inserting a record in dictionary format. The dictionary record consists of a *column qualifier-value pair*, wherein all the values should be byte strings or any type converted into bytes by the `encode('utf-8')` method. Also, `rowkey` should always be a byte string. The given `upsert_details()` inserts payment records into the `payments` table of the HBase database.

Aside from `put()`, the `Table` object has a `delete()` method that deletes a record using its `rowkey`. The following `delete_payment_details()` function of `PaymentRepository` highlights payment details being deleted from the `payments` table:

```python
def delete_payment_items(self, rowkey) -> bool:
    try:
        with self.pool.connection() as conn:
            tbl:Table = conn.table("payments")
            tbl.delete(rowkey.encode('utf-8'), columns=["items"])
        return True
    except Exception as e:
        print(e)
    return False
```

Aside from `rowkey`, the `delete()` method needs the name of the column family or families in its `columns` parameter, which means deleting the whole record. But sometimes, deletion requires only removing the column qualifier(s) or column(s) instead of the entire row so that only the column qualifier name(s) appear in the `columns` parameter.

The `Table` object has a `rows()` method that returns a `Tuple` value or list of tuples, each containing `rowkey` and the record in `bytes`. This method has two parameters, the *row key* and *column family or families* of the data records in the search. Here, `select_records_ids()` returns a list of payment records based on a selected list of row keys with some specified column families:

```python
def select_records_ids(self, rowkeys:List[str], cols:List[str] =
None):
    try:
        with self.pool.connection() as conn:
            tbl:Table = conn.table("payments")
            if cols == None or len(cols) == 0:
                rowkeys = tbl.rows(rowkeys)
                rows = [rec[1] for rec in rowkeys]
            else:
```

```
                    rowkeys = tbl.rows(rowkeys, cols)
                    rows = [rec[1] for rec in rowkeys]
            records = list()
            for r in rows:
                records.append({key.decode():value.decode() for key,
value in r.items()})
            return records
        except Exception as e:
            print(e)
        return None
```

The `rows()` method returns a `Tuple` value or tuples containing the *row key* as the first element and the *records* in dictionary format as the second element. Thus, we only need to shift the dictionary part using a list comprehension, as depicted in the code. Also, decoding each dictionary of fields will avoid JSON errors in Flask during its `Response` generation.

For its input, the `select_records_ids()` function can accept JSON requests containing the row keys of the records in search, as shown here:

```
{
    "rowkeys": ["1", "2", "101"],
    "cols": []
}
```

Alternatively, it can accept both the row keys and the column families, such as for the following request data:

```
{
    "rowkeys": ["1", "2", "101"],
    "cols": ["details"]
}
```

It can also accept specific column qualifiers that need to appear in the search output, as shown in the following code:

```
{
    "rowkeys": ["1", "2", "101"],
    "cols": ["details:stud_id", "details:tutor_id", "details:course_
code"]
}
```

Another way of retrieving data in the `happybase` module is through the `scan()` method, which returns a generator of tuples – similar tuples returned by `rows()`. Here, `select_all_records()` shows how to use `scan()` to retrieve all the payment records:

```
def select_all_records(self):
        records = []
        try:
            with self.pool.connection() as conn:
                tbl:Table = conn.table("payments")
                datalist = tbl.scan(columns=['details', 'items'])
                for key, data in datalist:
                    data_str = {k.decode(): v.decode() for k, v in
data.items()}
                    records.append(data_str)
                return records
        except Exception as e:
            print(e)
        return records
```

The method requires a `for` loop to extract all these records from the generator and decode all the details, which includes the column qualifier as the key and the value of each key, before adding them to a list. This retrieval consumes less running time than using lots of list and dictionary comprehensions with `rows()`.

Another advantage of using `scan()` instead of `rows()` is its advanced feature to filter records using the predicate conditions on columns, similar to a `WHERE` clause in a SQL statement. The following query transaction retrieves all payment records with a specific *tutor ID* specified by the client:

```
def select_records_tutor(self, tutor_id):
    records = []
    try:
        with self.pool.connection() as conn:
            tbl:Table = conn.table("payments")
            datalist = tbl.scan(columns=["details", "items"],
filter="SingleColumnValueFilter('details', 'tutor_id',
=,'binary:{}')".format(tutor_id))
            for key, data in datalist:
                data_str = {k.decode(): v.decode() for k, v in data.
items()}
                records.append(data_str)
        return records
    except Exception as e:
        print(e)
    return records
```

The scan() method has a filter parameter that accepts a *filter string* constituting the *filter class* and its *constructor arguments*, which will streamline the search. The filter parameter indicates what filter class to instantiate to build the appropriate search constraints. The given select_records_tutor() function uses SingleColumnValueFilter, which filters rows based on a value constraint given to the *column family*, *column qualifier*, *conditional operator*, and BinaryComparator (binary). Aside from SingleColumnValueFilter, here are some widely used types of filter classes that can create search conditions for the scan() method:

- RowFilter: Accepts a comparison operator and the preferred comparator (for example, ByteComparator, RegexStringComparator, and so on) needed to compare the indicated value with each row key.

- QualifierFilter: Accepts a conditional operator and the preferred comparator (for example, ByteComparator, RegexStringComparator, and so on) needed to compare the column qualifier name of each row with the given value.

- ColumnRangeFilter: Accepts the minimum range column and maximum range column and then checks if the indicated value falls between the range column values.

- ValueFilter: Accepts a conditional operator and the preferred comparator needed to compare the value to each field value.

Aside from BinaryComparator, other comparators that provide conversion and comparison methods for a filter class are BinaryPrefixComparator, RegexStringComparator, and SubStringComparator.

In the next section, we'll apply PaymentsRepository so that we can store and retrieve payment details in and from the payments table.

Applying a repository to API functions

The following API function uses upsert_details() from PaymentRepository to perform an *INSERT* transaction after receiving JSON request data from the client:

```
from modules import pool

@current_app.post('/ch07/payment/details/add')
def add_payment_details():
    data = request.get_json()
    repo = PaymentRepository(pool)
    result = repo.upsert_details(data['id'], data['tutor_id'],
data['stud_id'], data['ccode'], data['fee'])
    if result == False:
        return jsonify(message="error encountered in payment details
record insert"), 500
    return jsonify(message="inserted payment details record"), 201
```

The repository's `select_all_records()` provides the following `list_all_payments()` function to render all the records from the `payments` table:

```
from modules import pool

@current_app.get('/ch07/payment/list/all')
def list_all_payments():
    repo = PaymentRepository(pool)
    results = repo.select_all_records()
    return jsonify(records=results), 201
```

Here, `pool` is the `ConnectionPool` instance that was created in the `create_app()` factory from the `__init__.py` file of the `modules` package.

Now, for `happybase` to work, start up the *thrift server*. Let's showcase the Apache Thrift framework in the HBase platform.

Running the thrift server

Apache Thrift is an RPC-based framework that provides an interface for cross-language service development, such as building clients to HBase using Python, Java, C++, C#, or PHP to access HBase tables. To start the built-in thrift server of Apache HBase 2.5.5, run the `hbase thrift start` command. *Figure 7.8* shows the logs after starting up the thrift server:

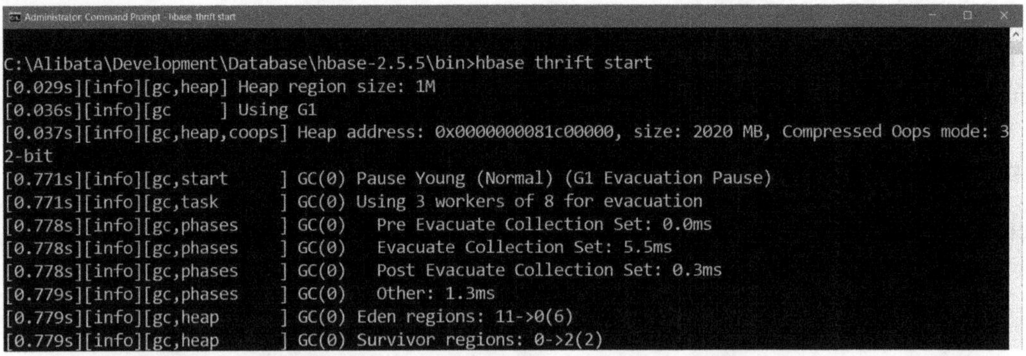
Administrator: Command Prompt - hbase thrift start
C:\Alibata\Development\Database\hbase-2.5.5\bin>hbase thrift start
[0.029s][info][gc,heap] Heap region size: 1M
[0.036s][info][gc   ] Using G1
[0.037s][info][gc,heap,coops] Heap address: 0x0000000081c00000, size: 2020 MB, Compressed Oops mode: 32-bit
[0.771s][info][gc,start   ] GC(0) Pause Young (Normal) (G1 Evacuation Pause)
[0.771s][info][gc,task   ] GC(0) Using 3 workers of 8 for evacuation
[0.778s][info][gc,phases   ] GC(0)   Pre Evacuate Collection Set: 0.0ms
[0.778s][info][gc,phases   ] GC(0)   Evacuate Collection Set: 5.5ms
[0.778s][info][gc,phases   ] GC(0)   Post Evacuate Collection Set: 0.3ms
[0.779s][info][gc,phases   ] GC(0)   Other: 1.3ms
[0.779s][info][gc,heap   ] GC(0) Eden regions: 11->0(6)
[0.779s][info][gc,heap   ] GC(0) Survivor regions: 0->2(2)

Figure 7.8 – Running a built-in HBase thrift server

Apache Thrift will only run if Apache HBase, Hadoop, and Zookeeper are all running.

The `happybase` module is non-Flask-specific, which means any Python client can use it to connect to the `HBase` server. The thrift server will always bridge between the client and HBase considering that the Python library uses the *Thrift 1* or *2* library to establish a connection. The `happybase` module uses the Thrift 1 library.

Now that we've created Flask repository transactions for the HBase database, let's explore a type of NoSQL storage that uses columns and rows for data storage.

Utilizing the column storage of Apache Cassandra

Apache Cassandra is a *column-family* NoSQL database that can also hold large amounts of data. HBase can share big data across its regions using auto-sharding, which makes HBase horizontally scalable. Likewise, Cassandra supports adding more nodes horizontally to improve server throughput, a characteristic of horizontal scaling. But there are also some differences between the two storages in terms of their architectures, table read and write performances, data modeling approaches, and query languages.

Let's start by designing our `course`, `degree_level`, `student`, and `student_perf` Cassandra tables.

Designing Cassandra tables

Unlike HBase, Cassandra stores its data per row, grouping all column fields, thus why the data model approach is a *column family*. Its database transactions are atomic, isolated, and durable, but with eventual or tunable consistency, so it doesn't offer an **Atomicity, Consistency, Isolation, Durability (ACID)** model like a **relational database management system (RDBMS)**. Sometimes, the Cassandra setup favors higher-availability performance over atomic and isolated transactions.

This project used *draw.io* to design the tables in Cassandra using UML class diagrams. *Figure 7.9* shows the data model for the project's Cassandra data storage:

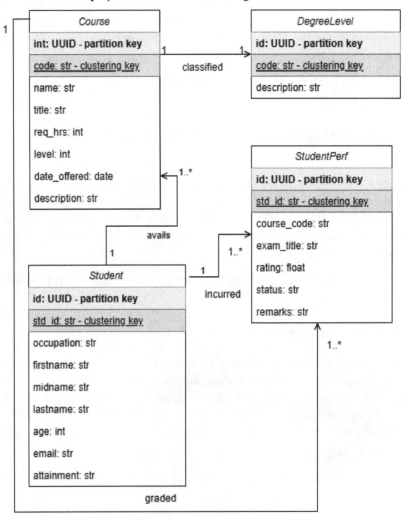

Figure 7.9 – Cassandra table designs using UML

Each Cassandra table must have a primary key. But, unlike in RDBMS, a primary key in *column-family* storage has at least one *partition key* and zero or more *clustering keys*. Since Cassandra storage runs on a distributed environment of clusters and nodes, the *partition key* evenly distributes the row data across the clustered storage. On the other hand, the *clustering key* sorts and manages the rows of data in a table. Also, the performance of the query transactions is the ultimate basis of the table design; the faster the query will be, the better the design.

Let's install Apache Cassandra so that we can realize our table design.

Installing and configuring Apache Cassandra

Download the ZIP file for Apache Cassandra at `https://cassandra.apache.org/_/download.html`. The **general availability (GA)** versions 3.x and below support Windows but not version 4.x. Since the project uses version 4.1.3, Windows PowerShell with WSL2 installed must be used to configure and run the server.

After unzipping the file, enable the Ubuntu firewall using the `sudo` command:

```
sudo ufw enable
```

Then, allow non-WSL clients to access ports *7000* (port for cluster communication), *9042* (default port for client access), and *7199* (port for JMX) using the following `sudo` commands:

```
sudo ufw allow 7000
sudo ufw allow 9042
sudo ufw allow 7199
```

Apache Cassandra 4.1.3 requires *Java 11* as its virtual machine, so run the following `sudo` command to install Java SDK 11 in the Ubuntu environment:

```
sudo apt install openjdk-11-jdk
```

After, go to the `/conf` directory of Cassandra's installation folder and open the `jvm11-server.options` file. Comment all the *CMS* GC option details and uncomment *G1GC*, the default GC option for Java 11.

Finally, run the following command from the `/conf` directory:

```
cassandra -f
```

To shut down the Cassandra server, use the `nodetool drain` command.

Now, let's open the Cassandra shell client to create the project's tables and learn **Cassandra Query Language (CQL)** commands.

Running the CQL shell client

Cassandra 4.1.3 has a query language called **Cassandra Query Language** (CQL) that supports row and column operations, similar to relational SQL. To run these CQL commands, spawn the CQL shell client by running the `cqlsh` command in the `/conf` directory. *Figure 7.10* shows the process of opening the *CQL shell* (`cqlsh`):

```
sjctrags@DESKTOP-56HNGC9:/mnt/c/Alibata/Development/Database/cassandra-4.1.3/bin $ cqlsh
Connected to Test Cluster at 127.0.0.1:9042
[cqlsh 6.1.0 | Cassandra 4.1.3 | CQL spec 3.4.6 | Native protocol v5]
Use HELP for help.
```

Figure 7.10 – Running the cqlsh command

CQL has **data definition languages** (DDLs), **data manipulation languages** (DML), queries, and general-purpose commands. For DDL, it has the `create|alter|drop keyspace`, `create |alter|drop table`, `use`, and `truncate` statements. For DML, it has `insert`, `delete`, `update`, and `batch` commands. For query transactions, it utilizes the `select` clause like in SQL. However, the `where` clause is limited only to partition, clustering, and composite keys. Some CQL commands end with a semicolon.

CQL has general-purpose commands such as `show version`, `expand`, and `describe`. To check all the clusters, run the `describe cluster` command. To check all the keyspaces, run the `describe keyspaces` command. To list all the tables in a keyspace, run the `describe tables` command. *Figure 7.11* shows a series of CQL commands that view Cassandra's data storage:

```
sjctrags@DESKTOP-56HNGC9: /mnt/c/Alibata/Development/Database/cassandra-4.1.3/bin
sjctrags@DESKTOP-56HNGC9:/mnt/c/Alibata/Development/Database/cassandra-4.1.3/bin $ cqlsh
Connected to Test Cluster at 127.0.0.1:9042
[cqlsh 6.1.0 | Cassandra 4.1.3 | CQL spec 3.4.6 | Native protocol v5]
Use HELP for help.
cqlsh> show version;
[cqlsh 6.1.0 | Cassandra 4.1.3 | CQL spec 3.4.6 | Native protocol v5]
cqlsh> describe cluster

Cluster: Test Cluster
Partitioner: Murmur3Partitioner
Snitch: DynamicEndpointSnitch

cqlsh> describe keyspaces;

packtspace  system_auth         system_schema   system_views
system      system_distributed  system_traces   system_virtual_schema

cqlsh> use packtspace;
cqlsh:packtspace> describe tables;

course  degree_level  student  student_perf
```

Figure 7.11 – Running CQL general-purpose commands

Running the `describe` command on a table returns the metadata description of the Cassandra table, as shown in *Figure 7.12*:

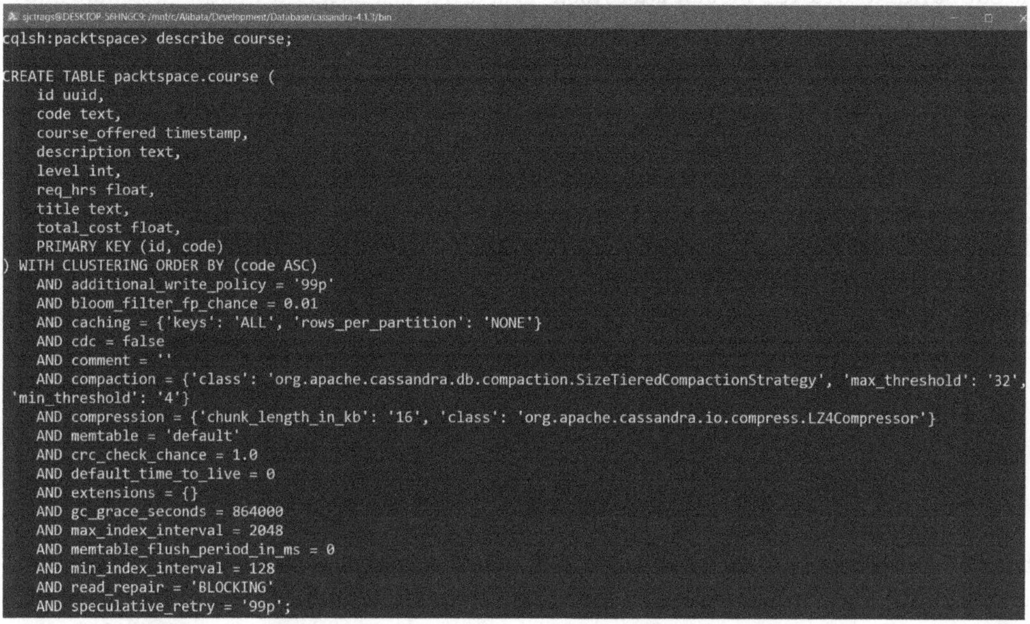

Figure 7.12 – Running the describe command on tables

A *cluster* contains more than one data center, and a *data center* can have more than one node. Each *node* must have a keyspace to hold all the tables, materialized views, user-defined types, functions, and aggregates. So, the first thing you must do with the CQL shell client is run the `create keyspace` command before building the project's tables. The following code creates `packtspace`, which holds the tables of our application:

```
CREATE KEYSPACE packtspace WITH replication = {'class':
'NetworkTopologyStrategy', 'datacenter1': '1'}  AND durable_writes =
false;
```

The *replication strategy* that's used is called `NetworkTopologyStrategy`. It makes `packtspace` open for replication and data storage expansion in the long run and also applicable for production deployment.

After creating `packspace`, you can manually create the `course`, `student`, `degree_level`, and `student_perf` tables in `keyspace` using the CQL command. However, DataStax has a `cassandra-driver` module that establishes a database connection to Cassandra and generates tables using the entity or model classes. Let's use this external module to build the application's model layer.

Establishing a database connection

The `cassandra-driver` module is a Python client driver that integrates Apache Cassandra into the Flask application. It contains classes and methods that will only be available after installing the module using the `pip` command:

```
pip install cassandra-driver
```

For our *Tutor Finder* to establish a Cassandra database connection, import `setup()` from the `cassandra.cqlengine.connection` module in the `__init__.py` file of the `modules` package and invoke `setup()` inside the `create_app()` factory method with the arguments for its `hosts`, `default_keyspace`, and `protocol_version` parameters. The following snippet shows the whole process:

```
from cassandra.cqlengine.connection import setup
def create_app(config_file):
    app = Flask(__name__)
    app.config.from_file(config_file, toml.load)
    setup(['127.0.0.1'], "packtspace", protocol_version=4)
```

The `hosts` parameter provides the initial set of IP addresses that will serve as the contact points for the clusters. The second parameter is `keyspace`, which was created beforehand with the CQL shell. The `protocol version` parameter refers to the native protocol that `cassandra-driver` uses to communicate with the server. It depicts the maximum number of requests a connection can handle during communication.

Next, we'll create the model layer.

Building the model layer

Instead of using the CQL shell to create the tables, we'll use the `cassandra-driver` module since it can create the tables programmatically using entity model classes that can translate into actual tables upon application server startup. These model classes are often referred to as *object mappers* since they also map to the metadata of the physical tables.

Unlike HBase, Cassandra recognizes data structures and data types for its tables. Thus, the driver has a `Model` class that subclasses entities for Cassandra table generation. It also provides helper classes, such as `UUID`, `Integer`, `Float`, and `DateTime`, that can define column metadata in an entity class. The following code shows the entity models that are created through the `cassandra-driver` module:

```
import uuid
from cassandra.cqlengine.columns import UUID, Text, Float, DateTime,
Integer, Blob
from cassandra.cqlengine.models import Model
from cassandra.cqlengine.management import sync_table
```

```
class Course(Model):
    id       = UUID(primary_key=True, default=uuid.uuid4)
    code     = Text(primary_key=True, max_length=20,
required=True,clustering_order="ASC")
    title    = Text(required=True, max_length=100)
    req_hrs = Float(required=True, default = 0)
    total_cost = Float(required=True, default = 0.0)
    course_offered  = DateTime()
    level = Integer(required=True, default=-1)
    description   = Text(required=False, max_length=200)

    def get_json(self):
        return {
            'id': str(self.id),
            'code': self.code,
            'title' : self.title,
            'req_hrs': self.req_hrs,
            'total_cost': self.total_cost,
            'course_offered': self.course_offered,
            'level': self.level,
            'description': self.description
    }
```

In the given `Course` entity, `id` and `code` are columns that are declared as *primary keys*; `id` is the *partition key*, while `code` is the *clustering key* that will manage and sort the records per node in ascending order. The `title`, `req_hrs`, `total_cost`, `course_offered`, `level`, and `descriptions` columns are typical columns that contain their respective metadata. On the other hand, the `get_json()` custom method is an optional mechanism that will serialize the model when `jsonify()` needs to render them as a JSON response.

The following model classes define the `degree_level`, `student`, and `student_perf` tables:

```
class DegreeLevel(Model):
    id = UUID(primary_key=True, default=uuid.uuid4)
    code = Integer(primary_key=True,required=True, clustering_
order="ASC")
    description = Text(required=True)
    ... ... ... ... ... ...

    ... ... ... ... ... ...
class Student(Model):
    id = UUID(primary_key=True, default=uuid.uuid4)
    std_id = Text(primary_key=True,required=True, max_length=12,
clustering_order="ASC")
    firstname = Text(required=True, max_length=60)
```

```
    midname = Text(required=True, max_length=60)
    ... ... ... ... ... ...
    ... ... ... ... ... ...
class StudentPerf(Model):
    id = UUID(primary_key=True, default=uuid.uuid4)
    std_id = Text(primary_key=True,required=True, max_length=12)
    course_code = Text(required=True, max_length=20)
    ... ... ... ... ... ...
    ... ... ... ... ... ...
sync_table(Course)
sync_table(DegreeLevel)
sync_table(Student)
sync_table(StudentPerf)
```

Here, `sync_table()` from `cassandra-driver` converts each model into a table and synchronizes any changes made in the model classes to the mapped table in `keyspace`. However, applying this method to the model class with too many changes may mess up the existing table's metadata. So, it is more acceptable to drop all old tables using the CQL shell before running `sync_table()` with the updated model classes.

After building the model layer, the subsequent procedure is to implement the repository transactions to access the data in Cassandra. So, let's access the keyspace and tables in our Cassandra platform so that we can perform CRUD operations.

Implementing the repository layer

Entity models inherit some attributes from the `Model` class, such as `__table_name__`, which accepts and replaces the default table name of the mapping, and `__keyspace__`, which replaces the default *keyspace* of the mapped table.

Moreover, entity models also inherit some other instance methods:

- `save()`: Persists the entity object in the database.
- `update(**kwargs)`: Updates the existing column fields based on the new column (`kwargs`) details.
- `delete()`: Removes the record from the database.
- `batch()`: Runs synchronized updates or inserts on replicas.
- `iff(**kwargs)`: Checks if the indicated `kwargs` matches the column values of the object before the update or delete happens.
- `if_exists()`/`if_not_exists()`: Verifies if the mapped record exists in the database.

Also, the entity classes derive the `objects` class variable from their `Model` superclass, which can provide query methods such as `filter()`, `allow_filtering()`, and `get()` for record retrieval. They also inherit the `create()` class method, which can insert records into the database, an option other than `save()`.

All these derived methods are the building blocks of our repository class. The following repository class shows how the `Course` model implements its CRUD transactions:

```python
from modules.models.db.cassandra_models import Course
from datetime import datetime
from typing import Dict, Any


class CourseRepository:
    def __init__(self):
        pass

    def insert_course(self, details:Dict[str, Any]):
        try:
            Course.create(**details)
            return True
        except Exception as e:
            print(e)
        return False
```

Gere, `insert_course()` uses the `create()` method to persist a `course` record instead of applying `save()`. For the update transaction, `update_course()` filters a `course` record by course code for updating:

```python
    def update_course(self, details:Dict[str, Any]):
        try:
            rec = Course.objects.filter( code=str(details['code'])) .
allow_filtering().get()
            del details['id']
            del details['code']
            rec.update(**details)
            return True
        except Exception as e:
            print(e)
        return False
```

In Cassandra, when querying records with constraints, the *partition key* must always be included in the constraint. However, `update_course()` uses the `allow_filtering()` method to allow data retrieval without the *partition key* and bypass the *Invalid Query Error* or *error code 2200*.

The following `delete_course_code()` transaction uses the `delete()` entity class method to remove the filtered record object. Again, the `allow_filtering()` method helps filter the record by code without messing up the *partition key*:

```
def delete_course_code(self, code):
    try:
        rec = Course.objects.filter(code=code) .allow_filtering().
get()
        rec.delete()
        return True
    except Exception as e:
        print(e)
    return False
```

Here, `search_by_code()` and `search_all_courses()` are the two query transactions of this `CourseRepository`. The former retrieves a single record based on a `course` code, while the latter filters all `course` records without any condition. The `get()` method of `objects` returns a non-JSONable `Course` object that `jsonify()` cannot process. But wrapping the object with `dict()` after converts it into a JSON serializable record. In `search_all_courses()`, the custom `get_json()` method helps generate a list of JSONable course records for easy `Response` generation:

```
def search_by_code(self, code:str):
    result = Course.objects.filter(code=code).allow_filtering().
get()
    records = dict(result)
    return records

def search_all_courses(self):
    result = Course.objects.all()
    records = [course.get_json() for course in result]
    return records
```

Cassandra is known for its faster write than read operations. It writes data to the commit log and then caches it simultaneously, preserving data from unexpected occurrences, damage, or downtime. But there is one form of NoSQL data storage that's popular for its faster reads operations: **Remote Dictionary Server (Redis)**.

Storing search data in Redis

Redis is a fast, open source, in-memory, *key-value* form of NoSQL storage that's popular in messaging and caching. In *Chapter 5*, we used it as the message broker of Celery, while in *Chapter 6*, we used it as a message queue for the SSE and WebSocket programming. However, this chapter will utilize Redis as a data cache to create a fast search mechanism.

First, let's install Redis on our system.

Installing the Redis server

For Windows, download the latest Redis TAR file from `https://redis.io/download/`, unzip it to an installation folder, and run `redis-server.exe` from the directory.

For WSL, run the following series of `sudo` commands:

```
sudo apt-add-repository ppa:redislabs/redis
sudo apt-get update
sudo apt-get upgrade
sudo apt-get install redis-server
```

Then, run `redis-cli -v` to check if the installation was successful. If so, run the `redis-server` command to start the Redis server. *Figure 7.13* shows the server log after the Redis server starts up on the WSL-Ubuntu platform:

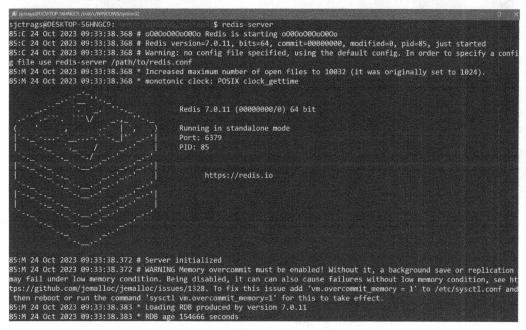

Figure 7.13 – Running the redis-server command

To stop the server, run the `redis-cli` shutdown command or press *Ctrl + C* on your keyboard.

Now, let's explore the Redis server using its client shell and understand its CLI commands so that we can run its CRUD operations.

Understanding the Redis database

Key-value storage uses a *hashtable* data structure model wherein its unique key value serves as a pointer to a corresponding value of any type. For Redis, the key is always a string that points to values of type strings, JSON, lists, sets, hashes, sorted set, streams, bitfields, geospatial, and time series. Since Redis is an in-memory storage type, it stores all its simple to complex key-value pairs of data in the host's RAM, which is volatile and cannot persist data permanently. However, in return, Redis can provide faster reads and access to its data than HBase and Cassandra.

To learn more about this storage, Redis has a built-in shell client that interacts with the database through some commands. *Figure 7.14* shows opening a client shell by running the `redis-cli` command and checking the number of databases the storage has using the `CONFIG GET databases` command:

```
PS C:\WINDOWS\system32> wsl
sjctrags@DESKTOP-56HNGC9:/mnt/c/WINDOWS/system32$ redis-cli
127.0.0.1:6379> CONFIG GET databases
1) "databases"
2) "16"
127.0.0.1:6379>
```

Figure 7.14 – Opening a Redis shell

The typical number of databases in Redis storage is *16*. Redis databases are named from *0* to *15*, like indexes of an array. The default database name is *0*, but it has a `select` command that chooses the preferred database other than 0 (for example, `select 1`).

Since Redis is a simple NoSQL database, it only has the following few commands, including CRUD, we need to consider:

- `set`: Adds a key-value pair to the database.
- `get`: Retrieves the value of a key.
- `hset`: Adds a hash with multiple key-value pairs.
- `hget`: Retrieves the value of a key in a hash.
- `hgetall`: Retrieves all the key-value pairs in a hash.
- `hkeys`: Retrieves all the keys in a hash.
- `hvals`: Retrieves all the values in a hash.
- `del`: Removes an existing key-value pair using the key or the whole hash.
- `hdel`: Removes single or multiple key-value pairs in a hash.

Redis hashes are records or structured types that can hold collections of field-value pairs with values of varying types. In a way, it can represent a Python object persisted in the database. *Figure 7.15* shows a list of Redis commands being run on the Redis shell:

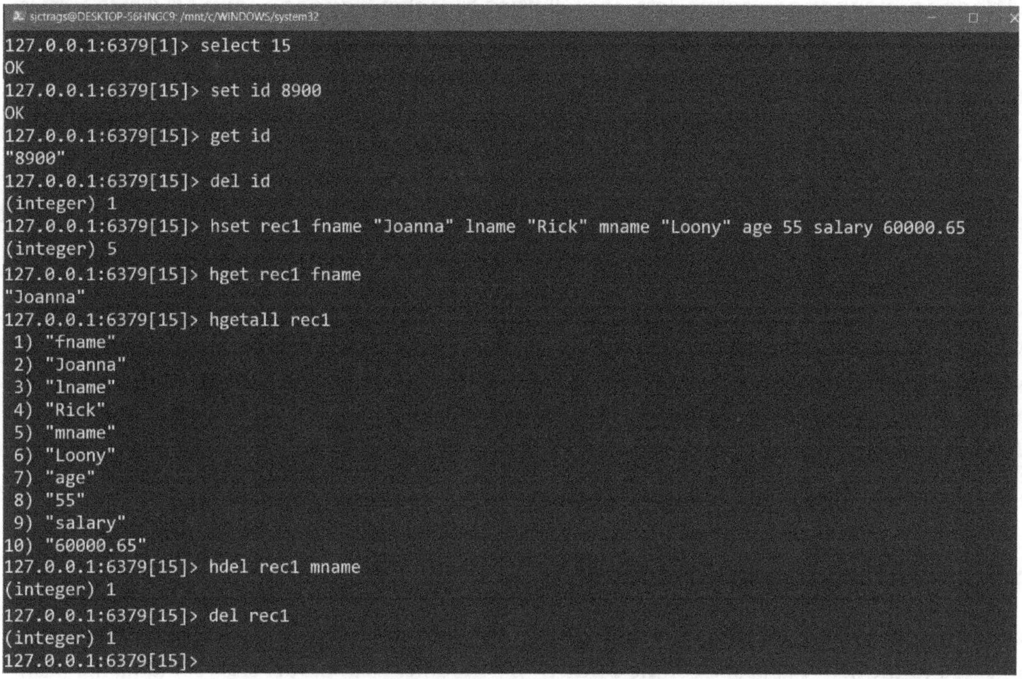

```
sjctrags@DESKTOP-S6HNGC9: /mnt/c/WINDOWS/system32                                              —  □  ×
127.0.0.1:6379[1]> select 15
OK
127.0.0.1:6379[15]> set id 8900
OK
127.0.0.1:6379[15]> get id
"8900"
127.0.0.1:6379[15]> del id
(integer) 1
127.0.0.1:6379[15]> hset rec1 fname "Joanna" lname "Rick" mname "Loony" age 55 salary 60000.65
(integer) 5
127.0.0.1:6379[15]> hget rec1 fname
"Joanna"
127.0.0.1:6379[15]> hgetall rec1
 1) "fname"
 2) "Joanna"
 3) "lname"
 4) "Rick"
 5) "mname"
 6) "Loony"
 7) "age"
 8) "55"
 9) "salary"
10) "60000.65"
127.0.0.1:6379[15]> hdel rec1 mname
(integer) 1
127.0.0.1:6379[15]> del rec1
(integer) 1
127.0.0.1:6379[15]>
```

Figure 7.15 – Running Redis commands on the Redis shell

But how can our *Tutor Finder* application connect to a Redis database as a client? We'll answer this question in the next section.

Establishing a database connection

In *Chapter 5* and *Chapter 6*, the *redis-py* module established a connection to Redis as a broker or message queue. This time, our application will connect to the Redis database for data storage and caching.

So far, the Redis OM module is the most efficient and convenient Redis database connector that can provide database connectivity and methods for CRUD operations, similar to an ORM. However, before accessing its utilities, install it using the `pip` command:

```
pip install redis-om
```

`redis-py` is the other library that's included in the installation of the `redis-om` module. The `redis` module has a Redis callable object that builds the database connectivity. The callable has a `from_url()` method that accepts the database URL and some parameter values for the `encoding` and `decode_responses` parameters. The following code shows `create_app()`, which creates the Redis connection to database 0:

```
import redis

def create_app(config_file):
    app = Flask(__name__)
    app.config.from_file(config_file, toml.load)
    redis.Redis.from_url("redis://localhost:6379/0", encoding="utf8",
decode_responses=True)
```

Here, `redis` in the URI indicates that the connection is a Redis standalone one to database 0 in localhost at port 6379. All responses of the `redis-om` transactions are decoded into strings because the `decode_responses` parameter is assigned a value of `True`. All these string results are in UTF-8 encoding.

At this point, the `redis-om` module is ready to build the application's model layer.

Implementing the model layer

Redis OM is a high-level and object-oriented library that uses entity classes to manage the Redis data compared to the native approach of the `redis-py` module. Each model class contains the hash fields with types validated by Pydantic validators. Once instantiated, the model object will hold the values of the keys before inserting them into the database with the auto-generated **hash value** or **pk**.

The `redis-om` module has the `HashModel` class, which will implement the entity classes of the application. The `HashModel` class is a representation of a Redis hash. It captures the key-value pairs and uses its instance methods to manage the data. It automatically generates the primary key or hash key for each model object. The following are the `HashModel` classes for the *course*, *student*, and *tutor* data:

```
from redis_om import  HashModel, Field, get_redis_connection

redis_conn = get_redis_connection(decode_responses=True)

class SearchCourse(HashModel):
    code: str   = Field(index=True)
    title: str
    description: str
    req_hrs: float
    total_cost: float
    level: int
```

```
class SearchStudent(HashModel):
    std_id: str
    firstname: str
    midname: str
    lastname: str
    … … … … … …


class SearchTutor(HashModel):
    firstname: str
    lastname: str
    midname: str
    … … … … … …
    class Meta:
        database = redis_conn
```

Here, CourseSearch, SearchStudent, and SearchTutor are model classes that have been created to cache incoming request data to the Redis database for fast search transactions. Each class has declared attributes that correspond to the keys of a record. After its instantiation, the model object will have a pk instance variable that contains the unique hash key of the data record.

Aside from relying on the Redis connection created by Redis.from_url(), a HashModel object can independently or directly connect to the Redis database by assigning a connection instance to its Meta object's database variable. In either of these connectivity approaches, the model object can still emit the methods that will operate the repository layer.

After establishing the Redis connection and creating the model classes, the next step is to build the repository layer.

Building the repository layer

Like in Cassandra's repository layer, the model object of the redis-om module implements the repository class. The HashModel entity emits methods that will implement the CRUD transactions. The following SearchCourseRepository class manages course details in the Redis database:

```
from modules.models.db.redis_models import SearchCourse
from typing import Dict, Any

class SearchCourseRepository:
    def __init__(self):
        pass

    def insert_course(self, details:Dict[str, Any]):
```

```
        try:
            course = SearchCourse(**details)
            course.save()
            return True
        except Exception as e:
            print(e)
        return False
```

The given `insert_course()` method uses the HashModel entity's `save()` instance method, which adds the course details as key-value pairs in database 0 of Redis. To update a record, retrieve the data object using its pk from the database and then invoke the `update()` method of the resulting model object with the new field values. The following `update_course()` method applies this `redis-om` approach:

```
    def update_course(self, details:Dict[str, Any]):
        try:
            record = SearchCourse.get(details['pk'])
            record.update(**details)
            return True
        except Exception as e:
            print(e)
        return False
```

When deleting a record, HashModel has a class method called `delete()` that removes a hashed object using its pk, similar to the following `delete_course()` method:

```
    def delete_course(self, pk):
        try:
            SearchCourse.delete(pk)
            return True
        except Exception as e:
            print(e)
        return False
```

When retrieving data from the database, the `get()` method is the only way to retrieve a single model object using an existing pk. Querying all the records requires a `for` loop to enumerate all pk values from the HashModel entity's `all_pks()` generator, which retrieves all pk from the database. The loop will fetch all the model objects using the enumerated pk. The following `select_course()` class retrieves all course details from the `search_course` table using the pk value of each record:

```
    def select_course(self, pk):
        try:
            record = SearchCourse.get(pk)
            return record.dict()
```

```
        except Exception as e:
            print(e)
        return None

    def select_all_course(self):
        records = list()
        for id in SearchCourse.all_pks():
            records.append(SearchCourse.get(id).dict())
        return records
```

All resulting objects from the query transactions are JSONable and don't need a JSON serializer. Running the given `select_all_course()` class will return the following sample Redis records:

```
{
    "records": [
        {
            "code": "PY-201",
            "description": "Advanced Python",
            "level": 3,
            "pk": "01HDH2VPZBGJJ16JKE3KE7RGPQ",
            "req_hrs": 50.0,
            "title": "Advanced Python Programming",
            "total_cost": 15000.0
        },
        {
            "code": "PY-101",
            "description": "Intro to Python programming",
            "level": 1,
            "pk": "01HDH2SVYR7AYMRD28RE6HSHYB",
            "req_hrs": 45.0,
            "title": "Python Basics",
            "total_cost": 5000.0
        },
```

Although Redis OM is perfectly compatible with FastAPI, it can also make any Flask application a client for Redis. Now, Redis OM cannot implement filtered queries. If Redis OM needs a filtered search with some constraints, it needs a *RediSearch* extension module that calibrates and provides more search constraints to query transactions. But *RediSearch* can only run with Redis OM if the application uses Redis Stack instead of the typical server.

The next section will highlight a *document-oriented* NoSQL database that's popular for enterprise application development: *MongoDB*.

Handling BSON-based documents with MongoDB

MongoDB is a NoSQL database that stores JSON-like documents of key-value pairs with a flexible and scalable schema, thus classified as a document-oriented database. It can store huge volumes of data with varying data structures, types, and formations.

These JSON-like documents use **Binary Javascript Object Notation** (**BSON**), a binary-encoded representation of JSON documents suitable for network-based data transport because of its compact nature. It has non-JSON-native data type support for date and binary data and recognizes embedded documents or an array of documents because of its traversable structure.

Next, we'll install MongoDB and compare its process to HBase, Cassandra, and Redis.

Installing and configuring the MongoDB server

First, download the preferred MongoDB community server from `https://www.mongodb.com/try/download/community`. Install it to your preferred drive and directory. Next, create a data directory where MongoDB can store its documents. If the data folder doesn't exist, MongoDB will resort to `C:\data\db` as its default data folder. Afterward, run the server by running the following command:

```
mongod.exe --dbpath="c:\data\db"
```

The default host of the server is localhost, and its port is `27017`.

Download MongoDB Shell from `https://www.mongodb.com/try/download/shell` to open the client console for the server. Also, download MongoDB Compass from `https://www.mongodb.com/try/download/compass`, the GUI administration tool for the database server.

So far, installing the MongoDB server and its tools takes less time than installing the other NoSQL databases. Next, we'll integrate MongoDB into our *Tutor Finder* application.

Establishing a database connection

To create database connectivity, MongoDB uses the `pymongo` module as its native driver, which is made from BSON utilities. However, the driver requires more codes to implement the CRUD transactions because it offers low-level utilities. A high-level and object-oriented module, such as `mongoengine`, can provide a better database connection than `pymongo`. The `mongoengine` library is a popular **object document mapper** (**ODM**) that can build a client application with a model and repository layers.

The `flask-mongoengine` library is written solely for Flask. However, since Flask 3.x, the `flask. json` module, on which the module is tightly dependent, was removed. This change affected the `MongoEngine` class of the `flask_mongoengine` library, which creates a MongoDB client. Until the library is updated so that it supports the latest version of Flask, the native `connect()` method from the native `mongoengine.connection` module will always be the solution to connect to the MongoDB database. The following snippet from the *Tutor Finder* application's `create_app()` factory method uses `connect()` to establish communication with the MongoDB server:

```
from mongoengine.connection import connect
def create_app(config_file):
    app = Flask(__name__)
    app.config.from_file(config_file, toml.load
    connect(host='localhost', port=27017, db='tfs',
uuidRepresentation='standard')
```

The `connect()` method requires `db` name, `host`, `port`, and the type of UUID the server will use to recognize the UUID primary key. This can be `unspecified`, `standard`, `pythonLegacy`, `javaLegacy`, or `csharpLegacy`.

On the other hand, the client can invoke the `disconnect()` method to close the connection.

Now, let's build the model layer using the helper classes from the `flask-mongoengine` module.

Building the model layer

The `flask-mongoengine` module has a **Document** base class that defines the structure of the MongoDB key-value pairs and their fields and properties. These model classes are translated afterward into BSON documents after the *INSERT* transactions. Here are the model classes that subclass the Document base class to build the login details of our application:

```
from mongoengine import Document, SequenceField, BooleanField,
EmbeddedDocumentField, BinaryField, IntField, StringField, DateField,
EmailField, EmbeddedDocumentListField, EmbeddedDocument

class Savings(EmbeddedDocument):
    acct_name = StringField(db_field='acct_name', max_length=100,
required=True)
    acct_number = StringField(db_field='acct_number', max_length=16,
required=True)
        ... ... ... ... ... ...


class Checking(EmbeddedDocument):
    acct_name = StringField(db_field='acct_name', max_length=100,
required=True)
    acct_number = StringField(db_field='acct_number', max_length=16,
required=True)
```

```
    bank =  StringField(db_field='bank', max_length=100,
required=True)
    … … … … … …


class PayPal(EmbeddedDocument):
    email = EmailField(db_field='email', max_length=20, required=True)
    address = StringField(db_field='address', max_length=200,
required=True)

class Tutor(EmbeddedDocument):
    firstname = StringField(db_field='firstname', max_length=50,
required=True)
    lastname = StringField(db_field='lastname', max_length=50,
required=True)
    … … …. … …
    savings = EmbeddedDocumentListField(Savings, required=False)
    checkings = EmbeddedDocumentListField(Checking, required=False)
    gcash = EmbeddedDocumentField(GCash, required=False)
    paypal = EmbeddedDocumentField(PayPal, required=False)

class TutorLogin(Document):
    id = SequenceField(required=True, primary_key=True)
    username = StringField(db_field='email',max_length=25,
required=True)
    password = StringField(db_field='password',maxent=25,
required=True)
    encpass = BinaryField(db_field='encpass', required=True)
    tutor = EmbeddedDocumentField(Tutor, required=False)
```

Like in SQLAlchemy, `flask-mongengine` offers helper classes, such as `StringField`, `BinaryField`, `DateField`, `IntField`, and `EmailField`, that build the metadata of a document. These helper classes have parameters, such as `db_field` and `required`, that will add details to the key-value pairs. Moreover, some parameters appear only in one helper class, such as `min_length` and `max_length` in `StringField`, because they control the number of characters in the string. Likewise, `ByteField` has a `max_bytes` parameter that will not appear in other helper classes. Note that, the `Document` base class' `BinaryField` translates to BSON's binary data and `DateField` to BSON's date type, not the common Python type.

Unlike Cassandra, Redis, and HBase, MongoDB allows relationships among structures. Although not normalized like in an RDBMS, MongoDB can link one document to its subdocuments using the `EmbeddedDocumentField` and `EmbeddedDocumentListField` helper classes. In the given model classes, the `TutorLogin` model will create a parent document collection called `tutor_login` that will reference a `tutor` sub-document because the sub-document's `Tutor` model is an `EmbeddedDocumentField` helper class of the parent `TutorLogin` model. The idea is similar to a one-to-one relationship concept in a relational ERD but not totally the same.

On the other hand, the relationship between `Tutor` and `Savings` is like a one-to-many relationship because `Savings` is the `EmbeddedDocumentListField` helper class of the `Tutor` model. In other words, the `tutor` document collections will reference a list of savings sub-documents. Here `EmbeddedDocumentField` does not have an `_id` field because it cannot construct an actual document collection, unlike in an independent `Document` base class.

Next, we'll create the repository layer from the `Tutor` document and its sub-documents.

Implementing the repository

The `Document` object emits utility methods that perform CRUD operations for the repository layer. Here is a `TutorLoginRepository` class that inserts, updates, deletes, and retrieves `tutor_login` documents:

```
from typing import Dict, Any
from modules.models.db.mongo_models import TutorLogin
import json


class TutorLoginRepository:
    def insert_login(self, details:Dict[str, Any]) -> bool:
        try:
            login = TutorLogin(**details)
            login.save()
        except Exception as e:
            print(e)
            return False
        return True
```

The `insert_login()` method uses the `save()` method of the `TutorLogin` model object for persistence. The `save()` method will persist all the `kwargs` data that's passed to the constructor of `TutorLogin`.

Like the Cassandra driver, the `Document` class has an `objects` class attribute that provides all query methods. Updating a document uses the `objects` attribute to filter the data using any document keys and then fetches the record using the attribute's `get()` method. If the document exists, the `update()` method of the filtered record object will update the given `kwargs` of fields that require updating. The following code shows `update_login()`, which updates a `TutorLogin` document:

```
    def update_login(self, id:int, details:Dict[str, Any]) -> bool:
        try:
            login = TutorLogin.objects(id=id).get()
            login.update(**details)
        except:
```

```
            return False
        return True
```

Deleting a document in MongoDB also uses the `objects` attribute to filter and extract the document that needs to be removed. The `delete()` method of the retrieved model object will delete the filtered record from the database once the repository invokes it. Here, `delete_login()` removes a filtered document from the database:

```
def delete_login(self, id:int) -> bool:
    try:
        login = TutorLogin.objects(id=id).get()
        login.delete()
    except:
        return False
    return True
```

The `objects` attribute is responsible for implementing all the query transactions. Here, `get_login()` fetches a single object identified by its unique `_id` value using the `get()` method, while the `get_login_username()` transaction retrieves a single record filtered by the tutor's username and password. On the other hand, `get_all_login()` retrieves all the `tutor_login` documents from the database:

```
def get_all_login(self):
    login = TutorLogin.objects()
    return json.loads(login.to_json())

def get_login(self, id:int):
    login = TutorLogin.objects(id=id).get()
    return login.to_json()

def get_login_username(self, username:str, password:str):
    login = TutorLogin.objects(username=username,
password=password).get()
    return login.to_json()
```

All these query transactions invoke the built-in `to_json()` method, which serializes and converts the BSON-based documents into JSON for the API's response generation process.

Embedded documents do not have dedicated collection storage because they are part of a parent document collection. Adding and removing embedded documents from the parent document requires using string queries and operators or typical object referencing in Python, such as setting to None when removing a sub-document. The following repository adds and removes a tutor's profile details from the login credentials:

```python
from typing import Dict, Any
from modules.models.db.mongo_models import TutorLogin, Tutor

class TutorProfileRepository:
    def add_tutor_profile(self, details:Dict[str, Any]) -> bool:
        try:
            login = TutorLogin.objects(id=details['id']).get()
            del details['id']
            profile = Tutor(**details)
            login.update(tutor=profile)
        except Exception as e:
            print(e)
            return False
        return True
```

Here, add_tutor_profile() embeds the TutorProfile document via the tutor key of the TutorLogin main document. Another solution is to pass the set_tutor=profile query parameter to the update() operation. The following transaction removes the tutor profile from the main document:

```python
    def delete_tutor_profile(self, id:int) -> bool:
        try:
            login = TutorLogin.objects(id=id).get()
            login.update(tutor=None)
        except Exception as e:
            print(e)
            return False
        return True
```

Then, delete_tutor_profile() unsets the profile document from the TutorLogin document by setting the tutor field to None. Another way to do this is to use the unset__tutor=True query parameter for the update() method.

The most effective way to manage a list of embedded documents is to use query strings or query parameters to avoid lengthy implementations. The following SavingsRepository class adds and removes a bank account from a list of savings accounts of a tutor. Its add_savings() method adds a new saving account to the tutor's list of saving accounts. It uses the update() method with the push__tutor__savings=savings query parameter, which pushes a new Savings instance to the list:

```python
from typing import Dict, Any
from modules.models.db.mongo_models import Savings, TutorLogin

class SavingsRepository:
    def add_savings(self, details:Dict[str, Any]):
        try:
            login = TutorLogin.objects(id=details['id']).get()
            del details['id']
            savings = Savings(**details)
            login.update(push__tutor__savings=savings)
        except Exception as e:
            print(e)
            return False
        return True
```

On the other hand, the delete_savings() method deletes an account using the pull__tutor__savings__acct_number= details['acct_number'] query parameter, which removes a savings account from the list:

```python
    def delete_savings(self, details:Dict[str, Any]):
        try:
            login = TutorLogin.objects(id=details['id']).get()
            login.update(pull__tutor__savings__acct_number=
details['acct_number'])
        except Exception as e:
            print(e)
            return False
        return True
```

Although MongoDB is popular and has the most support, it slows down when the number of users increases. When the datasets become massive, adding more replications and configurations becomes difficult due to its master-slave architecture. Adding caches is also part of the plan to improve data retrieval.

However, there is another document-oriented NoSQL database that's designed for distributed architecture and high availability with internal caching for datasets: **Couchbase**.

Managing key-based JSON documents with Couchbase

Couchbase is a NoSQL database that's designed for distributed architectures and offers high performance on concurrent, web-based, and cloud-based applications. It supports distributed **ACID** transactions and has a SQL-like language called **N1QL**. All documents stored in Couchbase databases are JSON-formatted.

Now, let's install and configure the Couchbase database server.

Installing and configuring the database instance

To begin, download Couchbase Community Edition from `https://www.couchbase.com/downloads/`. Once it's been installed, Couchbase will need cluster configuration details to be added, including the user profile for accessing the server dashboard at `http://localhost:8091/ui/index.html`. Accepting the user agreement for the configuration is also part of the process. After configuring the cluster, the URL will show us the login form to access the default server instance. *Figure 7.16* shows the login page of the Couchbase web portal:

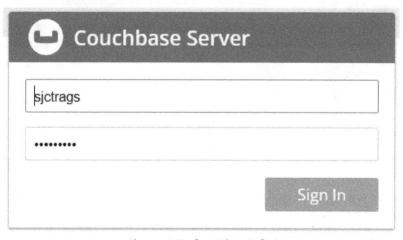

Chrome, Firefox, Edge, Safari

Figure 7.16 – Accessing the login page of Couchbase

After logging in to the portal, the next step is to create a *bucket*. A *bucket* is a named container that saves all the data in Couchbase. It groups all the keys and values based on collections and scopes. Somehow, it is similar to the concept of a database schema in a relational DBMS. *Figure 7.17* shows **packtbucket**, which has been created on the **Buckets** dashboard:

Figure 7.17 – Creating a bucket in the cluster

Afterward, create a scope that will hold the tables or document collections of the database instance. A bucket scope is a named mechanism that manages and organizes these collections. In some aspects, it is similar to a tablespace in a relational DBMS. To create these scopes, click the **Scopes & Collections** hyperlink to the right of the bucket name on the **Add Bucket** page. The **Add Scope** page will appear, as shown in *Figure 7.18*:

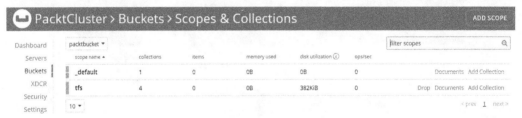

Figure 7.18 – Creating a scope in a bucket

On the **Add Scope** page, click the **Add Scope** hyperlink at the top-right portion and then enter the scope name – in our case, `tfs`.

Lastly, click the **Add Collection** hyperlink of a scope to add its collections. Clicking the scope's name will list all the document collections that have been created. *Figure 7.19* shows a list of all the `tfs` collections:

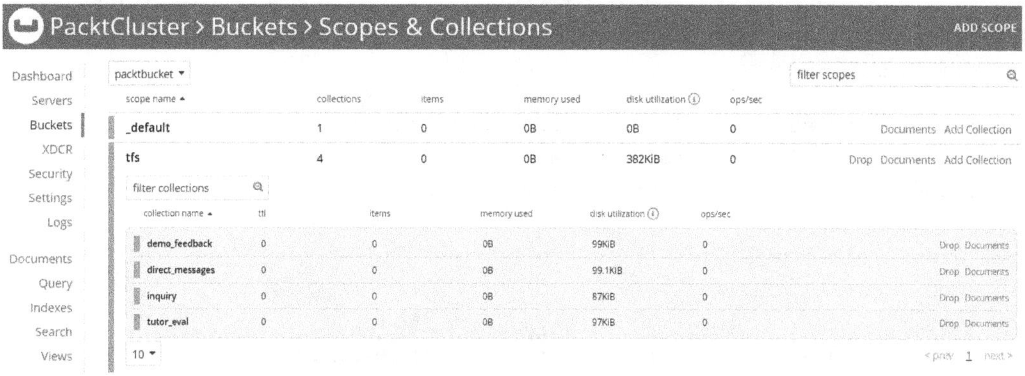

Figure 7.19 – List of collections in a tfs scope

Now, let's establish the bucket connection using the `couchbase` module.

Setting up the server connection

To create a client connection to Couchbase, install the `couchbase` module using the `pip` command:

```
pip install couchbase
```

In the `create_app()` factory function of the application, perform the following steps to access the Couchbase server instance:

1. Create a `PasswordAuthenticator` object with the correct user profile's credential to access the specified bucket.

2. Instantiate the `Cluster` class with its required constructor arguments, namely the Couchbase URL and some options, such as the `PasswordAuthenticator` object, wrapped in the `ClusterOptions` instance.

3. Access the preferred bucket by calling the `Cluster`'s `bucket()` instance method.

The following snippet shows how to implement these steps in our *Tutor Finder* application's `create_app()` method:

```python
from couchbase.auth import PasswordAuthenticator
from couchbase.cluster import Cluster
from couchbase.options import ClusterOptions

def create_app(config_file):
    app = Flask(__name__)
    app.config.from_file(config_file, toml.load)
    auth = PasswordAuthenticator("sjctrags", "packt2255",)
    cluster = Cluster('couchbase://localhost', ClusterOptions(auth))
    cluster.wait_until_ready(timedelta(seconds=5))
    global cb
    cb = cluster.bucket("packtbucket")
```

The `Cluster` object has a `wait_until_ready()` method that pings to the Couchbase services regarding the connection status and returns control to `create_app()` once the connection is ready. But calling this method slows down the startup of the Flask server. Our application has only invoked the method for experimentation purposes.

After a successful setup, we must ensure the `Bucket` object is ready for implementing the repository layer.

Creating the repository layer

The repository layer needs the `Bucket` object from `create_app()` to implement the CRUD transactions. The `Bucket` object has a `scope()` method that will access the container space that contains the collections. It returns a `Scope` object that emits `collection()`, which retrieves the preferred document collections. Here, `DirectMessageRepository` manages all the direct messages that students send to trainers and vice versa:

```python
class DirectMessageRepository:

    def insert_dm(self, details:Dict[str, Any]):
        try:
            cb_coll = cb.scope("tfs") .collection("direct_messages")
            key = "chat_" + str(details['id']) + '-' +
str(details["date_sent"])
            cb_coll.insert(key, details)
            return True
        except Exception as e:
            print(e)
        return False
```

The `dm_insert()` method gives us access to the `tfs` scope and its `direct_messages` document collections. Its main goal is to insert the details of a chat message between the tutor and trainer into the document collection using the given key through the collection's `insert()` method.

On the other hand, the `update_dm()` method uses the collection's `upsert()` method to update a JSON document using a key:

```python
def update_dm(self, details:Dict[str, Any]):
    try:
        cb_coll = cb.scope("tfs") .collection("direct_messages")
        key = "chat_" + str(details['id']) + '-' +
str(details["date_sent"])
        cb_coll.upsert(key, details)
        return True
    except Exception as e:
        print(e)
    return False
```

The collection's `remove()` method deletes a document from the collection. This can be seen in the following `delete_dm()` transaction, where it removes a chat message using its `key`:

```python
def delete_dm_key(self, details:Dict[str, Any]):
    try:
        cb_coll = cb.scope("tfs") .collection("direct_messages")
        key = "chat_" + str(details['id']) + '-' +
str(details["date_sent"])
        cb_coll.remove(key)
        return True
    except Exception as e:
        print(e)
    return False
```

Couchbase, unlike MongoDB, uses a SQL-like mechanism called *N1QL* to retrieve documents. The following *DELETE* transaction uses the N1QL query transaction instead of the collection's `delete()` method:

```python
def delete_dm_sender(self, sender):
    try:
        cb_scope = cb.scope("tfs")
        stmt = f"delete from `direct_messages` where `sender_id`
LIKE '{sender}'"
        cb_scope.query(stmt)
        return True
    except Exception as e:
        print(e)
    return False
```

The `Scope` instance, derived from the `Bucket` object's `scope()` method, has a `query()` method that executes a query statement in string form. The query statement should have its collection and field names enclosed in ticks (` `` `), while its string constraint values should be in single quotes. Thus, we have the `delete from `direct_messages` where `sender_id` LIKE '{sender},'` query statement in `delete_dm_sender()`, where `sender` is a parameter value.

The advantage of using N1QL queries in *DELETE* and *UPDATE* transactions is that the key is not the only basis for performing these operations. The *DELETE* operation can base its document removal on other fields, such as removing a chat message using the given sender ID:

```python
def delete_dm_sender(self, sender):
        try:
            cb_scope = cb.scope("tfs")
            stmt = f"delete from `direct_messages` where `sender_id`
LIKE '{sender}'"
            cb_scope.query(stmt)
            return True
        except Exception as e:
            print(e)
        return False
```

N1QL is popular in retrieving JSON documents from the keyspace with or without constraints. The following query transaction uses the *SELECT* query statement to retrieve all the documents in the `direct_messages` collections:

```python
def select_all_dm(self):
        cb_scope = cb.scope("tfs")
        raw_data = cb_scope.query('select * from `direct_messages`',
QueryOptions(read_only=True))
        records = [rec for rec in raw_data.rows()]
        return records
```

Couchbase can be a suitable form of backend storage for Flask applications when managing a dump of JSON data. Flask and Couchbase can build fast, scalable, and efficient microservices or distributed applications with rapid development and less database administration. However, compared to HBase, Redis, Cassandra, MongoDB, and Couchbase, Flask can integrate with graph databases, such as Neo4J, for graph-related algorithms.

Establishing a data relationship with Neo4J

Neo4J is a NoSQL database that focuses on relationships between data. Instead of documents, it stores nodes, relationships, and the properties that link these nodes. Neo4J is also known as a popular graph database because the concept relies on a graph model composed of nodes and lines directed between nodes.

Before integrating our application into the Neo4J database, we must install the current version of the Neo4J platform using Neo4J Desktop.

Installing Neo4J Desktop

Neo4J Desktop provides a local development environment and includes all the functionality needed to learn the database, from creating a custom local database to starting the Neo4J browser. Its installer can be found at `https://neo4j.com/download/`.

Once it's been installed, create a Neo4J project that will contain the local databases and configuration settings. Aside from the project's name, the process will also ask for a username and password for its authentication details. Once you've done this, delete its default Movie database and create the necessary graph database. *Figure 7.20* shows **Packt Flask Project** with a **Tutor** database:

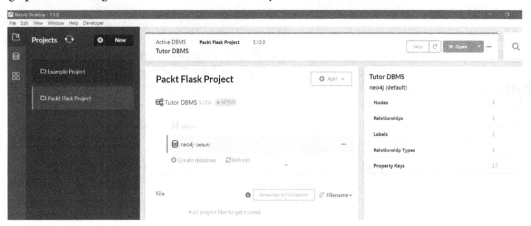

Figure 7.20 – The Neo4J Desktop dashboard

There are so many ways Flask can connect to a graph database, and one of them is through the py2neo library. We'll take a closer look at this in the next section.

Establishing a connection to the database

To start, install py2neo using the pip command:

```
pip install py2neo
```

Next, create a `neo4j_config.py` module in the main project folder with the following script to ensure database connectivity:

```
from py2neo import Graph
def db_auth():
    graph = Graph("bolt://127.0.0.1:7687", auth=("neo4j",
"packt2255"))
    return graph
```

Now, calling the given `db_auth()` method will initiate the bolts connection protocol with the host, port, and authentication details to open a connection for our *Tutor Finder* application through the `Graph` instance, the object responsible for repository layer implementation.

Implementing the repository

Cypher is the official query language of Neo4J and allows client applications to perform CRUD operations on the database. Python-Neo4J libraries provide various helper classes and methods to execute these Cypher commands at the Python level. In our case, the `Graph` instance has several utility methods to derive the building blocks of the module, namely `SubGraph`, `Node`, `NodeMatcher`, and `Relationship`. Here, `StudentNodeRepository` showcases the use of py2neo's API classes and methods in managing the student nodes:

```
from main import graph
from py2neo import Node, NodeMatcher, Subgraph, Transaction
from py2neo.cypher import Cursor
from typing import Any, Dict

class StudentNodeRepository:
    def __init__(self):
        pass

    def insert_student_node(self, details:Dict[str, Any]):
        try:
            tx:Transaction = graph.begin()
            node_trainer = Node("Tutor", **details)
            graph.create(node_trainer)
            graph.commit(tx)
            return True
        except Exception as e:
            print(e)
        return False
```

The `insert_student_node()` method creates a `Student` node and stores its details in the graph database. A node is the fundamental unit of data in Neo4J; it can be an independent node or connected to other nodes through a relationship.

There are two ways to create a node using the `py2neo` library:

- Running a query with Cypher's *CREATE* transaction using Graph's `query()` or `run()` methods.
- Persisting the `Node` object using the `Graph` object's `create()` method.

Creating a node requires transaction management, so we must start a `Transaction` context to commit all the data manipulation operations. Here, `insert_student_node()` creates a `Transaction` object to create a logical context for the node creation operation for the `Graph` object's `commit()` method to commit:

```python
def update_student_node(self, details:Dict[str, Any]):
    try:
        tx = graph.begin()
        matcher = NodeMatcher(graph)
        student_node:Node  = matcher.match('Student', student_
id=details['student_id']).first()
        if not student_node == None:
            del details['student_id']
            student_node.update(**details)
            graph.push(student_node)
            graph.commit(tx)
            return True
        else:
            return False
    except Exception as e:
        print(e)
    return False
```

NodeManager can locate a specific node given the criteria in key-value pairs. Here, `update_student_node()` uses the `match()` method from `NodeManager` to single out a `Node` object with the specific `student_id` value. After retrieving a graph node, if there is one, you must call the `Node` object's `update()` method with the `kwargs` value of the new data. To merge the updated `Node` object with its committed version, invoke the `Graph` object's `push()` method and perform a commit.

Another way of searching and retrieving a Node match is through the Graph object's query() method. It can execute *CREATE* and other Cipher manipulation commands because it has auto-commit features. But in most cases, it is applied in node retrieval transactions. Here, delete_student_node() uses the query() method with the MATCH command to retrieve a specific node for deletion:

```
def delete_student_node(self, student_id:str):
    try:
        tx = graph.begin()
        student_cur:Cursor = graph.query(f"MATCH (st:Student)
WHERE st.student_id = '{student_id}' Return st")
        student_sg:Subgraph = student_cur.to_subgraph()
        graph.delete(student_sg)
        graph.commit(tx)
        return True
    except Exception as e:
        print(e)
    return False
```

The Graph object's query() method returns Cursor, which is a navigator for streams of nodes. The Graph object has a delete() method that can delete any nodes retrieved by query(), but the nodes should be in *SubGraph* form. To delete the retrieved nodes, convert the Cursor object into a *SubGraph* by calling it from the to_subgraph() method. Then, call commit() to handle the whole delete transaction.

Retrieving nodes in py2neo can utilize either NodeManager or the Graph object's query() method. Here, get_student_node() retrieves a specific Student node filtered by student ID using NodeMatcher, while select_student_nodes() uses query() to retrieve a list of Student nodes:

```
def get_student_node(self, student_id:str):
    matcher = NodeMatcher(graph)
    student_node:Node  = matcher.match('Student', student_
id=student_id).first()
    record = dict(student_node)
    return record


def select_student_nodes(self):
    student_cur:Cursor = graph.query(f"MATCH (st:Student) Return
st")
    records = student_cur.data()
    return records
```

The `dict()` function converts a `Node` object into a dictionary, thus wrapping a `Student` node with the `dict()` function in the given `get_student_node()`. On the other hand, `Cursor` has a `data()` function to convert the streams of `Node` objects into a list of dictionary elements. So, `select_student_nodes()` returns the stream of `Student` nodes as a list of `Student` records.

Summary

There are lots of NoSQL databases that can store non-relational data for big data applications built with Flask 3.x. Flask can PUT, GET, and SCAN data in HBase using HDFS, access the Cassandra database, execute HGET an HSET with Redis, perform CRUD operations in Couchbase and MongoDB, and manage nodes with Neo4J. Although there are changes in some support modules, such as in `flask-mongoengine`, because of the transformations in the Flask internal modules (for example, the removal of `flask.json`), Flask can still adapt to other Python module extensions and workarounds to connect to and manage its data, such as using the FastAPI-compatible Redis OM.

In general, this chapter showcased Flask's compatibility with almost all the efficient, popular, and widely used NoSQL databases. It is also a Python framework that's fit for building big data applications for many enterprises and scientific development because it supports many NoSQL storages.

The next chapter is about using Flask to implement task management with workflows.

8

Building Workflows with Flask

Workflows are sequences or groups of repetitive tasks, activities, or small processes that require a complete start-to-end execution to satisfy a particular business process. Each task is equivalent to routinary transactions such as sending emails, running scripts or terminal commands, data transformation and serialization, database transactions, and other highly computational operations. These tasks can be simple sequence, parallel, and complex types.

Several tools and platforms can provide best practices, rules, and technical specifications to build workflows for industry, enterprise, and scientific problems. However, most of these solutions require Java more than Python as their core language. Now, the main goal of this chapter is to prove that Python, particularly the Flask framework, can simulate workflows that utilize **Business Process Modeling Notation (BPMN)** and also **non-BPMN** workflows using popular and modern platforms such as *Zeebe/Camunda*, *Airflow 2.0*, and *Temporal*. Moreover, the chapter will also showcase the use of *Celery tasks* in building custom workflows for Flask applications.

This chapter will cover the following topics that will discuss the different mechanisms and procedures in implementing workflow activities with the Flask framework:

- Building workflows with Celery tasks
- Creating BPMN and non-BPMN workflows with SpiffWorkflow
- Building service tasks with the Zeebe/Camunda platforms
- Using Airflow 2.x in orchestrating API endpoints
- Implementing workflows using Temporal.io

Technical requirements

This chapter aims to implement *Doctor's Appointment Management Software* that uses workflows to implement its business processes. It has the following five different Flask projects, showcasing the varying workflow solutions to build the Flask application:

- `ch08-celery-redis`, which focuses on designing dynamic workflows with Celery tasks.
- `ch08-spiff-web`, which implements a web application for the appointment system using the SpiffWorkflow library.
- `ch08-temporal`, which uses the Temporal platform to build distributed architecture.
- `ch08-zeebe`, which utilizes the Zeebe/Camunda platform for BPMN workflows.
- `ch08-airflow`, which integrates with the Airflow 2.x workflow engine to manage API services.

Although with different workflow solutions, each of these projects targets the practical and optimal process performance for the *user's login transactions*, *appointment processes*, *doctor engagement*, *billing processes*, and *releasing transactions*. All database transactions are relational and use PostgreSQL as their database. All these projects, on the other hand, are available at `https://github.com/ PacktPublishing/Mastering-Flask-Web-Development/tree/main/ch08`.

Building workflows with Celery tasks

Utilizing **Celery** as a task queue manager and **Redis** as its broker was part of our *Chapter 5* content. The chapter explicitly discussed all setups and installations to build Flask-Celery-Redis integration. It also expounded on how Celery can run background processes asynchronously outside the context of Flask's request-response transaction. Additionally, this chapter will show us another feature of Celery that can solve business process optimization.

Celery has a mechanism to build dynamic workflows, types of workflows that run outside the bounds of some schema definitions and rules from start to end of workflow activities. Its first requirement is to wrap all tasks in *signatures*.

Creating task signatures

In a typical scenario, calling Celery tasks requires invoking directly its `delay()` method to run the underlying process the standard way or `apply_async()` to run it asynchronously. But to manage Celery tasks to build custom dynamic workflows, individual tasks must invoke the `signature()` or `s()` method first. This allows passing the Celery task invocation to workflow operations, linking a Celery task to another task as callbacks after its successful execution, and also helps manage its inputs, arguments, and execution options. A signature is like a wrapper to a task ready to be passed as an argument to Celery's workflow operations.

The following `add_login_task_wrapper()` task, for instance, can be wrapped inside a signature just by calling its `signature()` or `s()` method:

```
@shared_task(ignore_result=False)
def add_login_task_wrapper(details):
    async def add_login_task(details):
        try:
            async with db_session() as sess:
                async with sess.begin():
                    repo = LoginRepository(sess)
                    details_dict = loads(details)
                    print(details_dict)
                    login = Login(**details_dict)
                    result = await repo.insert_login(login)
                    if result:
                        return str(True)
                    else:
                        return str(False)
        except Exception as e:
            print(e)
            return str(False)
    return run(add_login_task(details))
```

Since the given task has one required parameter, emitting its `signature()` method includes having a tuple argument, as in the following snippet:

```
add_login_task_wrapper.signature((login_str))
```

Or, a shorter way is to call the `s()` equivalent with a typical parameter list containing the arguments, as in the following snippet:

```
add_login_task_wrapper.s(login_str)
```

Invoke `delay()`, `apply_async()`, or simply `()` right after the `signature()` call to run the task, if necessary. Now, let us explore Celery's built-in signatures called *primitives*, used in building simple and complex workflows.

Utilizing Celery primitives

Now, Celery provides the following core workflow operations called primitives, which are also signature objects themselves that take a list of task signatures to build dynamic workflow transactions:

- `chain()` – A Celery function that takes a series of signatures that are linked together to form a chain of callbacks executed from left to right.

- `group()` – A Celery operator that takes a list of signatures that will execute in parallel.

- `chord()` – A Celery operator that takes a list of signatures that will execute in parallel but with a callback that will consolidate their results.

Let us first, in the next section, showcase Celery's chained workflow execution.

Implementing a sequential workflow

Celery primitives are the components of building dynamic Celery workflows. The most commonly used primitive is the *chain* primitive, which can establish a pipeline of tasks with results passed from one task to another in a left-to-right manner. Since it is dynamic, it can follow any specific sequence based on the software specification, but it prefers smaller and straightforward tasks to avoid unwanted performance degradation. *Figure 8.1* shows a workflow diagram that the `ch08-celery-redis` project implemented for an efficient user signup transaction:

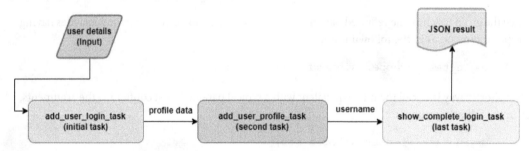

Figure 8.1 – Task signatures in a chain operation

Similar to the `add_user_login_task_wrapper()` task, `add_user_profile_task_wrapper()` and `show_complete_profile_task_wrapper()` are asynchronous Celery tasks that can emit their respective signature to establish a dynamic workflow. The following endpoint function calls the signatures of these tasks in sequence using the `chain()` primitive:

```
from modules.login.services.workflow_tasks import add_user_login_task_
wrapper, add_user_profile_task_wrapper, show_complete_login_task_
wrapper

@login_bp.post('/login/user/add')
```

```
async def add_user_workflow():
    user_json = request.get_json()
    user_str = dumps(user_json)
    task = chain(add_user_login_task_wrapper.s(user_str), add_user_
profile_task_wrapper.s(), show_complete_login_task_wrapper.s())()
    result = task.get()
    records = loads(result)
    return jsonify(profile=records), 201
```

The presence of () at the end of the chain() primitive means the execution of the chained sequence since chain() is also a signature but a predefined one. Now, the purpose of the add_user_ workflow() endpoint is to merge the *INSERT* transaction of the login credentials and the login profile details of the user instead of accessing two separate endpoints for the whole process. Also, it's there to render the login credentials to the user after a successful workflow execution. So, all three tasks are in one execution frame with one JSON input of combined user profile and login details to the initial task, add_user_login_task_wrapper(). But what if tasks need arguments? Does the signature() method accept parameter(s) for its task? Let's take a look in the next section.

Passing inputs to signatures

As mentioned earlier in this chapter, the required arguments for the Celery tasks can be passed to the s() or signature() function. In the given chained tasks, the add_user_login_task_ wrapper() is the only task among the three that needs input from the API, as depicted in its code here:

```
@shared_task(ignore_result=False)
def add_user_login_task_wrapper(details):
    async def add_user_task(details):
        try:
            async with db_session() as sess:
                async with sess.begin():
                    repo = LoginRepository(sess)
                    details_dict = loads(details)
                    … … … … … …
                    login = Login(**user_dict)
                    result = await repo.insert_login(login)

                    if result:
                        profile_details = dumps(details_dict)
                        return profile_details
                    else:
                        return ""
        except Exception as e:
            print(e)
```

```
                return ""
    return run(add_user_task(details))
```

The `details` parameter is the complete JSON details passed from the endpoint function to the `s()` method so that the task will retrieve only the *login credentials* for the *INSERT* login transaction. Now, the task will return the remaining details, the user profile information, as input to the next task in the sequence, `add_user_profile_task_wrapper()`. The following code shows the presence of a local parameter in the `add_user_profile_task_wrapper()` task that will receive the result of the previous task:

```
@shared_task(ignore_result=False)
def add_user_profile_task_wrapper(details):
    async def add_user_profile_task(details):
        try:
            async with db_session() as sess:
                async with sess.begin():
                    ... ... ... ... ... ...
                    role = profile_dict['role']
                    result = False
                    if role == 0:
                        repo = AdminRepository(sess)
                        admin = Administrator(**profile_dict)
                        result = await repo.insert_admin(admin)
                    elif role == 1:
                        repo = DoctorRepository(sess)
                        doc = Doctor(**profile_dict)
                        result = await repo.insert_doctor(doc)
                    elif role == 2:
                        repo = PatientRepository(sess)
                        patient = Patient(**profile_dict)
                        result = await repo.insert_patient(patient)
                    ... ... ... ... ... ...
                    ... ... ... ... ... ...
    return run(add_user_profile_task(details))
```

If the preceding task of a chained workflow returns a value, the next task must have a first local parameter to receive that result. In the given code of `add_user_profile_task_wrapper()`, the `details` parameter pertains to the returned value of `add_user_login_task_wrapper()`. The first parameter will always receive the result of the preceding tasks. Now, the `add_user_profile_task_wrapper()` task will check the role to determine what table to insert the profile information in. Then, it will return the *username* as input to the final task, `show_complete_login_task_wrapper()`, which will render the user credentials.

The dynamic workflow must have strict exception handling from the inside of the tasks and from the outside of the Celery workflow execution to establish a continuous and blockage-free passing of results or input from the initial task to the end.

On the other hand, running independent Celery tasks requires a different Celery primitive operation called group(). Let us now scrutinize some parallel tasks from our application.

Running independent and parallel tasks

The group() primitive can run tasks concurrently and even return consolidated results from functional tasks. Our sample grouped workflow, shown in *Figure 8.2*, focuses only on void tasks that serialize a list of records to CSV files, so no consolidation of results is needed:

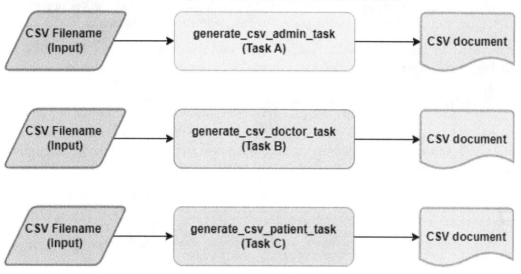

Figure 8.2 – Task signatures in grouped workflow

The group() operation can accept varying Celery tasks with different arguments but prefers those that *read from and write to files*, *perform database transactions*, *extract resources from API endpoints*, *download files from external storages*, or *perform any I/O operations*. Our create_reports() endpoint function performs the grouped workflow presented in *Figure 8.2*, which aims to back up the list of user administrators, patients, and doctors to their respective CSV files. The following is the code of the endpoint function:

```
from modules.admin.services.reports_tasks import generate_csv_admin_
task_wrapper, generate_csv_doctor_task_wrapper, generate_csv_patient_
task_wrapper

@admin_bp.get('/admin/reports/create')
async def create_reports():
```

```
    admin_csv_filename = os.getcwd() + "/files/dams_admin.csv"
    patient_csv_filename = os.getcwd() + "/files/dams_patient.csv"
    doctor_csv_filename = os.getcwd() + "/files/dams_doc.csv"
    workflow = group( generate_csv_admin_task_wrapper.s(admin_csv_
filename), generate_csv_doctor_task_wrapper.s( doctor_csv_filename),
generate_csv_patient_task_wrapper.s( patient_csv_filename))()
    workflow.get()
    return jsonify(message="done backup"), 201
```

The `create_reports()` endpoint passes different filenames to the three tasks. The `generate_csv_admin_task_wrapper()` method will back up all administrator records to `dams_admin.csv`, `generate_csv_patient_task_wrapper()` will dump all patient records to `dams_patient.csv`, and `generate_csv_doctor_task_wrapper()` will save all doctor profiles to `dams_doctor.csv`. All three will concurrently execute after running the `group()` operation.

But if the concern is to manage all the results of these concurrently running tasks, the `chord()` workflow operation, as shown in the next section, will be the best option for this scenario.

Using callbacks to manage task results

The `chord()` primitive works like the `group()` operation except for its callback task requirement, which will handle and manage all results of the independent tasks. The following API endpoint generates a report on a doctor's appointments and laboratory requests:

```
from modules.admin.services.doctor_stats_tasks import count_patients_
doctor_task_wrapper, count_request_doctor_task_wrapper, create_doctor_
stats_task_wrapper

@admin_bp.get('/admin/doc/stats')
async def derive_doctor_stats():
    docid = request.args.get("docid")
    workflow = chord((count_patients_doctor_task_wrapper.s(docid),
count_request_doctor_task_wrapper.s(docid)), create_doctor_stats_task_
wrapper.s(docid))()
    result = workflow.get()
    return jsonify(message=result), 201
```

The `derive_doctor_stats()` method aims to execute the workflow shown in *Figure 8.3*, which uses the `chord()` operation to run `count_patients_doctor_task_wrapper()` to determine the number of patients of a particular doctor and `count_request_doctor_task_wrapper()` to extract the total number of laboratory requests of the same doctor. The results of the tasks are stored in a list according to the order of their executions before passing it to the callback task, `create_doctor_stats_task_wrapper()`, for processing. Unlike in the `group()` primitive, the results are managed by a callback task before returning the final result to the API function:

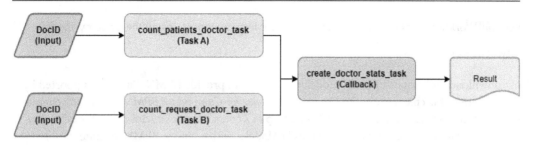

Figure 8.3 – Task signatures in chord() primitive

A sample output of the `create_doctor_stats_task_wrapper()` task will be like this: *"Doctor HSP-200 has 2 patients and 0 lab requests."*

There are lots of ways to build complex dynamic workflows using combinations of `chain()`, `group()`, and `chord()`, which will implement the workflows that the Flask applications need to optimize some business processes. It is possible for a chained task to call the `group()` primitive from the inside to spawn and run a group of independent tasks. It is also feasible to use Celery's *subtasks* to implement conditional task executions. There are also miscellaneous primitives such as `map()`, `starmap()`, and `chunks()` that can manage arguments of tasks in the workflow. A Celery workflow is flexible and open to any implementation using its primitives and signatures since it targets dynamic workflows. Celery workflows can read and execute workflows from XML files, such as BPMN workflows. However, there is a workflow solution that can work on both dynamic and BPMN workflows: SpiffWorkflow.

Creating BPMN and non-BPMN workflows with SpiffWorkflow

SpiffWorkflow is a flexible Python execution engine for workflow activities. Its latest installment focuses more on BPMN models, but it always has strong support classes to build and run non-BPMN workflows translated into Python and JSON. The library has a *BPMN interpreter* that can execute tasks indicated in BPMN diagrams created by BPMN modeling tools and *serializers* to run JSON-based workflows.

To start SpiffWorkflow, we need to install some required dependencies.

Setting up the development environment

No broker or server is needed to run workflows with SpiffWorkflow. However, installing the main plugin using the `pip` command is a requirement:

```
pip install spiffworkflow
```

Then, for serialization and parsing purposes, install the `lxml` dependency:

```
pip install lxml
```

Since SpiffWorkflow uses the Celery client library for legacy support, install the `celery` module:

```
pip install celery
```

Now, download and install a BPMN modeler tool that can provide BPMN diagrams supported by SpiffWorkflow. This chapter uses the *Camunda Modeler for Camunda 7 BPMN* version to generate BPMN diagrams, which we can download from `https://camunda.com/download/modeler/`. *Figure 8.4* provides a screenshot of the Camunda Modeler with a sample BPMN diagram:

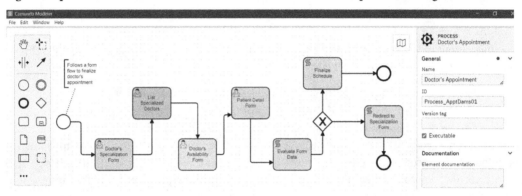

Figure 8.4 – Camunda Modeler with BPMN model for Camunda 7

The version of SpiffWorkflow used by this chapter can only parse and execute the BPMN model for the Camunda 7 platform. Hopefully, its future releases can support Camunda 8 or higher versions of BPMN diagrams.

Let us now create our workflow using the BPMN modeler tool.

Creating a BPMN diagram

BPMN is an open standard for business process diagrams. It is a graphical mechanism to visualize and simulate a systematic set of activities in one process flow that goals a successful result. A BPMN diagram has a set of graphical elements, called *flow objects*, composed of *activities*, *events*, *sequence flows*, and *gateways*.

An activity represents work that needs execution inside a workflow process. A work can be simple and atomic, such as a *task*, or complex, such as a *sub-process*. When an activity is atomic and cannot break down further, despite the complexity of the process, then that is considered a task. A task in BPMN is denoted as a *rounded-corner rectangle shape* component. There are several types of tasks, but SpiffWorkflow only supports the following:

- **Manual task** – A non-automated task that a human can perform outside the context of the workflow.

- **Script task** – A task that runs a modeler-defined script.

- **User task** – A typical task that a human actor can carry out using some application-related operation, such as clicking a button.

The tasks presented in the BPMN diagram of *Figure 8.4*, namely **Doctor's Specialization Form**, **List Specialized Doctors**, **Doctor's Availability Form**, and **Patient Detail Form**, are *user tasks*. Usually, user tasks can represent actions such as web form handling, console-based transactions with user inputs, or transactions in applications involving editing and submitting form data. On the other hand, the **Evaluate Form Data** and **Finalize Schedule** tasks are considered *script tasks*.

A *sequence flow* is a one-directional line connector between activities or tasks. The BPMN standard allows adding descriptions or labels to sequence flows to determine which paths to take from one activity to another.

Now, the workflow will not work without *start* and *stop events*. An *event* is an occurrence along the workflow required to execute due to some triggers to produce some result. The start event, represented by a *small and open circle with a thin-lined boundary*, triggers the start of the workflow. The stop event, defined by a *small, open circle with a single thick-lined boundary*, ends the workflow activities. Other than these two, there are *cancel, signal, error, message, timer,* and *escalation* events supported by SpiffWorkflow, and all these are represented as circles.

The *diamond-shaped component* in *Figure 8.4* is a *gateway* component. It diverges or converges its incoming or outgoing process flows. It can control multiple incoming and multiple outgoing process flows. SpiffWorkflow supports the following types of gateways:

- **Exclusive gateway** – Caters to multiple incoming flows and will emit only one output flow based on some evaluation.

- **Parallel gateway** – Emits an independent process flow that will execute tasks without order but will wait for all the tasks to finish.

- **Event gateway** – Emits an outgoing flow based on some events from an outside source.

- **Inclusive gateway** – Caters to multiple incoming flows and can emit more than one output flow based on some complex evaluation.

The gateway in *Figure 8.4* is an example of an exclusive gateway because it will allow **the Finalize Schedule** task execution to proceed if, and only if, the form data is complete. Otherwise, it will redirect the sequence flow to the **Doctor's Specialization Form** web form task again for data re-entry.

Now, let us start the showcase on how SpiffWorkflow can interpret a BPMN diagram for **business process management (BPM)**.

Implementing the BPMN workflow

SpiffWorkflow can translate mainly the *user*, *manual*, and *script* tasks of a BPMN diagram. So, it can best handle business process optimization involving sophisticated web flows in a web application.

Since there is nothing to configure in the `create_app()` factory or `main.py` module for SpiffWorkflow, the next step after dependency module installations and the BPMN diagram design is the view function implementation for the BPMN diagram simulation. The view functions must initiate and execute SpiffWorkflow tasks to run the entire BPMN workflow.

The first support class to call in the module script is `CamundaParser`, a support class found in the `SpiffWorkflow.camunda.parser.CamundaParser` module of SpiffWorkflow. The `CamundaParser` class will parse the BPMN tags of the BPMN file based on the Camunda 7 standards. The BPMN file is an XML document with tags corresponding to the *flow objects* of the workflow. Now, the `CamundaParser` class will need the name or ID of the BPMN definition to load the document and verify if the XML schema of the BPMN document is well formed and valid. The following is the first portion of the `/view/appointment.py` module of the `doctor` Blueprint module that instantiates the `CamundaParser` class that will load our `dams_appointment.bpmn` file, the workflow design depicted in the BPMN workflow diagram of *Figure 8.4*:

```
from SpiffWorkflow.bpmn.workflow import BpmnWorkflow
from SpiffWorkflow.camunda.parser.CamundaParser import CamundaParser
from SpiffWorkflow.bpmn.specs.defaults import ScriptTask
from SpiffWorkflow.camunda.specs.user_task import UserTask
from SpiffWorkflow.task import Task, TaskState
from SpiffWorkflow.util.deep_merge import DeepMerge

parser = CamundaParser()
filepath = os.path.join("bpmn/dams_appointment.bpmn")
parser.add_bpmn_file(filepath)
spec = parser.get_spec('Process_ApptDams01')
```

The `add_bpmn_file()` function of the API will load the BPMN file, while the `get_spec()` function will parse the document starting with the process definition ID call. Now, *Figure 8.5* shows a snapshot of the BPMN file with the process definition ID:

```
bpmn >  dams_appointment.bpmn
  1   <?xml version="1.0" encoding="UTF-8"?>
  2   <bpmn:definitions xmlns:bpmn="http://www.omg.org/spec/BPMN/20100524/MODEL" xmlns:bpmndi="http://www.omg.org/spec/
  3   <bpmn:process id="Process_ApptDams01" name="Doctor&#39;s Appointment" isExecutable="true" camunda:historyTimeTo
  4     <bpmn:startEvent id="StartEvent_1">
  5       <bpmn:outgoing>Flow_0fr3hj5</bpmn:outgoing>
  6     </bpmn:startEvent>
  7     <bpmn:sequenceFlow id="Flow_0fr3hj5" sourceRef="StartEvent_1" targetRef="Activity_SpecialDoc" />
  8     <bpmn:userTask id="Activity_SpecialDoc" name="Doctor&#39;s Specialization Form">
  9       <bpmn:extensionElements>
 10         <camunda:formData>
 11           <camunda:formField id="specialization" label="Specialization" />
 12         </camunda:formData>
 13       </bpmn:extensionElements>
 14       <bpmn:incoming>Flow_0fr3hj5</bpmn:incoming>
 15       <bpmn:outgoing>Flow_0h8baw4</bpmn:outgoing>
 16     </bpmn:userTask>
 17     <bpmn:sequenceFlow id="Flow_0h8baw4" sourceRef="Activity_SpecialDoc" targetRef="Activity_SelectDoc" />
 18     <bpmn:userTask id="Activity_SelectDoc" name="List Specialized Doctors">
```

Figure 8.5 – A snapshot of a BPMN file containing the process definition ID

After activating SpiffWorkflow with its parser, the next step is to build web flows through the view functions. The view implementations will be a series of page redirections that will gather all the necessary form data values for the *user tasks* of the BPMN workflow. The following `choose_specialization()` view will be the first web form to start since it will simulate the **Doctor's Specialization Form** task:

```
@doc_bp.route("/doctor/expertise", methods = ["GET", "POST"])
async def choose_specialization():
    if request.method == "GET":
      return render_template("doc_specialization_form.html")
    session['specialization'] = request.form['specialization']
    return redirect(url_for("doc_bp.select_doctor"))
```

This view will redirect the user to `select_doctor()` to list all doctors with the specialization indicated by `choose_specialization()`. The following snippet presents the code for the `select_doctor()` view:

```
@doc_bp.route("/doctor/select", methods = ["GET", "POST"])
async def select_doctor():
    if request.method == "GET":
        return render_template("doc_doctors_form.html")
    session['docid'] = request.form['docid']
    return redirect(url_for("doc_bp.reserve_schedule") )
```

After the `select_doctor()` view, the user will choose a date and time for the appointment through the `reserve_schedule()` view. The last view of the web flow is `provide_patient_details()`, which will ask for the patient details needed for the diagnosis and payment. The following code presents the implementation of the `reserve_schedule()` view:

```python
@doc_bp.route("/doctor/schedule", methods = ["GET", "POST"])
async def reserve_schedule():
    if request.method == "GET":
        return render_template("doc_schedule_form.html"), 201
    session['appt_date'] = request.form['appt_date']
    session['appt_time'] = request.form['appt_time']
    return redirect( url_for("doc_bp.provide_patient_details"))
```

Now, the last view function, `provide_patient_details()`, will trigger the workflow execution besides its goal to extract the patient information required for the appointment scheduling and consolidate it with the other details from the previous views. The following is the code for the `provide_patient_details()` view:

```python
from SpiffWorkflow.bpmn.workflow import BpmnWorkflow
from SpiffWorkflow.camunda.parser.CamundaParser import CamundaParser
from SpiffWorkflow.bpmn.specs.bpmn_task_spec import TaskSpec
from SpiffWorkflow.camunda.specs.user_task import UserTask
from SpiffWorkflow.task import Task, TaskState

@doc_bp.route("/doctor/patient", methods = ["GET", "POST"])
async def provide_patient_details():
    if request.method == "GET":
        return render_template("doc_patient_form.html"), 201
    form_data = dict()
    form_data['specialization'] = session['specialization']
    form_data['docid'] = session['docid']
    form_data['appt_date'] = session['appt_date']
    form_data['appt_time'] = session['appt_time']
    form_data['ticketid'] = request.form['ticketid']
    form_data['patientid'] = request.form['patientid']
    form_data['priority_level'] = request.form['priority_level']
    workflow = BpmnWorkflow(spec)
    workflow.do_engine_steps()
    ready_tasks: List[Task] = workflow.get_tasks(TaskState.READY)
    while len(ready_tasks) > 0:
        for task in ready_tasks:
            if isinstance(task.task_spec, UserTask):
                upload_login_form_data(task, form_data)
            workflow.run_task_from_id(task_id=task.id)
```

```
        else:
            task_details:TaskSpec = task.task_spec
            print("Complete Task ", task_details.name)
     workflow.do_engine_steps()
     ready_tasks = workflow.get_tasks(TaskState.READY)
   dashboard_page = workflow.data['finalize_sched']
   if dashboard_page:
     return render_template("doc_dashboard.html"), 201
   else:
     return redirect(url_for("doc_bp.choose_specialization"))
```

Session handling provides the `provide_patient_details()` view with the ability to gather all appointment details from the previous web views. As depicted in the given code, all session data, including the patient details from its form, are placed in its `form_data` dictionary. Utilizing the session is a workaround because it is not feasible to fuse the workflow loops required by the SpiffWorkflow library and the web flows. The last redirected page must initiate the workflow with the `BpmnWorkflow` class. But what is the difference between the `CamundaParser` and `BpmnWorkflow` API classes? We answer this question in the next section.

Distinguishing between workflow specifications and instances

There are two categories of components in SpiffWorkflow: *specification* and *instance* objects. `CamundaParser`, through its `get_spec()` method, returns a `WorkflowSpec` instance object, a specification or model object that defines the BPMN workflow. On the other hand, `BpmnWorkflow` creates a `Workflow` instance object, which tracks and returns actual workflow activities. However, `BpmnWorkflow` requires the workflow specification object as its constructor parameter before instantiation.

The workflow instance will provide all sequence flows from the start until the stop event and the tasks with their corresponding state. All the states, such as READY, CANCELLED, COMPLETED, and FUTURE, are indicated in the `TaskState` API coupled with hook methods found in the `Task` instance object. But how does SpiffWorkflow determine a BPMN task? We will see that in the next section.

Identifying between task specifications and instances

As with the workflow, each SpiffWorkflow task has a specification object called `TaskSpec`, which provides details such as the *name of the task definition* and *task type*, such as `UserTask` or `ScriptTask`. On the other hand, the task instance object is named `Task`. The workflow instance object provides `get_tasks()` overrides that give all tasks based on a specific state or `TaskSpec` instance. Moreover, it has `get_task_from_id()` to extract the `Task` instance object based on *task ID*, `get_task_spec_from_name()` to retrieve the `TaskSpec` name based on its indicated BPMN name, and `get_tasks_from_spec_name()` to retrieve all tasks based on a `TaskSpec` definition name.

To traverse and track every `UserTask`, `ManualTask`, or `Gateway` task and their trailing `ScriptTask` task(s) based on the BPMN diagram starting from `StartEvent`, invoke the `do_engine_steps()` of the workflow instance. A loop must call the `do_engine_steps()` method to track every activity in the workflow, including events and `ScriptTask` tasks until it reaches `EndEvent`. Thus, `provide_patient_details()` has a `while` loop in the POST transaction to traverse the workflow and execute every `Task` object with the `run_task_from_id()` method of the workflow instance.

But running tasks, specifically `UserTask` and `ScriptTask`, is not only concerned with the fulfillment of the workflow activity but also the passing of some task data.

Passing form data to UserTask

`UserTask`'s form fields are among the several sources of BPMN workflow data. The Camunda Modeler allows the BPMN designer to create form variables for each `UserTask` task. *Figure 8.6* shows the three form fields, namely `patientid`, `ticketid`, and `priority_level`, of the **Patient Detail Form** task and the portion of the Camunda Modeler where to add form variables:

Figure 8.6 – Adding form fields to UserTask

The presence of form fields in a custom-generated form requires data passing to these form variables through the view function. Form fields without values will yield exceptions that can halt workflow executions, eventually ruining the Flask application. The `while` loop in the following code snippet of the `provide_patient_details()` view calls an `upload_login_form_data()` custom method that assigns values from the `form_data` dictionary to each `UserTask` form variable:

```python
from SpiffWorkflow.util.deep_merge import DeepMerge

def upload_login_form_data(task: UserTask, form_data):
    form = task.task_spec.form
    data = {}
    if task.data is None:
        task.data = {}

    for field in form.fields:
        if field.id == "specialization":
            process_data = form_data["specialization"]
        elif field.id == "docid":
            process_data = form_data["docid"]
        elif field.id == "date_scheduled":
            process_data = form_data["appt_date"]
        ... ... ... ... ... ...
        update_data(data, field.id,  process_data)
        DeepMerge.merge(task.data, data)

@doc_bp.route("/doctor/patient", methods = ["GET", "POST"])
async def provide_patient_details():
    ... ... ... ... ... ...
    while len(ready_tasks) > 0:
        for task in ready_tasks:
            if isinstance(task.task_spec, UserTask):
                upload_login_form_data(task, form_data)
            else:
                task_details:TaskSpec = task.task_spec
                print("Complete Task ", task_details.name)
            workflow.run_task_from_id(task_id=task.id)
        ... ... ... ... ... ...
        return redirect(url_for("doc_bp.choose_specialization"))
```

The `upload_login_form_data()` method determines each form field through its *ID* and extracts its appropriate value from the `form_data` dictionary. Then, the custom method, shown in the following snippet, assigns the value to the form field and uploads the field-value pair as *workflow data* using the `DeepMerge` utility class of SpiffWorkflow:

```
def update_data(dct, name, value):
    path = name.split('.')
    current = dct
    for component in path[:-1]:
        if component not in current:
            current[component] = {}
        current = current[component]
    current[path[-1]] = value
```

Technically, `update_data()` creates a dictionary object containing the field name as the key and its corresponding `form_data` value.

But how about `ScriptTask`? Can it have form variables, too? Let's explore that in the next section.

Adding input variables to ScriptTask

`ScriptTask` can also have input variables but not form fields. These input variables also need values from the view function because these are essential parts of its expressions. Sometimes, `ScriptTask` does not need inputs from views because it can extract existing workflow data to build its conditional expression. But for sure, it must emit output variable(s) that the succeeding `Gateway`, `ScriptTask`, or `UserTask` task needs to pursue their execution. *Figure 8.7* shows the **Evaluate Form Data** task with its `proceed` output variable and how it extracts and uses the profile information from the workflow data:

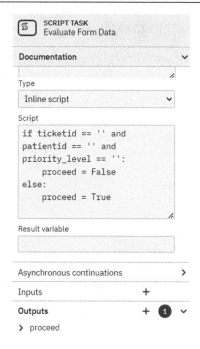

Figure 8.7 – Utilizing variables in ScriptTask

After running all the tasks and uploading all the values to the different variables in the workflow, the result of the workflow must be variables that will decide the result of the view function; in our case, the `provide_patient_details()` view. Let us now retrieve these results to determine the type of responses our views will render.

Managing the result of the workflow

The goal of our workflow through SpiffWorkflow is to determine the view page a route function will render. Together with this is the execution of the required backend transactions, such as saving the scheduled appointment into the database, sending notifications to the doctors for the newly created appointment, and generating the necessary documents for the schedule. The workflow's generated data will determine the resulting processes of the view. In our appointment workflow, when the generated `finalize_sched` variable is `True`, the view will redirect the user to the doctor's dashboard page. Otherwise, the user will see the first page of the data-gathering process.

Let us now explore the capability of SpiffWorkflow to implement non-BPMN workflows.

Implementing a non-BPMN workflow

SpiffWorkflow can implement workflows in JSON or Python configurations. In our `ch08-spiff-web` project, we have the following Python class that implements a prototype of a payment process workflow:

```python
from SpiffWorkflow.specs.WorkflowSpec import WorkflowSpec
from SpiffWorkflow.specs.ExclusiveChoice import ExclusiveChoice
from SpiffWorkflow.specs.Simple import Simple
from SpiffWorkflow.operators import Equal, Attrib


class PaymentWorkflowSpec(WorkflowSpec):
    def __init__(self):
        super().__init__()
        patient_pay = Simple(wf_spec=self, name='dams_patient_pay')
        patient_pay.ready_event.connect( callback=tx_patient_pay)
        self.start.connect(taskspec=patient_pay)
        payment_verify = ExclusiveChoice(wf_spec=self, name='payment_
check')
        patient_pay.connect(taskspec=payment_verify)
        patient_release = Simple(wf_spec=self, name='dams_patient_
release')
        cond = Equal(Attrib(name='amount'), Attrib(name='charge'))
        payment_verify.connect_if(condition=cond, task_spec=patient_
release)
        patient_release.completed_event.connect( callback=tx_patient_
release)
        patient_hold = Simple(wf_spec=self, name='dams_patient_
onhold')
        payment_verify.connect(task_spec=patient_hold)
        patient_hold.completed_event.connect( callback=tx_patient_
onhold)
```

`WorkflowSpec` is responsible for the non-BPMN workflow implementation in Python format. The constructor of the `WorkflowSpec` sub-class creates generic, simple, and atomic tasks using the `Simple` API of the `SpiffWorkflow.specs.Simple` module. The task can have more than one input and any number of output task variables. There is also an `ExclusiveChoice` sub-class that works like a gateway for the workflow.

Moreover, each task has a `connect()` method to establish sequence flows. It also has event variables, such as `ready_event`, `cancelled_event`, `completed_event`, and `reached_event`, that run their respective callback method, such as our `tx_patient_pay()`, `tx_patient_release()`, and `tx_patient_onhold()` methods. Calling these event objects marks a transition from one task's current state to another.

The `Attrib` helper class recognizes a task variable and retrieves its data for comparison performed by internal API classes, such as `Equal`, `NotEqual`, and `LessThan`, of the `SpiffWorkflow.operators` module.

Let us now run our `PaymentWorkflowSpec` workflow using a view function.

Running a non-BPMN workflow

Since this is not a Camunda-based workflow, running the workflow does not need a parser. Immediately wrap and instantiate the custom `WorkflowSpec` sub-class inside the `Workflow` class and call `get_tasks()` inside the view function to prepare the non-BPMN workflow for the task traversal and executions. But the following `start_payment_form()` function opts for individual access of tasks using the workflow instance's `get_tasks_from_spec_name()` function instead of using a `while` loop for task traversal:

```
@payment_bp.route("/payment/start", methods = ["GET", "POST"])
async def start_payment_form():
    if request.method == "GET":
        return render_template("payment_form.html"), 201
    ... ... ... ... ... ...
    workflow_instance = Workflow(workflow_spec=PaymentWorkflowSpec())
    workflow_instance.get_tasks()
```

The following `Task` list will start the workflow:

```
    start_tasks: list[Task] = workflow_instance.get_tasks_from_spec_
name( name='Start')
    for task in start_tasks:
        if task.state == TaskState.READY:
            workflow_instance.run_task_from_id( task_id=task.id)
```

This `Task` list will load all payment data to the workflow and execute the `tx_patient_pay()` callback method to process payment transactions:

```
    patient_pay_task: list[Task] = workflow_instance.get_tasks_from_
spec_name( name='dams_patient_pay')
    for task in patient_pay_task:
        if task.state == TaskState.READY:
            task.set_data(ticketid=ticketid, patientid=patientid,
charge=charge, amount=amount, discount=discount, status=status, date_
released=date_released)
            workflow_instance.run_task_from_id( task_id=task.id)
```

This part of the workflow will execute the `ExclusiveChoice` event to compare the payment amount paid by the patient against the patient's total charges:

```
    payment_check_task: list[Task] = workflow_instance.get_tasks_from_
spec_name( name='payment_check')
    for task in payment_check_task:
        if task.state == TaskState.READY:
            workflow_instance.run_task_from_id( task_id=task.id)
```

If the patient fully paid the charges, the following tasks will execute the `tx_patient_release()` callback method to clear and issue release notifications to the patient:

```
    for_releasing = False
    patient_release_task: list[Task] = workflow_instance.get_tasks_
from_spec_name( name='dams_patient_release')
    for task in patient_release_task:
        if task.state == TaskState.READY:
            for_releasing = True
            workflow_instance.run_task_from_id( task_id=task.id)
```

If the patient has partially paid the charges, the following tasks will execute the `tx_patient_onhold()` callback method:

```
    patient_onhold_task: list[Task] = workflow_instance.get_tasks_
from_spec_name( name='dams_patient_onhold')
    for task in patient_onhold_task:
        if task.state == TaskState.READY:
            workflow_instance.run_task_from_id( task_id=task.id)

    if for_releasing == True:
        return redirect(url_for('payment_bp.release_
patient'), code=307)
    else:
        return redirect(url_for('payment_bp.hold_patient'), code=307)
```

The result of the workflow will decide on what page the view will redirect the user to, whether the *releasing* or *on-hold* page.

Now, SpiffWorkflow will lessen the coding effort in building the workflow because it has defined API classes that support both BPMN and non-BPMN workflow implementation. But what if the need is to trigger workflows through API endpoints that SpiffWorkflow can hardly handle?

The next topic will focus on using a BPMN workflow engine that the Camunda platform uses in running tasks through API endpoints.

Building service tasks with the Zeebe/Camunda platforms

Camunda is a popular lightweight workflow and decision automation engine with built-in powerful tools, such as the *Camunda Modeler*, *Cawemo*, and the *Zeebe* broker. But this chapter is not about Camunda but about using Camunda's *Zeebe server* to deploy, run, and execute workflow tasks built by the Flask framework. The goal is to create a Flask client application that will deploy and run BPMN workflows designed by the Camunda Modeler using the Zeebe workflow engine.

Let us start with the setup and configurations needed to integrate Flask with the Zeebe server.

Setting up the Zeebe server

The easiest way to run the Zeebe server is to use Docker to run its `camunda/zeebe` image. So, read first the updated *Docker Subscription Service Agreement* before downloading and installing Docker Desktop, available from `https://docs.docker.com/desktop/install/windows-install/`.

After the installation, start the Docker engine, open a terminal, and run the following Docker command:

```
docker run --name zeebe --rm -p 26500-26502:26500-26502 -d
--network=ch08-network camunda/zeebe:latest
```

A *Docker network*, as with our `ch08-network`, is needed to expose the ports to the development platform. Zeebe's port `26500` is where the Flask client application will communicate to the server's gateway API. After using Zeebe, run the `docker stop` command with *Zeebe's container ID* to shut down the broker.

Now, the next step is to install the suitable Python Zeebe client for the application.

Installing the pyzeebe library

Lots of effective and popular Zeebe client libraries are Java-based. However, `pyzeebe` is one of the few Python external modules that are simple, easy to use, lightweight, and effective in establishing connectivity to the Zeebe server. It is a *gRPC*-based client library for Zeebe, typically designed to manage workflows that involve RESTful services.

> **Important note**
>
> gRPC is a flexible and high-performance RPC framework that can run in any environment and easily connect to any cluster, with support for access authentication, API health checking, load balancing, and open tracing. All Zeebe client libraries use gRPC to communicate with the server.

Let us now install the `pyzeebe` library using the `pip` command:

```
pip install pyzeebe
```

After the installation and setup, it is time to create a BPMN workflow diagram using the Camunda Modeler.

Creating a BPMN diagram for pyzeebe

The `pyzeebe` module can load and parse BPMN files used by *Camunda version 8.0*. Since it is a small library, it can only read and execute `ServiceTask` tasks. *Figure 8.8* shows a BPMN diagram with two `ServiceTask` tasks: the **Get Diagnostics** task, which retrieves all patients' diagnoses, and the **Get Analysis** task, which returns the doctor's resolutions or prescriptions to the diagnoses:

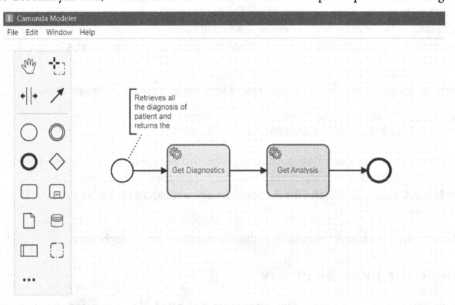

Figure 8.8 – A BPMN diagram with two ServiceTask tasks

The next step is to load and run the final BPMN document using the `pyzeebe` client library. Running the workflow activities from the BPMN diagram is impossible without a `pyzeebe` *worker* and *client*. But implementation of the worker must come first.

Creating a pyzeebe worker

A `pyzeebe` worker or a `ZeebeWorker` worker is a typical Zeebe worker that handles all `ServiceTask` tasks. It runs asynchronously in the background using `asyncio`. `pyzeebe`, as an asynchronous library, prefers a `Flask[async]` platform with `asyncio` utilities. But it requires `grpc.aio.Channel` as a constructor parameter before instantiation.

The library provides three methods to create the needed channel, namely create_insecure_ channel(), create_secure_channel(), and create_camunda_cloud_channel(). All three instantiate the channel, but w create_insecure_channel() disregards the TLS protocol, and create_camunda_cloud_channel() considers the connection to the Camunda cloud. Our ch08-zeebe application uses the insecure one to instantiate the ZeebeWorker worker and eventually manage the ServiceTask tasks indicated in our BPMN file. The following worker-tasks module script shows an independent Python application that contains the ZeebeWorker instantiation and its tasks or jobs:

```python
from pyzeebe import ZeebeWorker, create_insecure_channel
import asyncio
from modules.models.config import db_session, init_db
from modules.doctors.repository.diagnosis import DiagnosisRepository


print('starting the Zeebe worker...')
print('initialize database connectivity...')
init_db()


channel = create_insecure_channel()
worker = ZeebeWorker(channel, max_connection_retries= Zeebe.ZEEBE_MAX_
CONNECTION_RETRIES)
```

This portion of the script shows the creation of the ZeebeWorker worker with its constructor parameters. The initdb() call is included in the module because our tasks will need CRUD transactions:

```python
@worker.task(task_type="select_diagnosis", **Zeebe.TASK_DEFAULT_
PARAMS)
async def select_diagnosis(docid, patientid):
    async with db_session() as sess:
        async with sess.begin():
            try:
                repo = DiagnosisRepository(sess)
                records = await repo.  select_diag_doc_patient(docid,
patientid)
                diagnosis_rec = [rec.to_json() for rec in records]
                diagnosis_str = json.dumps(diagnosis_rec, default=json_
date_serializer)
                return {"data": diagnosis_str}
            except Exception as e:
                print(e)
                return {"data": json.dumps([])}
```

The `select_diagnosis()` method is a `pyzeebe` worker decorated with the `@worker.task()` annotation. The `task_type` attribute of the `@worker.task()` annotation indicates its `ServiceTask` name in the BPMN model. The decorator can also include other attributes, such as `exception_handler` and `timeout_ms`. Now, `select_diagnosis()` looks for all patients' diagnoses from the database with `docid` and `patientid` parameters as filters to the search. It returns a dictionary with a key named `data` handling the result:

```
@worker.task(task_type="retrieve_analysis", **Zeebe.TASK_DEFAULT_
PARAMS)
async def retrieve_analysis(records):
    try:
        records_diagnosis = json.loads(records)
        diagnosis_text = [dt['resolution'] for dt in records_diagnosis]
        return {"result": diagnosis_text}
    except Exception as e:
        print(e)
    return {"result": []}
```

On the other hand, this `retrieve_analysis()` task takes records from `select_diagnosis()` in string form but is serialized back to the list form with `json.loads()`. This task will extract only all resolutions from the patients' records and return them to the caller. The task returns a dictionary also.

The *local parameter names* and the *dictionary keys* returned by the worker's tasks must be *BPMN variable names* because the client will also fetch these local parameters to assign values and dictionary keys for the output extraction for the preceding `ServiceTask` task.

Since our Flask client application uses its event loop, our worker must run on a separate event loop using `asyncio` to avoid exceptions. The following `worker_tasks.py` snippet shows how to run the worker on an `asyncio` environment:

```
if __name__ == "__main__":
    loop = asyncio.get_event_loop()
    loop.run_until_complete(worker.work())
```

The `ZeebeWorker` instance has a `work()` coroutine that must be running asynchronously at the back using an independent event, disconnected from Flask operations. Always run the module with the Python command, such as `python worker-tasks.py`.

Let us now implement the `pyzeebe` client.

Implementing the pyzeebe client

The Flask application needs to instantiate the `ZeebeClient` class to connect to Zeebe. As with the `ZeebeWorker`, it also requires the same `grpc.aio.Channel` parameter as a constructor parameter before its instantiation. Since `ZeebeClient` behaves asynchronously like `ZeebeWorker`, all its operations must run asynchronously in the background as Celery tasks. But, unlike the worker, `ZeebeClient` appears in every Blueprint module as part of its Celery service tasks. The following is the `diagnosis_tasks` module script of the *doctor* Blueprint module that instantiates `ZeebeClient` with the Celery tasks:

```
from celery import shared_task
import asyncio
from pyzeebe import ZeebeClient, create_insecure_channel

channel = create_insecure_channel(hostname="localhost", port=26500)
client = ZeebeClient(channel)
```

The given code snippet creates a `ZeebeClient` instance. The port to connect the Zeebe client is 26500:

```
@shared_task(ignore_result=False)
def deploy_zeebe_wf(bpmn_file):
    async def zeebe_wf(bpmn_file):
        try:
            await client.deploy_process(bpmn_file)
            return True
        except Exception as e:
            print(e)
        return False
    loop = asyncio.get_event_loop()
    return loop.run_until_complete(zeebe_wf(bpmn_file))
```

The `deploy_zeebe_wf()` task is the first process to run before anything else. The API endpoint calling this will load, parse, and deploy the BPMN file with the workflow to the Zeebe server using the asynchronous `deploy_process()` method of `ZeebeClient`. The task will throw an exception if the BPMN file has schema problems, is not well formed, or is invalid:

```
@shared_task(ignore_result=False)
def run_zeebe_task(docid, patientid):
    async def zeebe_task(docid, patientid):
        try:
            process_instance_key, result = await client.run_process_
with_result(
```

```
                     bpmn_process_id= "Process_
Diagnostics", variables={"docid": docid, "patientid":patientid},
variables_to_fetch =["result"], timeout=10000)
            return result
        except Exception as e:
            print(e)
            return {}
    loop = asyncio.get_event_loop()
    return loop.run_until_complete(zeebe_task(docid, patientid))
```

ZeebeClient has two asynchronous methods that can execute process definitions in the BPMN file, and these are run_process() and run_process_with_result(). Both methods pass values to the first task of the workflow, but only run_process_with_result() returns an output value. The given run_zeebe_task() method will execute the first ServiceTask task, the worker's select_diagnosis() task, pass values to its docid and patientid parameters, and retrieve the dictionary output of the last ServiceTask task, retrieve_analysis(), indicated by the result key. A ServiceTask task's parameters are considered BPMN variables that the BPMN file or the ZeebeClient operations can fetch at any time. Likewise, the key of the dictionary returned by ServiceTask becomes a BPMN variable, too. So, the variables parameter of the run_process_with_result() method fetches the local parameters of the first worker's task, and its variables_to_fetch property retrieves the returned dictionary of any ServiceTask task indicated by the key name.

To enable the ZeebeClient operations, run Celery and the Redis broker. Let us now implement API endpoints that will simulate the diagnosis workflow.

Building API endpoints

The following API endpoint passes the filename of the BPMN file to the pyzeebe client by calling the deploy_zeebe_wf() Celery task:

```
@doc_bp.get("/diagnosis/bpmn/deploy")
async def deploy_diagnosis_analysis_bpmn():
    try:
        filepath = os.path.join(Zeebe.BPMN_DUMP_PATH, "dams_diagnosis.
bpmn")
        task = deploy_zeebe_wf.apply_async(args=[filepath])
        result = task.get()
        return jsonify(data=result), 201
    except Exception as e:
            print(e)
    return jsonify(data="error"), 500
```

Afterward, the following `extract_analysis_text()` endpoint can run the workflow by calling the `run_zeebe_task()` Celery task:

```
@doc_bp.post("/diagnosis/analysis/text")
async def extract_analysis_text():
        try:
            data = request.get_json()
            docid = data['docid']
            patientid = int(data['patientid'])
            task = run_zeebe_task.apply_async(args=[docid, patientid])
            result = task.get()
            return jsonify(result), 201
        except Exception as e:
            print(e)
        return jsonify(data="error"), 500
```

The given endpoint will also pass the `docid` and `patientid` values to the client task.

The `pyzeebe` library has many limitations, such as supporting `UserTask` and web flows and implementing workflows that call API endpoints for results. Although connecting our Flask application to the enterprise Camunda platform can address these problems with `pyzeebe`, it is a practical and clever approach to use the Airflow 2.x platform instead.

Using Airflow 2.x in orchestrating API endpoints

Airflow 2.x is an open source platform that provides workflow authorization, monitoring, scheduling, and maintenance with its easy-to-use UI dashboard. It can manage **extract, transform, load** (ETL) workflows and data analytics.

Airflow uses Flask Blueprints internally and allows customization just by adding custom Blueprints in its Airflow directory. However, the main goal of this chapter is to use Airflow as an API orchestration tool to run sets of workflow activities that consume API services for resources.

Let us begin with the installation of the Airflow 2.x platform.

Installing and configuring Airflow 2.x

There is no direct Airflow 2.x installation for the Windows platform yet. But there is a Docker image that can run Airflow on Windows and operating systems with low memory resources. Our approach was to install Airflow directly on WSL2 (Ubuntu) through Windows PowerShell and also use Ubuntu to implement our Flask application for this topic.

Now, follow the next procedures:

1. For Windows users, run the `wsl` command on PowerShell and log in to its home account using the *WSL credentials*.

2. Then, run the `cd ~` Linux command to ensure all installations happen in the home directory.

3. After installing Python 11.x and all its required Ubuntu libraries, create a virtual environment (for example, `ch08-airflow-env`) using the `python3 -m venv` command for the `airflow` module installation.

4. Activate the virtual environment by running the `source <venv_folder>/bin/activate` command.

5. Next, find a directory in the system that can be the Airflow core directory where all Airflow configurations and customizations happen. In our case, it is the `/mnt/c/Alibata/Development/Server/Airflow` folder.

6. Open the `bashrc` configuration file and add the `AIRFLOW_HOME` variable with the Airflow core directory path. The following is a sample of registering the variable:

    ```
    export AIRFLOW_HOME=/mnt/c/Alibata/Development/Server/Airflow
    ```

7. Now, install the `airflow` module using the `pip` command:

    ```
    pip install apache-airflow
    ```

8. Initialize its metadata database and generate configuration files in the `AIRFLOW_HOME` directory using the `airflow db migrate` command.

9. Create an administrator account for its UI dashboard using the following command: `airflow users create --username <user> --password <pass> --firstname <fname> --lastname <lname> --role Admin --email <xxxx@yyyy.com>`. The role value should be `Admin`.

10. Verify if the user account is added to its database using the `airflow users list` command.

11. At this point, log in to the *root account* and activate the virtual environment using `root`. Run the scheduler using the `airflow scheduler` command.

12. With the root account, run the server using the `airflow webserver --port 8080` command. Port `8080` is its default port.

13. Lastly, access the Airflow portal at `http://localhost:8080` and use your `Admin` account to log in to the dashboard.

Figure 8.9 shows the home dashboard of Airflow 2.x:

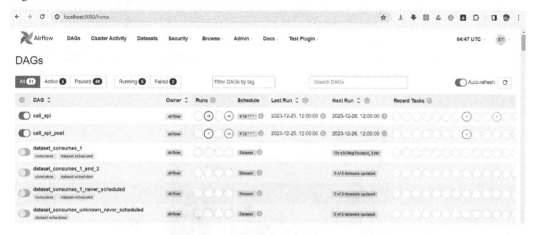

Figure 8.9 – The home page of the Airflow 2.x UI

An Airflow architecture is composed of the following components:

- **Web server** – Runs the UI management dashboard and executes and monitors tasks.

- **Scheduler** – Checks the status of tasks, updates tasks' state details in the metadata database, and queues the next task for executions.

- **Metadata database** – Stores the states of a task, **cross-communications** (**XComs**) data, and **directed acyclic graph** (**DAG**) variables; processes perform read and write operations in this database.

- **Executor** – Executes tasks and updates the metadata database.

Next, let us create workflow tasks.

Creating tasks

Airflow uses DAG files to implement tasks and their sequence flows. A DAG is a high-level design of the workflow and exclusive tasks based on their task definitions, schedules, relationships, and dependencies. Airflow provides the API classes that implement a DAG in Python code. But, before creating DAG files, open the `AIRFLOW_HOME` directory and create a `dags` sub-folder inside it. *Figure 8.10* shows our Airflow core directory with the created `dags` folder:

Figure 8.10 – Custom dags folder in AIRFLOW_HOME

One of the files in our `$AIRFLOW_HOME/dag` directory is `report_login_count_dag.py`, which builds a sequence flow composed of two orchestrated API executions, each with service tasks. *Figure 8.11* provides an overview of the workflow design:

Figure 8.11 – An overview of an Airflow DAG

DAG is an API class from the `airflow` module that implements an entire workflow activity. It is composed of different *operators* that represent tasks. A DAG file can implement more than one DAG if needed. The following code is the DAG script in the `report_login_count_dag.py` file that implements the workflow depicted in *Figure 8.11*:

```
from airflow import DAG
from airflow.operators.python import PythonOperator
from airflow.providers.http.operators.http import SimpleHttpOperator
from datetime import datetime
with DAG(dag_id="report_login_count",
    description="Report the number of login accounts",
    start_date=datetime(2023, 12, 27),
    schedule_interval="0 12 * * *",
    ) as dag:
```

Every DAG has a unique ID or dag_id value. Aside from `description`, DAG has parameters, such as `start_date` and `schedule_interval`, that work like a Cron (time) scheduler for the workflow. The `schedule_interval` parameter can have the `@hourly`, `@daily`, `@weekly`, `@monthly`, or `@yearly` Cron preset options run periodically or a Cron-based expression, such as `*/15 * * * *`, that schedules the workflow to run every *15 minutes*. Setting the parameter to None will disable the periodic execution, requiring a trigger to run the tasks:

```
    task1 = SimpleHttpOperator(
        task_id="list_all_login",
        method="GET",
        http_conn_id="packt_dag",
        endpoint="/ch08/login/list/all",
        headers={"Content-Type": "application/json"},
        response_check=lambda response: handle_response(response),
        dag=dag
    )
    task2 = PythonOperator(
        task_id='count_login',
        python_callable=count_login,
        provide_context=True,
        do_xcom_push=True,
        dag=dag
    )
```

An *Airflow operator* implements a task. But, there are many types of operators to choose from depending on what kind of task the DAG requires. Some widely used operators in training and workplaces are the following:

- `EmptyOperator` – Initiates a built-in execution.
- `PythonOperator` – Calls a Python function that implements a business logic.
- `BashOperator` – Aims to run `bash` commands.
- `EmailOperator` – Sends an email through a protocol.
- `SimpleHttpOperator` – Sends an HTTP request.

Other operators may require installing the needed modules. For example, the `PostgresOperator` operator used for executing PostgreSQL commands requires installing the `apache-airflow[postgres]` module through the `pip` command.

Each task must have a unique `task_id` value for Airflow identification. Our `Task1` task is a `SimpleHTTPOperator` operator that sends a `GET` request to an HTTP `GET` API endpoint expected to return a JSON resource. It has an ID named `list_all_login` and connects to Airflow's HTTP connection object named `packt_dag`. All `SimpleHTTPOperator` needs is a `Connection` object, which stores the HTTP details of the external server resource that the operation will need to establish a connection. Accessing the **Admin** > **Connection** page of Airflow's UI dashboard will provide the necessary form pages for creating a `Connection` object. *Figure 8.12* shows the form that accepts HTTP details of the connection and creates the object:

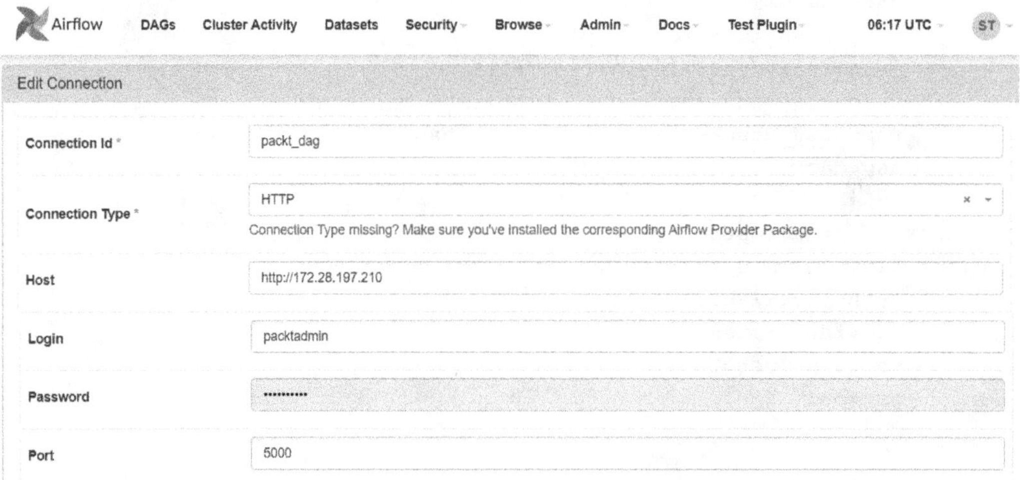

Figure 8.12 – Creating an HTTP Connection object

Also, a `SimpleHTTPOperator` operator provides a callback method indicated by its `response_check` parameter. The callback method accesses the response and other related data and can perform evaluation and logging on the API response. The following is the implementation of the callback method of `Task1`:

```python
def handle_response(response, **context):
    if response.status_code == 201:
        print("executed API successfully...")
        return True
    else:
        print("executed with errors...")
        return False
```

On the other hand, `Task2` is a `PythonOperator` operator that runs a Python function, `count_login()`, for retrieving the JSON data from the API executed in `Task1` and counting the number of records from the JSON resource. Setting its `provide_context` parameter to `True` allows its `python_callable` method to access the `taskInstance` object that pulls the API resource from `Task1`. The `count_login()` function can also set an `xcom` variable, a form of workflow data, because the value of `Task2`'s `do_xcom_push` parameter is `True`. The following snippet is the implementation of `count_login()`:

```python
def count_login(ti, **context):
    data = ti.xcom_pull(task_ids=['list_all_login'])
    if not len(data):
        raise ValueError('Data is empty')
    records_dict = json.loads(data[0])
    count = len(records_dict["records"])
    ti.xcom_push(key="records", value=count)
    return count
    task3 = SimpleHttpOperator(
        task_id='report_count',
        method="GET",
        http_conn_id="packt_dag",
        endpoint="/ch08/login/report/count",
        data={"login_count": "{{ task_instance.xcom_pull( task_
ids=['list_all_login','count_login'], key='records')[0] }}"},
        headers={"Content-Type": "application/json"},
        dag=dag
    )

    ... ... ... ... ... ...
task1 >> task2 >> task3 >> task4
```

Task3 is also a `SimpleHTTPOperator` operator, but its goal is to call an HTTP GET API and pass a request parameter, `login_count`, with a value derived from XCom data. Operators can access Airflow built-in objects, such as dag_run and `task_instance`, using the { { } } Jinja2 delimiter. In Task3, `task_instance`, using its xcom_pull() function, retrieves from the list of tasks the XCom variable records. The result of xcom_pull() is always a list with the value of the XCom variable at its *0 index*.

The last portion of the DAG file is where to place the sequence flow of the DAG's task. There are two ways to establish dependency from one task to another. `>>`, or the *upstream dependency*, connects a flow from left to right, which means the execution of the task from the right depends on the success of the left task. The other one, `<<` or the *downstream dependency*, follows the reverse flow. If two or more tasks depend on the same task, brackets enclose those dependent tasks, such as the `task1 >>` `[task2, task3]` flow, where task2 and task3 are dependent tasks of task1. In the given DAG file, it is just a sequential flow from task1 to task4.

What executes our tasks are called *executors*. The default executor is `SequentialExecutor`, which runs the task flows one task at a time. `LocalExecutor` runs the workflow sequentially, but the tasks may run in parallel mode. There is `CeleryExecutor`, which runs workflows composed of Celery tasks, and `KubernetesExecutor`, which runs tasks on a cluster.

To deploy and re-deploy the DAG files, *restart* the scheduler and the web server. Let us now implement an API endpoint function that will run the DAG deployed in the Airflow server.

Utilizing Airflow built-in REST endpoints

To trigger the DAG is to run the workflow. Running a DAG requires using the Airflow UI's DAG page, applying Airflow APIs for console-based triggers, or consuming Airflow's built-in REST API with the Flask application or Postman. This chapter implemented the ch08-airflow project to provide the report_login_count DAG with API endpoints for Task1 and Task3 executions and also to trigger the workflow using some Airflow REST endpoints. The following is a custom endpoint function that triggers the report_login_count DAG with a dag_run_id value of a UUID type:

```
@login_bp.get("/login/dag/report/login/count")
async def trigger_report_login_count():
    token = "cGFja3RhZG1pbjpwYWNrdGFkbWlu"
    dag_id = "report_login_count"
    deployment_url = "localhost:8080"
    response = requests.post(
        url=f"http://{deployment_url} /api/v1/dags/{dag_id}/dagRuns",
        headers={
            "Authorization": f"Basic {token}",
            "Content-Type": "application/json",
            "Accept": "*/*",
            "Connection": "keep-alive",
```

```
                    "Accept-Encoding": "gzip, deflate, br"
        },
        data = '{"dag_run_id": "d08a62c6-ed71-49fc-81a4-
47991221aea5"}'
    )
    result = response.content.decode(encoding="utf-8")
    return jsonify(message=result), 201
```

Airflow requires *Basic authentication* before consuming its REST endpoints. Any REST access must include an `Authorization` header with the generated token of a valid username and password. Also, install a REST client module, such as `requests`, to consume the API libraries. Running `/api/v1/dags/report_login_count/dagRuns` with an HTTP `POST` request will give us a JSON response like this:

```
{
    "conf": {},
    "dag_id": "report_login_count",
    "dag_run_id": "01c04a4b-a3d9-4dc5-b0c3-e4e59e2db554",
    "data_interval_end": "2023-12-27T12:00:00+00:00",
    "data_interval_start": "2023-12-26T12:00:00+00:00",
    "end_date": null,
    "execution_date": "2023-12-27T13:55:44.910773+00:00",
    "external_trigger": true,
    "last_scheduling_decision": null,
    "logical_date": "2023-12-27T13:55:44.910773+00:00",
    "note": null,
    "run_type": "manual",
    "start_date": null,
    "state": "queued"
}
```

Then, running the following Airflow REST endpoint using the same `dag_run_id` value will provide us with the result of the workflow:

```
@login_bp.get("/login/dag/xcom/values")
async def extract_xcom_count():
    try:
        token = "cGFja3RhZG1pbjpwYWNrdGFkbWlu"
        dag_id = "report_login_count"
        task_id = "return_report"
        dag_run_id = "d08a62c6-ed71-49fc-81a4-47991221aea5"
        deployment_url = "localhost:8080"
        response = requests.get(
```

```
            url=f"http://{deployment_url} /api/v1/dags/{dag_id}/
    dagRuns /{dag_run_id}/taskInstances/{task_id} /xcomEntries/{'report_
    msg'}",
            headers={
                "Authorization": f"Basic {token}",
                ... ... ... ... ... ...
            }
        )
        result = response.json()
        message = result['value']
        return jsonify(message=message)
    except Exception as e:
        print(e)
    return jsonify(message="")
```

The given HTTP GET request API will provide us a JSON result like so:

```
{
    "message": "There are 20 users as of 2023-12-28 00:38:17.592287."
}
```

Airflow is a big platform that can offer us many solutions, especially in building pipelines of tasks for data transformation, batch processing, and data analytics. Its strength is also in implementing API orchestration for microservices. But for complex, long-running, and distributed workflow transactions, it is Temporal.io that can provide durable, reliable, and scalable solutions.

Implementing workflows using Temporal.io

The **Temporal.io** server manages loosely coupled workflows and activities, those not limited by the architecture of the Temporal.io platform. Thus, all workflow components are coded from the ground up without hooks and callable methods appearing in the implementation. The server expects the execution of activities rather than tasks. In BPMN, an activity is more complex than a task. The server is responsible for building a fault-tolerant workflow because it can recover failed activity execution by restarting its execution from the start.

So, let us begin this topic with the Temporal.io server setup.

Setting up the environment

The Temporal.io server has an installer for *macOS*, *Windows*, and *Linux* platforms. For Windows users, download the ZIP file from the https://temporal.download/cli/archive/latest?platform=windows&arch=amd64 link. Then, unzip the file to the local machine. Start the server using the temporal server start-dev command.

Now, to integrate our Flask application with the server, install the `temporalio` module to the virtual environment using the `pip` command. Establish a server connection in the `main.py` module of the application using the `Client` class of the `temporalio` module. The following `main.py` script shows how to instantiate a `Client` instance:

```python
from temporalio.client import Client
from modules import create_app
import asyncio


app, celery_app= create_app("../config_dev.toml")


async def connect_temporal(app):
    client = await Client.connect("localhost:7233")
    app.temporal_client = client


if __name__ == "__main__":
    asyncio.run(connect_temporal(app))
    app.run(debug=True)
```

The `connect_temporal()` method instantiates the Client API class and creates a `temporal_client` environment variable in the Flask platform for the API endpoints to run the workflow. Since `main.py` is the entry point module, an event loop will execute the method during the Flask server startup.

After setting up the Temporal.io server and its connection to the Flask application, let us discuss the distinct approach to workflow implementation.

Implementing activities and a workflow

Temporal uses the code-first approach of implementing a workflow and its activities. Activities in a Temporal platform must be *idempotent*, meaning its parameters and results are non-changing through the course or history of its executions. The following is an example of a complex but idempotent activity:

```python
from temporalio import activity
@activity.defn
async def reserve_schedule(appointmentwf: AppointmentWf) -> str:
    try:
        async with db_session() as sess:
            async with sess.begin():
                repo = AppointmentRepository(sess)
                … … … … … … …
                result = await repo.insert_appt(appt)
                if result == False:
                    … … … … … …
```

```
                return "failure"
        ... ... ... ... ... ...
                return "success"
    except Exception as e:
        print(e)
    ... ... ... ... ... ...
    return "failure"
```

A *Temporal activity* is any function decorated by the `@activity.defn` annotation, with a workflow data class as a local parameter, and returns a non-varying and non-changing value. The returned value can be a fixed string, number, or any string with a non-varying length. Avoid returning collection or model objects with varying property values. Our `reserve_schedule()` activity accepts an `AppointmentWf` object containing appointment details and saves the record of information into the database. It returns only either `"successful"` or `"failure"`.

An activity is where access to external services, such as databases, emails, or APIs, is permitted by Temporal and not in the workflow implementation. The following code is a *Temporal workflow* that runs the `reserve_schedule()` activity:

```
@workflow.defn(sandboxed=False)
class ReserveAppointmentWorkflow():
    def __init__(self) -> None:
        self.appointmentwf = AppointmentWf()
```

The Temporal workflow is the main unit of execution of the application. A typical Python class with a `@workflow.defn` decorator implements a workflow. An example is our `ReserveAppointmentWorkflow` class.

It maintains the same execution state starting from the beginning, making it a deterministic workflow. It also manages all its states through replays to determine some exceptions and provide recovery after a non-deterministic state.

Moreover, Temporal workflows are designed to run continuously without time limits but with proper scheduling to handle long-running and complex activities. However, using threads for concurrency is not allowed in Temporal workflows. It must have an instance method decorated by `@workflow.run` to create a continuous loop for its activities. The following `run()` method accepts a request model with appointment details from the user and loops until the cancellation of the appointment, where `appointmentwf.status` becomes `False`:

```
@workflow.run
async def run(self, data: ReqAppointment) -> None:
    duration = 12
    self.appointmentwf.ticketid = data.ticketid
    self.appointmentwf.patientid = data.patientid
    ... ... ... ... ... ...
```

```
        while self.appointmentwf.status:
            self.appointmentwf.remarks = "Doctor reservation being
processed...."
            try:
                await workflow.execute_activity(
                    reserve_schedule,
                    self.appointmentwf,
                    start_to_close_timeout=timedelta( seconds=10),
                )
                await asyncio.sleep(duration)
            except asyncio.CancelledError as err:
                self.appointmentwf.status = False
                self.appointmentwf.remarks = "Appointment with doctor
done."
                await workflow.execute_activity(
                    close_schedule,
                    self.appointmentwf,
                    start_to_close_timeout=timedelta( seconds=10),
                )
                raise err
```

As part of the fault-tolerant platform, the workflow allows implementing responses to *completion*, *cancellation*, and *timeout* events. Our `ReserveAppointmentWorkflow` instance, when canceled, will throw a `CancelledError` exception that will trigger the exception clause that sets `appointmentwf.status` to `False` and executes the `start_to_close()` activity.

Aside from the loop and the constructor, a workflow implementation can emit `resultset` instances or information about the workflow. To carry this out, implement an instance method and decorate it with `@workflow.query`. The following method returns an appointment record:

```
@workflow.query
def details(self) -> AppointmentWf:
    return self.appointmentwf
```

Unlike in Zeebe/Camunda, where the server executes and manages the workflow, the Temporal. io server does not run any workflow instance but the worker. We learn more about workers in the following section.

Building a worker

A **worker** is responsible for managing the **task queue**. The Worker class from the temporalio. worker module requires the client connection, task_queue, workflows, and activities as constructor parameters before its instantiation. Our worker should be outside the context of Flask, so we added the workflow_runner parameter to the parameters. The following code is our implementation of the Temporal worker:

```
from temporalio.client import Client
from temporalio.worker import Worker, UnsandboxedWorkflowRunner
import asyncio

from modules.admin.activities.workflow import reserve_schedule, close_
schedule
from modules.models.workflow import appt_queue_id
from modules.workflows.transactions import ReserveAppointmentWorkflow

async def main():
    client = await Client.connect("localhost:7233")
    worker = Worker(
        client,
        task_queue=appt_queue_id,
        workflows=[ReserveAppointmentWorkflow],
        activities=[reserve_schedule, close_schedule],
        workflow_runner=UnsandboxedWorkflowRunner,
    )
    await worker.run()

if __name__ == "__main__":
    print("Temporal worker started...")
    asyncio.run(main())
```

Workflows and activities will not execute unless registered in the worker's settings. The Worker instance needs to know what workflows to queue and activities to run before the client application triggers their executions. Now, passing the UnsandboxedWorkflowRunner object to the workflow_runner parameter indicates that our worker will be running as an independent Python application outside the context of our Flask platform or any sandbox environment, thus the setting of the sandboxed parameter in the @workflow.defn decorator of every workflow class to False. To run the worker, call and await the run() method of the Worker instance.

Lastly, after implementing the workflows, activities, and the worker, it is time to trigger the workflow for execution.

Running activities

The ch08-temporal project is a RESTful application in Flask, so to run a workflow, an API endpoint must import and use app.temporal_client to connect to the server and to invoke the start_workflow() method that will trigger the workflow execution.

The start_workflow() method requires the workflow's *"run"* method, the single model object parameter, the unique workflow ID, and task_queue. The following API endpoint triggers the execution of our ReserveAppointmentWorkflow class:

```python
@admin_bp.route("/appointment/doctor", methods=["POST"])
async def request_appointment():
    client = get_client()
    appt_json = request.get_json()
    appointment = ReqAppointment(**appt_json)
    await client.start_workflow(
        ReserveAppointmentWorkflow.run,
        appointment,
        id=appointment.ticketid,
        task_queue=appt_queue_id,
    )
    message = jsonify({"message": "Appointment for doctor
requested...."})
    response = make_response(message, 201)
    return response
```

After a successful workflow trigger, another API can query the details or results of the workflow by extracting the workflow's WorkflowHandler class from the client using its *workflow ID*. The following endpoint function shows how to retrieve the result of the completed workflow:

```python
@admin_bp.route("/appointment/details", methods=["GET"])
async def get_appointment_details():
    client = get_client()
    ticketid = request.args.get("ticketid")
    print(ticketid)
    handle = client.get_workflow_handle_for(
ReserveAppointmentWorkflow.run, ticketid)
    results = await handle.query( ReserveAppointmentWorkflow.details)
    message = jsonify({
            "ticketid": results.ticketid,
            "patientid": results.patientid,
            "docid": results.docid,
            "date_scheduled": results.date_scheduled,
            "time_scheduled": results.time_scheduled,
        }
```

```
    )
    response = make_response(message, 200)
    return response
```

To prove that Temporal workflows can respond to cancellation events, the following API invokes the `cancel()` method from the `WorkflowHandler` class for its workflow to throw a `CancelledError` exception, leading to the execution of the `close_schedule()` activity:

```
@admin_bp.route("/appointment/close", methods=["DELETE"])
async def end_subscription():
    client = get_client()
    ticketid = request.args.get("ticketid")
    handle = client.get_workflow_handle(ticketid,)
    await handle.cancel()
    message = jsonify({"message": "Requesting cancellation"})
    response = make_response(message, 202)
    return response
```

Although there is still a lot to discuss about the architecture and the behavior of these big-time workflow solutions, the main goal is to highlight the feasibility of integrating different workflow engines into the asynchronous Flask platform and take into consideration workarounds for integrations to work with Flask applications.

Summary

This chapter proved that `Flask[async]` can work with different workflow engines, starting with Celery tasks. `Flask[async]`, combined with the workflows created by Celery's signatures and primitives, works well in building chained, grouped, and chorded processes.

Then, `Flask[async]` was proven to work with SpiffWorkflow for some BPMN serialization that focuses on `UserTask` and `ScriptTask` tasks. Also, this chapter even considered solving BPMN enterprise problems using the Zeebe/Camunda platform that showcases `ServiceTask` tasks.

Moreover, `Flask[async]` created an environment with Airflow 2.x to implement pipelines of tasks building an API orchestration. In the last part, the chapter established the integration between `Flask[async]` and Temporal.io and demonstrated the implementation of deterministic and distributed workflows.

This chapter provided a clear picture of the extensibility, usability, and scalability of the Flask framework in building scientific and big data applications and even BPMN-related and ETL-involved business processes.

The next chapter will discuss the different authentication and authorization mechanisms to secure Flask applications.

9

Securing Flask Applications

Like any web application, Flask applications have vulnerabilities that require protection from external attacks, which exploit these software defects. These cyber-attacks are mainly due to broken access control problems, **Cross-Site Scripting (XSS)**, **Cross-Site Request Forgery (CSRF)**, **Server-Side Request Forgery (SSRF)**, **SQL Injection**, and **Denial-of-Service (DoS)**, as well as outdated modules and libraries.

Implementing security measures must be an utmost priority of any Flask application, especially if it is more dependent on external modules when building its models, repository layers, and workflow-related transactions. Using third-party libraries can inflict risks to the Flask applications because some library codes can contain coding errors or vulnerabilities. This is especially true for codes sourced from outdated third-party modules and libraries with unreliable sources.

It is easier to build Flask components and features with external modules, such as implementing the authentication and authorization measures using the **Authlib** module instead of composing it from the ground up. To decrease the chance of, if not avoid, web attacks, one should devise a security plan that will employ only reliable and updated modules. This will protect the application from outside attackers.

The main goal of this chapter is to provide possible security solutions for Flask applications to avoid some of the well-known web attacks using Flask's built-in components, as well as some up-to-date and reliable third-party libraries.

Here are the topics that we will cover in the context of helping us secure our Flask applications:

- Adding protection from web vulnerabilities
- Securing response data
- Managing user credentials
- Implementing web form authentication
- Preventing CSRF attacks
- Implementing user authentication and authorization
- Controlling the view or API access

Technical requirements

This chapter is about a **vaccine reporting and management system** that showcases security solutions for preventing attacks from outside the system. Its web forms explain how to impose sanitation and validation to avoid injection attacks and XSS. Specifically, its login form discusses how to save credentials to the browser. Also, it highlights views that render unsanitized data from user requests. Most importantly, this application provides options on authentication and authorization procedures applicable and accepted by Flask to secure and manage the access control of its web forms and API resources. By the way, the application has several versions that serve as specimens in depicting various security issues and the solutions on how to manage them. All the projects use `Flask[async]` features, including asynchronous `Flask-SQLAlchemy` transactions. They are available at `https://github.com/PacktPublishing/Mastering-Flask-Web-Development/tree/main/ch09`.

Adding protection from web vulnerabilities

SQL injection, SSRF, and XSS attacks are the most common web vulnerabilities that corrupt many web applications. They also affect any applications that use HTTP-based transactions, such as `POST`, `PUT`, `PATCH`, and `DELETE`. SQL injection occurs when an attacker infiltrates the backend datastore that manages the content of the trusted application. Embedded malicious SQL code can tamper with the data, rendering unwanted pages or destroying the database. XSS attacks commonly insert malicious scripts into the pages of the application to steal cookies, session data, and sensitive credentials from the system. On the other hand, CSRF occurs inside an authenticated environment. It happens when a valid user performs an HTTP transaction and a malicious script lurking in the browser usurps valid credentials with bogus and invalid ones to lead transactions to untrusted systems.

Applying form validation to request data

One solution to avoid these attacks is to design a form validation that will not cost several lines of code or add more performance overhead to the view or API functions. The **Flask-WTF** module, for instance, can provide a `FlaskForm` class that sub-classes form models with attributes mapped to the appropriate field classes. Each field class (e.g., `StringField`, `BooleanField`, `DateField`, or `TimeField`) has properties and built-in validators (e.g., `Length()`, `Email()`, or `DataRequired()`) with support on custom validators when the validation procedure requires intricate conditions. With the validators in place, it can more or less protect the application from exploits. Further discussion on using Flask-WTF is included in *Chapter 4*.

If the application is non-web or web-based and does not need form model classes, **Flask-Gladiator** is the suitable form validation tool to check incoming request data. Flask-Gladiator is a general server-side data validation module that can impose basic validation rules on incoming request parameters, such as type checking, string length evaluation, pattern matching, input masking, and checking whether values fall within a specific range. In *Chapter 6*, we used Flask-Validator for validating symbolic equations and mathematical expressions as form inputs. To use flask-gladiator, install the module using the following `pip` command:

```
pip install flask-gladiator
```

After the installation, the module does not need further setup. It can be used readily when building the validation rules, such as in the following validation rules that scrutinize the incoming administrator's profile:

```
import gladiator as glv
from gladiator.core import ValidationResult

def validate_form(form_data):
    field_validations = (
        ('adminid', glv.required, glv.length_max(12)),
        ('username', glv.required, glv.type_(str)),
        ('firstname', glv.required, glv.length_max(50), glv.regex_
('[a-zA-Z][a-zA-Z ]+')),
        ('midname', glv.required, glv.length_max(50), glv.regex_
('[a-zA-Z][a-zA-Z ]+')),
        ('lastname', glv.required, glv.length_max(50), glv.regex_
('[a-zA-Z][a-zA-Z ]+')),
        ('email', glv.required, glv.length_max(25), glv.format_email),
        ('mobile', glv.required, glv.length_max(15)),
        ('position', glv.required,  glv.length_max(100)),
        ('status', glv.required, glv.in_(['true', 'false'])),
        ('gender', glv.required, glv.in_(['male', 'female'])),
    )
    result:ValidationResult = glv.validate(field_validations, form_
data)
    return result.success
```

The most essential component of the `gladiator` module is the `validate()` method, which has two required parameters: `form_data` and `validators`. The validators are placed in a tuple of tuples, as shown in the preceding code, wherein each tuple contains the request parameter name followed by all its validators. Our `ch09-web-passphrase` project uses the following validators:

- `required()`: Requires the parameter to have a value.
- `length_max()`: Checks whether the given string length is lower than or equal to a maximum value.

- `type_()`: Checks the type of the request data (e.g., a form parameter is always a string).
- `regex_()`: Matches the string to a regular expression.
- `format_email()`: Checks whether the request data follows the email regex.
- `in_()`: Checks whether the value is within the list of options.

The list shows only a few of the many validator functions that the `gladiator` module can provide to establish the validation rules. Now, the `validate()` method returns a `ValidationResult` object, which has a boolean `success` variable that yields `True` if all the validators have no hits. Otherwise, it yields `False`. The following code shows how the `ch09-web-passphrase`'s `add_admin_profile()` method utilizes the given `validate_form()` view function:

```
@current_app.route('/admin/profile/add', methods=['GET', 'POST'])
async def add_admin_profile():
    if not session.get("user"):
        return redirect('/login/auth')
    … … … … … …
    if request.method == 'GET':
        return render_template('admin/add_admin_profile.html',
admin=admin_rec), 200
    else:
        result = validate_form(request.form)
        if result == False:
            flash(f'Validation problem.', 'error')
            return render_template('admin/add_admin_profile.html',
admin=admin_rec), 200

        … … … … … …
        return render_template('admin/add_admin_profile.html',
admin=admin_rec), 200
```

Now, filtering malicious text can be effective if we combine the validation and sanitation of this form data. Sanitizing inputs means encoding special characters that might trigger the execution of malicious scripts from the browser.

Sanitizing form inputs

Aside from validation, view or API functions must also sanitize incoming request data by converting special characters and suspicious symbols to purely text so that XML- and HTML-based templates can render them without side effects. This process is known as **escaping**. The `markupsafe` module has an `escape()` method that can normalize request data with query strings that intend to control the JavaScript codes, modify the UI experience, or tamper browser cookies when Jinja2 templates render them. The following snippet is a portion of the `add_admin_profile()` view function that sanitizes the form data after `gladiator` validation:

```
@current_app.route('/admin/profile/add', methods=['GET', 'POST'])
async def add_admin_profile():
    ... ... ... ... ... ...
    result = validate_form(request.form)
    if result == False:
        flash(f'Validation problem.', 'error')
        return render_template('admin/add_admin_profile.html',
admin=admin_rec), 200
    username = request.form['username']
    ... ... ... ... ... ...
    admin_details = {
        "adminid": escape(request.form['adminid'].strip()),
        "username": escape(request.form['username'].strip()),
        "firstname": escape(request.form['firstname'].strip()),
        ... ... ... ... ... ...
        "gender": escape(request.form['gender'].strip())
    }
    admin = Administrator(**admin_details)
    result = await repo.insert_admin(admin)
    if result == False:
        flash(f'Error adding ... profile.', 'error')
    else:
        flash(f'Successfully added a user ... )
    return render_template('admin/add_admin_profile.html',
admin=admin_rec), 200
```

Removing leading and trailing whitespaces or defined suspicious characters using Python's `strip()` method with the escaping process may lower the risk of injection and XSS attacks. However, be sure that the validation rules and sanitation techniques combined will neither ruin the performance of the view or API function nor change the actual request data. Also, tight validation rules can affect the overall runtime performance, so choose the appropriate number and types of validators for every form.

To avoid SQL injection, use an ORM such as **SQLAlchemy**, **Pony**, or **Peewee** that can provide a more abstract form of SQL transactions and even escape utilities to sanitize column values before persistence. Avoid using native and dynamic queries where the field values are concatenated to the query string because they are prone to manipulation and exploitation.

Sanitation can also be applied to response data to avoid another type of attack called the **Server-Side Template Injection** (**SSTI**). Let us now discuss how to protect the application from SSTIs by managing the response data.

Securing response data

Jinja2 has a built-in escaping mechanism to avoid SSTIs. SSTIs allow attackers to inject malicious template scripts or fragments that can run in the background. These then ruin the response or perform unwanted executions that can ruin server-side operations. Thus, applying the `safe` filter in Jinja templates to perform dynamic content augmentation is not a good practice. The `safe` filter turns off the Jinja2's escaping mechanism and allows for running these malicious attacks. In connection with this, avoid using **dynamic hypertext links** using the `<a>` tag in templates (e.g., `<a href="{{ var_link }}">Click Me</a>`). Instead, utilize the `url_for()` utility method to call dynamic view functions because it validates and checks whether the Jinja variable in the expression is a valid view name. *Chapter 1* discusses how to apply `url_for()` for hyperlinks.

On the other hand, there are also issues in Flask that need handling to prevent injection attacks on the Jinja templates, such as managing how the view functions will render the context data and add security response headers.

Rendering Jinja2 variables

There is no ultimate solution to avoid injection but to apply escaping to context data before rendering them to Jinja2 templates. Moreover, avoid using `render_template_string()` even if this is part of the Flask framework. Rendering HTML page-generated content may accidentally run malicious data from inputs overlooked by filtering and escaping. It is always good practice to place all HTML content in a file with an `.html` extension, or XML content in a `.xml` file, to enable Jinja2's default escaping feature. Then, render them using the `render_template()` method with or without the escaped and validated context data. All our projects use `render_template()` in rendering Jinja2 templates.

Security response headers must also be part of the response object when rendering every view template. Let us explore these security response headers and learn where to build them.

Adding security response headers

HTTP security response headers are directives used by many web applications to mitigate vulnerability attacks, such as XXS and public exposure of user details. They are headers added in the response object during the rendition of the Jinja2 templates or JSON results. Some of these headers include the following:

- **Content-Type**: This indicates the original media type of the resource rendered; by default, it is HTML. If it is HTML, it is necessary to indicate the UTF-8 charset to avoid XSS.

- **X-Content-Type-Options**: This tells the browser to follow the indicated content-type. It also blocks the browser's media-type sniffing, so its value should be nosniff.

- **X-Frame-Options**: This indicates whether the browser is allowed to load the page in a <frame>, <iframe>, <embed>, or <objects>. Possible values include DENY and SAMEORIGIN. The DENY option disallows rendering pages on a frame, while SAMEORIGIN allows rendering a page on a frame with the same URL site as the page.

- **Strict-Transport-Security**: This indicates that the browser can only access the page through the HTTPS protocol.

In our ch09-web-passphrase project, the global @after_request function creates a list of security response headers for every view function call. The following code snippet in the main.py module shows this function implementation:

```python
@app.after_request
def create_sec_resp_headers(response):
    response.headers['Content-Type'] = 'text/html; charset=UTF-8'
    response.headers['X-Content-Type-Options'] = 'nosniff'
    response.headers['X-Frame-Options'] = 'SAMEORIGIN'
    response.headers['Strict-Transport-Security'] = 'Strict-Transport-
Security: max-age=63072000; includeSubDomains; preload'
    return response
```

Here, `Content-Type`, `X-Content-Type-Options`, `X-Frame-Options`, and `Strict-Transport-Security` are the most essential response headers for web applications. By the way, `SAMEORIGIN` is the ideal value for `X-Frame-Options` because it prevents view pages from displaying outside the site domain of the project, mitigating **clickjacking** attacks. *Figure 9.1* shows the response header tracked down by the browser after rendering the `/admin/profile/add` view.

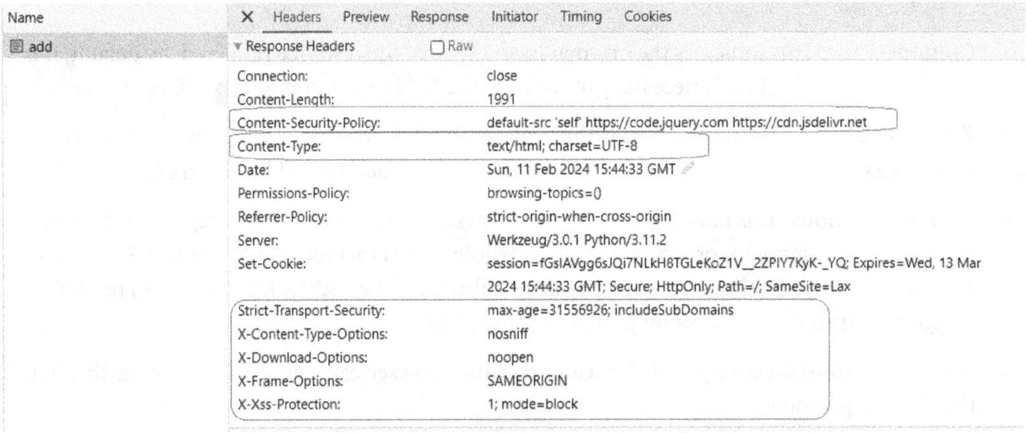

Name						
	✕ Headers	Preview	Response	Initiator	Timing	Cookies
🗎 add	▼ Response Headers		☐ Raw			
	Connection:		close			
	Content-Length:		1991			
	Content-Security-Policy:		default-src 'self' https://code.jquery.com https://cdn.jsdelivr.net			
	Content-Type:		text/html; charset=UTF-8			
	Date:		Sun, 11 Feb 2024 15:44:33 GMT			
	Permissions-Policy:		browsing-topics=()			
	Referrer-Policy:		strict-origin-when-cross-origin			
	Server:		Werkzeug/3.0.1 Python/3.11.2			
	Set-Cookie:		session=fGsIAVgg6sJQi7NLkH8TGLeKoZ1V__2ZPIY7KyK-_YQ; Expires=Wed, 13 Mar 2024 15:44:33 GMT; Secure; HttpOnly; Path=/; SameSite=Lax			
	Strict-Transport-Security:		max-age=31556926; includeSubDomains			
	X-Content-Type-Options:		nosniff			
	X-Download-Options:		noopen			
	X-Frame-Options:		SAMEORIGIN			
	X-Xss-Protection:		1; mode=block			

Figure 9.1 – The response headers when running the view function

On the other hand, another way to manage security response headers is through the Flask module **Flask-Talisman**, or **Talisman** for short. To use it, install the `flask-talisman` module using the following `pip` command:

```
pip install flask-talisman
```

Afterward, instantiate the `Talisman` class in the `create_app()` method and integrate the module into the Flask application by adding and configuring the web application's security response headers using Talisman libraries, as shown in the following snippet:

```
from flask_talisman import Talisman

def create_app(config_file):
    app = Flask(__name__,template_folder= '../modules/pages', static_
folder= '../modules/resources')
    app.config.from_file(config_file, toml.load)
    … … … … … …
    talisman = Talisman(app)
    csp = {
        'default-src': [
            '\'self\'',
            'https://code.jquery.com',
            'https://cdnjs.com',
```

```
            'https://cdn.jsdelivr.net',
    ]
}
hsts = {
    'max-age': 31536000,
    'includeSubDomains': True
}
talisman.force_https = True
talisman.force_file_save = True
talisman.x_xss_protection = True
talisman.session_cookie_secure = True
talisman.frame_options_allow_from = 'https://www.google.com'
talisman.content_security_policy = csp
talisman.strict_transport_security = hsts
```

Talisman provides an easier and Pythonic way to set up the **Content Security Policy (CSP)**. The CSP header restricts which web resources and sites to load on the browser through its policy directives (e.g., `default-src`, `image-src`, `style-src`, `media-src`, `object-src`). In our configuration, JS files must only come from `https://code.jquery.com`, `https://cdnjs.com`, `https://cdn.jsdelivr.net`, and the localhost, while both CSS and images must be fetched from the localhost as indicated in `default-src`, the fallback resources for each view page. Specifying `script-src` with specific JS sources, `style-src` with CSS resources, and `image-src` with the targeted images will bypass the `default-src` setting.

Aside from CSP, the Talisman can add `X-XSS-Protection`, `Referrer-Policy`, and `Set-Cookie`, as well as the headers previously included in the response by the `@after_request` function. Caution is needed in combining the two approaches because overlapping of header settings may happen.

Adding the `Strict-Transport-Security` header in the response and setting the `force_https` of Talisman's property to `True` requires running the application in HTTPS mode. Let us explore the latest and easiest way to enable HTTPS for a Flask application.

Using HTTPS to run request/response transactions

HTTPS is a TLS-encrypted HTTP protocol. It establishes secured communication between the transmitter and receiver of data, protecting the cookies, URLs, and sensitive information that flows during the exchange. It also guards the integrity of the data and the user's authenticity since it requires the user's private key to allow access. With that, to enable the HTTPS protocol, the WSGI server must run with a public and private key certificate generated by an SSL key generator. By convention, the certificate must be saved inside the project directory or somewhere safe in the host server. This chapter utilizes the **OpenSSL** tool to produce the certificate.

Install the latest `pyopenssl` using the following `pip` command:

```
pip install pyopenssl
```

Now, to run the application, include the private and public keys in `run()` through its `ssl_context` parameter. The following `main.py` snippet shows how to run the application using HTTPS on a development server:

```
app, celery_app, auth = create_app('../config_dev.toml')
… … … … … …
if __name__ == '__main__':
    app.run(ssl_context=('cert.pem', 'key.pem'))
```

Running the `python main.py` command with the `ssl_context` parameter will show a log on the terminal console, as shown in *Figure 9.2*:

* Serving Flask app 'modules'
* Debug mode: off
WARNING: This is a development server. Do not use it in a production deployment. Use a production WS
GI server instead.
* Running on https://127.0.0.1:5000
Press CTRL+C to quit

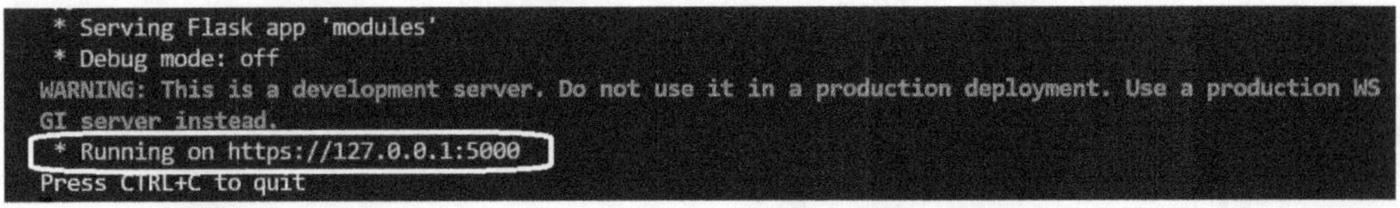

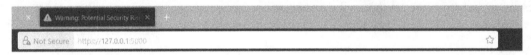

Figure 9.2 – The server log when running on HTTPS

When opening the `https://127.0.0.1:5000/` link on a browser, a warning page will pop up on the screen, such as the one depicted in *Figure 9.3*, indicating that we are entering a secured page from a non-secured browser.

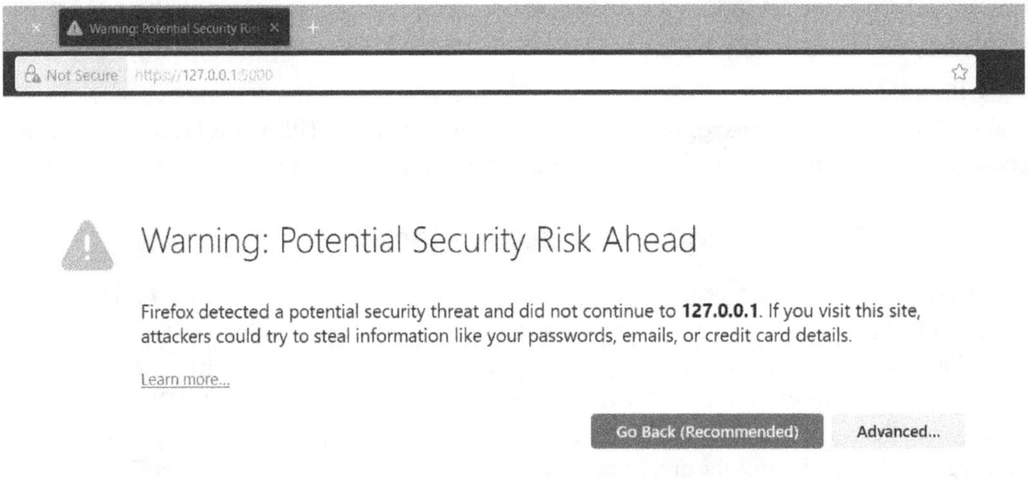

Warning: Potential Security Risk Ahead
Not Secure
Firefox detected a potential security threat and did not continue to 127.0.0.1. If you visit this site, attackers could try to steal information like your passwords, emails, or credit card details.
Learn more...
Go Back (Recommended)
Advanced...

Figure 9.3 – A warning page on opening secured links

Another way to run Flask applications on an HTTP protocol is to include the key files in the command line, such as `python main.py --cert=cert.pem --key=key.pem`. In the production environment, we run Flask applications according to the procedure followed by the secured production server.

Encryption does not apply only when establishing an HTTP connection but also when securing sensitive user information such as usernames and passwords. In the next section, we will discuss the different ways of **hashing** and encrypting user credentials.

Managing user credentials

The most common procedure for protecting any application from attacks is to control access to the user's sensitive details, such as their username and password. Direct use of saved raw user credentials for login validation will not protect the application from attacks unless the application derives passphrases from the passwords, saves them into the database, and applies them for user validation instead.

This topic will cover password hashing using **Hashlib** and **Bcrypt**, password encryption using **symmetric cryptography**, and utilizing the `sqlalchemy_utils` module for the seamless and automatic encryption of sensitive data.

Encrypting user passwords

Generating a passphrase from the username and password of the user is the typical and easiest way to protect the application from attackers who want to crack down or hack a user account. In Flask, there are two ways to generate a passphrase from user credentials:

- **The hashing process**: A one-way approach that involves generating a fixed-length passphrase of the original text.

- **The encryption process**: A two-way approach that involves generating a variable-length text using random symbols that can be traced back to its original text.

The `ch09-api-bcrypt` and `ch09-auth-basic` projects use hashing to manage the passwords of a user. The `ch09-auth-basic` project utilizes Hashlib as its primary hashing library for passphrase generation. Flask has the `werkzeug.security` module that provides `generate_password_hash()`, a function that uses Hashlib's `scrypt` algorithm to generate a passphrase from a text. The project's `add_signup()` API endpoint function that utilizes the `werkzeug.security` module in generating the passphrase from the user's password is as follows:

```
from werkzeug.security import generate_password_hash

@current_app.post('/login/signup')
async def add_signup():
    login_json = request.get_json()
```

```
    password = login_json["password"]
    passphrase = generate_password_hash(password)
    async with db_session() as sess:
        async with sess.begin():
            repo = LoginRepository(sess)
            login = Login(username=login_json["username"],
password=passphrase, role=login_json["role"])
            result = await repo.insert_login(login)
            if result == False:
               return jsonify(message="error in insert"), 201
            return jsonify(record=login_json), 200
```

The generate_password_hash() method has three parameters:

- The actual password is the first parameter.

- The hashing method is the second parameter with a default value of scrypt.

- The **salt** length is the third parameter.

The salt length will determine the number of alphanumerics that the method will use to generate a salt. A salt is the additional alphanumerics with a fixed length that are added to the end of the password to make the passphrase more unbreachable or uncrackable. The process of adding salt to hashing is called **salting**.

On the other hand, the werkzeug.security module also supports pbkdf2 as an option for the hashing method parameter. However, it is less secure than the Scrypt algorithm. Scrypt is a simple and effective hashing algorithm that requires salt to hash a password. The generate_password_hash() method defaults the salt length to 16, which can be replaced anytime by passing any preferred length. Moreover, Scrypt is memory intensive, since it needs storage to temporarily hold all the initial salted random alphanumerics until it returns the final passphrase.

Since there is no way to re-assemble the passphrase to extract the original text, the werkzeug.security module has a check_password_hash() method that returns True if the given text value matches the hashed value. The following snippet validates the password of an authenticated user if it matches an account in the database with the same username but a hashed password:

```
from werkzeug.security import check_password_hash

@auth.verify_password
def verify_password(username, password):
    task = get_user_task_wrapper.apply_async( args=[username])
    login:Login = task.get()
    if login == None:
        abort(403)
    if check_password_hash(login.password, password):
```

```
        return login.username
    else:
        abort(403)
```

The `check_password_hash()` method requires two parameters, namely the passphrase as the first and the original password as the second. If the `werkzeug.security` module is not the option for your requirement due to its slowness, the **Hashlib** module, using its **SHA-256** algorithm, is a better replacement. However, do not forget to install the `hashlib` module using the following `pip` command before applying it:

```
pip install hashlib
```

On the other hand, `ch09-api-bcrypt` uses the Bcrypt algorithm to generate a passphrase for a password. Since **Bcrypt** is an extension module, the initial step is to install it using the following `pip` command:

```
pip install bcrypt
```

Afterward, instantiate the `Bcrypt` class container in the `create_app()` factory method and integrate the module into the Flask application through the `app` instance. The following snippet shows the setup of the Bcrypt module in the Flask application:

```
from flask_bcrypt import Bcrypt

bcrypt = Bcrypt()
def create_app(config_file):
    app = Flask(__name__, template_folder='../modules/pages', static_
folder='../modules/resources')
    app.config.from_file(config_file, toml.load)
    app.config.from_prefixed_env()
    … … … … … …
    bcrypt.init_app(app)
    … … … … … …
```

The `bcrypt` object will provide every module component with the utility methods to hash credential details such as passwords. The following `ch09-api-bcrypt`'s version of the `add_signup()` endpoint hashes the password of an account using the imported `bcrypt` object before saving the user's credentials into the database:

```
from modules import bcrypt

@current_app.post('/login/signup')
async def add_signup():
    login_json = request.get_json()
    password = login_json["password"]
```

```
        passphrase = bcrypt.generate_password_hash(password) .
decode('utf-8')
      async with db_session() as sess:
        async with sess.begin():
          repo = LoginRepository(sess)
          … … … … … …
          result = await repo.insert_login(login)
          if result == False:
            return jsonify(message="error in insert"), 201
          return jsonify(record=login_json), 200
```

Like Hashlib algorithms (e.g., `scrypt` or `pbkdf2`), Bcrypt is not capable of extracting the original password from the passphrase. However, it also has a `check_password_hash()` method, which validates whether a password has the correct passphrase. However, compared to Hashlib, Bcrypt is more secure and modern because it uses the **Blowfish Cipher** algorithm. Its only drawback is its slow hashing process, which may affect the application's overall performance.

Aside from hashing, the encryption algorithms can also help secure the internal data of any Flask application, especially passwords. A well-known module that can provide reliable encryption methods is the `cryptography` module. So, let us first install the module using the following `pip` command before using its cryptographic recipes and utilities:

```
pip install cryptography
```

The `cryptography` module offers both symmetric and asymmetric cryptography. The former uses one key to initiate the encryption and decryption algorithms, while the latter uses two keys: the public and private keys. Since our application only needs one key to encrypt user credentials, it will use symmetric cryptography through the `Fernet` class, the utility class that implements symmetric cryptography for the module. Now, after the installation, call `Fernet` in the `create_app()` method to generate a key through its `generate_key()` class method. The following snippet in the factory method shows how the application created and kept the key for the entire runtime:

```
from cryptography.fernet import Fernet

def create_app(config_file):
    app = Flask(__name__, template_folder='../modules/pages', static_
folder='../modules/resources')
    app.config.from_file(config_file, toml.load)
    … … … … … …
    with open("enc_key.txt", mode="w") as file:
        file.write(Fernet.generate_key().decode())
```

The `Fernet` token or secret key is a URL-safe base64-encoded alphanumeric that will instigate the encryption and decryption algorithms. The application should store the key in a safe and inaccessible location during startup, such as saving it in a file inside a secured directory. Missing the key will lead to `cryptography.fernet.InvalidToken` and `cryptography.exceptions.InvalidSignature` errors.

After generating the secret key, instantiate the `Fernet` class with the key as the constructor argument to emit the `encrypt()` method. The following `ch09-auth-digest`'s version of `add_signup()` encrypts the user password using `Fernet`:

```python
from cryptography.fernet import Fernet

@current_app.post('/login/signup')
async def add_signup():
    … … … … … …
    password = login_json["password"]
    with open("enc_key.txt", mode="r") as file:
        enc_key = bytes(file.read(), "utf-8")
    fernet = Fernet(enc_key)
    passphrase = fernet.encrypt(bytes(password, 'utf-8'))
    async with db_session() as sess:
        async with sess.begin():
            … … … … … …
            result = await repo.insert_login(login)
            … … … … … …
            return jsonify(record=login_json), 200
```

To instantiate `Fernet`, `add_signup()` must extract the token from the file, convert it into bytes, and pass it as a constructor argument to the `Fernet` class. The `Fernet` instance provides an `encrypt()` method that uses the **128-bit AES in CBC mode** and **HMAC with SHA-256** algorithms to encode the password in bytes. It also has a `decrypt()` method to extract the original password from the encrypted message. The following is `ch09-auth-digest`'s password validation scheme that retrieves the user credentials from the database with the encoded password and decrypts the encoded message to extract the actual password:

```python
@auth.get_password
def get_passwd(username):
    task = get_user_task_wrapper.apply_async( args=[username])
    login:Login = task.get()
    with open("enc_key.txt", mode="r") as file:
        enc_key = bytes(file.read(), "utf-8")
    fernet = Fernet(enc_key)
    password = fernet.decrypt(login.password) .decode('utf-8')
    if login == None:
```

```
        return None
    else:
        return password
```

Again, `get_passwd()` needs the token from the file to instantiate `Fernet`. Using the `Fernet` instance, `get_passwd()` can emit the `decrypt()` method to reassemble the encrypted message and extract the actual password in the `UTF-8` format. Compared to hashing, encryption involves reassembling plain text into an unreadable and uncrackable ciphertext using a token and reverting that ciphertext into its original readable form. So, it's a two-way process, unlike in hashing.

If the goal is to persist encoded data into the database without adding unnecessary cryptographic blunders that can slow down software performance, the solution is to use `sqlalchemy_utils`.

Using sqlalchemy_utils for encrypted columns

The `sqlalchemy_utils` module provides additional utility methods and column types to SQLAlchemy model classes, which include `StringEncryptedType`. Since the module utilizes cryptographic recipes of the cryptography module, be sure to install the latter before `sqlalchemy_utils` using the following `pip` command:

```
pip install cryptography sqlalchemy_utils
```

Afterward, design your model classes by applying `StringEncryptedType` to table columns that need `Fernet`'s encryption, such as the `username` and `password` columns. The following is the `Login` model class of the `ch09-web-passphrase` project with `username` and `password` columns of `StringEncryptedType`:

```
from sqlalchemy_utils import StringEncryptedType

enc_key = "packt_pazzword"
class Login(Base):
    __tablename__ = 'login'
    id = Column(Integer, Sequence('login_id_seq', increment=1),
primary_key = True)
    username = Column(StringEncryptedType(String(20), enc_key),
nullable=False, unique=True)
    password = Column(StringEncryptedType(String(50), enc_key),
nullable=False)
    role = Column(Integer, nullable=False)
    … … … … … …
```

`StringEncryptedType` automatically encrypts the column data during the `INSERT` transaction and decrypts the encoded field value during the `SELECT` statements. To apply the utility class to the column, map it to the column field enclosing the actual SQLAlchemy column type with the custom-generated `Fernet` token. It may look like a column field wrapper that will filter and encrypt the inserted field value and decrypt it during retrieval. No other additional coding from the `view` functions or repository layer is needed to perform the encryption and decryption processes on these field values.

When using `Flask-Migrate`, add the `import sqlalchemy_utils` statement to the generated `env.py` and `script.py.mako` files inside the `migrations` folder after running the `db init` command and before running the `db migrate` and `db upgrade` operations. The following are the modified `ch09-web-passphrase` migration files with the imported `sqlalchemy_utils` module:

```
(env.py)
import logging
from logging.config import fileConfig

from flask import current_app

from alembic import context
import sqlalchemy_utils
... ... ... ... ... ...


(script.py.mako)
"""${message}

Revision ID: ${up_revision}
Revises: ${down_revision | comma,n}
Create Date: ${create_date}

"""
from alembic import op
import sqlalchemy as sa
import sqlalchemy_utils
${imports if imports else ""}
... ... ... ... ... ...
```

The highlighted lines provided in the given migration files are the proper places to add extra imports used in the SQLAlchemy model classes. This includes not only the `sqlalchemy_util` classes but also other libraries that may help to establish the desired model layer.

When customizing user authentication, the application utilizes the default Flask session to store user information, such as username. This session saves information to the browser. To protect the app from broken access control attacks, you can use a reliable authentication and authorization mechanism or apply server-side session handling through the **Flask-Session** module if custom session-based authentication suits your requirements.

Utilizing the server-side sessions

In *Chapter 4*, the **Flask-Session** module was integrated into the Flask application to replace the browser-based session with a server-side one. Storing important user details in the browser exposes the user details, as well as the application in general, to other user profiles without consent and permission. This is a perfect example of a broken access control scenario. So, server-side session management is an acceptable solution to protect the application from exposing user information, such as `username`, to the public.

After setting up the `Session` class of the module through the `app` instance, Flask's built-in `session` dictionary object can readily store session data on the server side. The following `login_user()` view function stores the `username` of the credential to the server-side session after user credential confirmation:

```
@current_app.route('/login/auth', methods=['GET', 'POST'])
async def login_user():
    if request.method == 'GET':
        return render_template('login/authenticate.html'), 200
    username = request.form['username'].strip()
    password = request.form['password'].strip()
    async with db_session() as sess:
        async with sess.begin():
            repo = LoginRepository(sess)
            records = await repo.select_login_username_
passwd(username, password)
            login_rec = [rec.to_json() for rec in records]
            if len(login_rec) >= 1:
                session["user"] = username
                return redirect('/menu')
            else:
                ... ... ... ... ... ...
                return render_template('login/authenticate.html'), 200
```

All session data needs to be cleared when logging out. Removing all session data will **close** the session, protecting the application from exposing user information to others. The following snippet shows the `logout()` view function of the `ch09-web-paraphrase` project:

```
@current_app.route('/logout', methods=['GET'])
async def logout():
    session["user"] = None
    return redirect('/login/auth')
```

Aside from setting the session attribute to None, the `pop()` method of the `session` object can also help remove the session data. Removing all session data is the same as invalidating the current session.

Now, if custom web login implementation does not apply to your requirement, the **Flask-Login** module can offer built-in utilities for user login and logout. Let us now discuss how to use Flask-Login for a Flask application.

Implementing web form authentication

Flask-Login is an extension module that provides utility methods to manage user sessions, supports Bcrypt hashing, and equips helper classes to build the Flask-Login model and callback methods to secure view functions with user credentials. To use `flask-login`, install it first using the following `pip` command:

```
pip install flask-login
```

Also, install and set up the Flask-Session module for Flask-Login to store its user session in the filesystem.

Then, to integrate Flask-login into the Flask application, instantiate its `LoginManager` class in the `create_app()` method and set it up through the app instance. Define some of its properties such as `session_protection`, which requires the installation of Flask-Bcrypt, and `login_view`, which designates the `login` `view` function. The following snippet shows the setup of Flask-Login for our `ch09-web-login` project:

```
from flask_login import LoginManager

def create_app(config_file):
    app = Flask(__name__,template_folder= '../modules/pages', static_
folder=  '../modules/resources')
    app.config.from_file(config_file, toml.load)
    app.config.from_prefixed_env()
    ... ... ... ... ... ...
    login_auth = LoginManager()
    login_auth.init_app(app)
```

```
login_auth.session_protection = "strong"
login_auth.login_view = "modules.views.login.login_valid_user"
```

After the setup, add a Login model class to the model layer through your desired ORM and sub-class it with the UserMixin helper class of the flask-login module. The following is the Login model of our project that will persist the user's id, username, passphrase, and role:

```
from flask_login import UserMixin
from sqlalchemy_utils import StringEncryptedType
enc_key = "packt_pazzword"


class Login(UserMixin, Base):
    __tablename__ = 'login'
    id = Column(Integer, Sequence('login_id_seq', increment=1),
primary_key = True)
    username = Column(StringEncryptedType(String(20), enc_key),
nullable=False, unique=True)
    password = Column(StringEncryptedType(String(50), enc_key),
nullable=False)
    role = Column(Integer, nullable=False)
    ... ... ... ... ... ...
```

Instead of utilizing the Flask-Bcrypt module, our application uses the built-in StringEncryptedType hashing mechanism from the sqlalchemy_utils module. Now, the use of the UserMixin superclass allows the use of some properties, such as is_authenticated, is_active, and is_anonymous, as well as some utility methods, such as get_id (), provided by the current_ user object from the flask_login module.

Flask-Login stores the id of a user in the session after a successful login authentication. With Flask-Session, it will store the id somewhere that has been secured. The id is vital to the **user loader callback function**, which is the Flask-Login component that is responsible for retrieving the user's details from the database. The function automatically fetches the id from the session, retrieves the object using its id parameter and a repository class, and returns the Login object to the application. Here is ch09-web-login's implementation for the user loader:

```
@login_auth.user_loader
def load_user(id):
    task = get_user_task_wrapper.apply_async(args=[id])
    result = task.get()
    return result
```

In our project, the function is part of the `main.py` module. Now, using the `get_user_task_wrapper()` Celery task, the `load_user()` uses the `select_login()` transaction of the `LoginRepository` to retrieve a `Login` record based on the given `id` parameter. The application automatically calls `load_user()` for every request to access the views. The continuous call to the callback function checks the validity of the user. The returned `Login` object serves as a token that permits the user to access the application. To declare a user loader callback function, create a function with one local `id` parameter and decorate it with the `userloader()` decorator of the `LoginManager` instance.

The `login` view function caches the `Login` object, saves the `Login`'s id to the session, and maps it to the `current_user` proxy object of the `flask_login` module. The following snippet shows the `login` view function indicated in our setup:

```python
from flask_login import login_user

@current_app.route('/login/auth', methods=['GET', 'POST'])
async def login_valid_user():
    if request.method == 'GET':
        return render_template('login/authenticate.html'), 200
    username = request.form['username'].strip()
    password = request.form['password'].strip()
    async with db_session() as sess:
        async with sess.begin():
            repo = LoginRepository(sess)
            records = await repo.select_login_username_passwd( username,
password)
            if(len(records) >= 1):
                login_user(records[0])
                return render_template('login/signup.html'), 200
            else:
                … … … … … …
                return render_template( 'login/authenticate.html'), 200
```

If the POST-submitted user credentials are correct based on the database validation, the `login` view function, namely our `login_valid_user()`, should invoke the `login_user()` method of the `flask_login` module. The view must pass the queried `Login` object containing the user's login credentials to the `login_user()` function. Aside from the `Login` object, the method has other options, such as the following:

- `remember`: A boolean parameter that enables the `remember_me` feature, which allows the user session to be alive even after a browser's accidental exit.

- `fresh`: A boolean parameter that sets a user as `not fresh` if their session is valid right after the closing of a browser.

- `force`: A boolean parameter that forces a user to be logged in.

- `duration`: The amount of time before the `remember_me` cookie expires.

After the successful authentication, users can now access the restricted views or APIs that are off-limits to non-authenticated users: those views with `@login_required` decorator. The following is a sample view function of our `ch09-web-login` that needs authenticated user access:

```python
from flask_login import login_required


@current_app.route('/doctor/profile/add', methods=['GET', 'POST'])
@login_required
async def add_doctor_profile():
    if request.method == 'GET':
        async with db_session() as sess:
            async with sess.begin():
                repo = LoginRepository(sess)
                records = await repo.select_all_doctor()
                doc_rec = [rec.to_json() for rec in records]
                return render_template('doctor/add_doctor_profile.
html', docs=doc_rec), 200
    else:
        username = request.form['username']
        … … … … … …
        return render_template('doctor/add_doctor_profile.html',
doctors=doc_rec), 200
```

Aside from the decorator, the `current_user`'s `is_authenticated` property can also restrict the execution of some code fragments in the views and Jinja templates.

Lastly, to complete the integration of Flask-Login, implement a **logout view** that calls the `logout_user()` utility of the `flask_login` module. The following code is the logout view implementation for our project:

```python
from flask_login import logout_user


@current_app.route("/logout")
async def logout():
    logout_user()
    return redirect(url_for('login_valid_user'))
```

Be sure that the logout view redirects the user to the login view page instead of just rendering the login page to avoid **HTTP status code 405** (*Method Not Allowed*) during re-login.

Does having a secured web form authentication prevent CSRF attacks from happening? Let us focus on protecting our applications from attackers who want to divert transactions to other suspicious sites.

Preventing CSRF attacks

CSRF is an attack whereby authenticated users are duped into diverting sensitive data to hidden and malicious sites. This attack happens when users perform POST, DELETE, PUT, or PATCH transactions, whereby form data are retrieved and submitted to the application. In Flask, the most common solution is to use Flask-WTF because it has a built-in CSRFProtect class that globally protects every form transaction of the application. Once enabled, CSRFProtect allows the generation of unique tokens for every form transaction. Those form submissions that will not generate a token will cause CSRFProtect to trigger an error message, detecting a CSRF attack.

Chapter 4 highlights the setup of the Flask-WTF module in a Flask application. After its installation, import CSRFProtect and instantiate it in create_app(), as shown in the following code snippet:

```
from flask_wtf.csrf import CSRFProtect

def create_app(config_file):
    app = Flask(__name__,template_folder= '../modules/pages', static_
folder= '../modules/resources')
    ... ... ... ... ... ...
    csrf = CSRFProtect(app)
```

However, this setup will need a SECRET_KEY or WTF_CSRF_SECRET_KEY to be defined in the configuration file. Now, after integrating it into the application through the app instance, all <form> in Jinja templates must have the {{ csrf_token() }} component or a <input type="hidden" name="csrf_token" value = "{{ csrf_token() }}" /> component to impose CSRF protection.

If there is no plan to use the entire Flask-WTF, another option is to apply **Flask-Seasurf** instead.

After showcasing the web-based authentication strategies, it is now time to discuss the different authentication types for API-based applications.

Implementing user authentication and authorization

There is a strong foundation of extension modules that can secure API services from unwanted access, such as the **Flask-HTTPAuth** module. This has the updated utility classes for **Basic**, **Digest**, and **Bearer token** authentication implementation and the Authlib module for building OAuth2 authorization servers based on various grants. These two libraries are updated and reliable, and when combined with other safety procedures, such as password hashing and encryption, can provide baseline protection for Flask applications. Let us start identifying the steps in employing the Flask-HTTPAuth module in our application.

Utilizing the Flask-HTTPAuth module

After you have installed the `Flask-HTTPAuth` module and its extensions, it can provide its `HTTPBasicAuth` class to build Basic authentication, the `HTTPDigestAuth` class to implement Digest authentication, and the `HTTPTokenAuth` class for the Bearer token authentication scheme.

Basic authentication

Basic authentication requires an unencrypted base64 format of the user's `username` and `password` credentials through the `Authorization` request header. To implement this authentication type in Flask, instantiate the module's `HTTPBasicAuth` in `create_app()` and register the instance to the Flask app instance, as shown in the following snippet:

```
from flask_httpauth import HTTPBasicAuth

def create_app(config_file):
    app = Flask(__name__,template_folder= '../modules/pages', static_
folder= '../modules/resources')
    app.config.from_file(config_file, toml.load)
    … … … … … …
    global auth
    auth = HTTPBasicAuth()
```

Also, the `HTTPBasicAuth` implementation needs a callback function that will retrieve the username and password from the client, query the database to check the user's records, and return the valid username to the application if it exists, like in the following code:

```
app, celery_app, auth = create_app('../config_dev.toml')
… … … … … …
@auth.verify_password
def verify_password(username, password):
    task = get_user_task_wrapper.apply_async( args= [username])
    login:Login = task.get()
    if login == None:
        abort(403)
    if check_password_hash(login.password,password) == True:
        return login.username
    else:
        abort(403)
```

To be `HTTPBasicAuth`'s callback function, the given `check_password()` must have the username and password parameters and should be annotated with `HTTPBasicAuth`'s `verify_password()` decorator. Our callback uses the Celery task to search and retrieve the `Login` object containing the `username` and `password` details and raises **HTTP status code 403** if the record does not exist. If the authentication is successful, users will have permission to access all API methods decorated by `HTTPBasicAuth`'s `login_required()` decorator.

The `Flask-HTTPAuth` module has built-in authorization support. If the basic authentication needs a role-based authorization, the application only needs to have a separate callback function decorated by `get_user_roles()` from the `HTTPBasicAuth` class. The following is `ch09-auth-basic`'s callback function for retrieving user roles from its users:

```
from werkzeug.datastructures.auth import Authorization

app, celery_app, auth = create_app('../config_dev.toml')
... ... ... ... ... ...
@auth.get_user_roles
def get_scope(user:Authorization):
    task = get_user_task_wrapper.apply_async( args=[user.username])
    login:Login = task.get()
    return str(login.role)
```

The given `get_scope()` automatically retrieves the `werkzeug.datastructures.auth.Authorization` object from the request. The `Authorization` object contains the `username` on which the `get_user_task_wrapper()` Celery task will base its search for the `Login` record object of the user from the database. The return value of the callback function can be a single role in string format or a list of roles attributed to the user. The given `del_doctor_profile_id()` from the `ch09-auth-digest` project does not permit any authenticated users except for doctors whose `role` is equivalent to the `1` code:

```
@current_app.delete('/doctor/profile/delete/<int:id>')
@auth.login_required(role="1")
async def del_doctor_profile_id(id:int):
    async with db_session() as sess:
      async with sess.begin():
          repo = DoctorRepository(sess)
          ... ... ... ... ... ...
          return jsonify(record="deleted record"), 200
```

Here, `del_doctor_profile_id()` is an API function that deletes a doctor's profile information in the database. No role can perform the transaction but the doctor (`role=1`) himself/herself.

Digest authentication

On the other hand, the module's `HTTPDigestAuth` builds a digest authentication scheme for the API-based applications, which encrypts the credentials and some of its additional headers to the applications such as `realm`, `nonce`, `opaque`, and `nonce count`. Thus, it is more secure than the basic authentication scheme. The following snippet shows how to set up digest authentication in the `create_app()` factory:

```
from flask_httpauth import HTTPDigestAuth


def create_app(config_file):
    app = Flask(__name__,template_folder= '../modules/pages', static_
folder= '../modules/resources')
    app.config.from_file(config_file, toml.load)
    … … … … … …
    global auth
    auth = HTTPDigestAuth()
```

The `HTTPDigestAuth`'s constructors have five parameters, two of which have default values, namely qop and `algorithm`. The **Quality of Protection (qop)** option has the `auth` value, which means that the application is at the basic protection level of the digest scheme. So far, the highest protection level is `auth-int`, which is, at the time of writing this book, not yet functional in the `Flask-HTTPAuth` module. The other parameter, `algorithm`, has the md5 default value for the encryption, but the requirement can change it to `md5-sess` or any supported encryption method. Now, the three other optional parameters are the following:

- `realm`: This contains the `username` of the user and the name of the application's host.

- `scheme`: This is a replacement to the default value of the `Digest scheme` header in the `WWW-Authenticate` response.

- `use_hw1_pw`: If this is set to `True`, the `get_password()` callback function must return a hashed password.

In digest authentication, the user must submit their username, password, nonce, opaque, and nonce count to the application for verification. *Figure 9.4* shows a postman client submitting the header information to the `ch09-auth-digest` app:

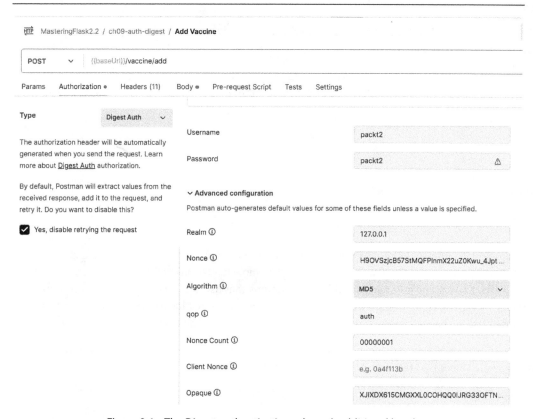

Figure 9.4 – The Digest authentication scheme's additional headers

A nonce is a unique base64 or hexadecimal string that the server generates for every **HTTP status code 401** response. The content of the compressed string, usually the estimated timestamp when the client received the response, must be unique to every access.

Also specified by the server is **opaque**, a base64 or hexadecimal string value that the client needs to return to the server for validation if it is the same value as generated before.

The nonce count value or **nc-value** is a non-repeating hexadecimal value, usually starting at 0000001, that checks the integrity of the user credentials and protects data from playback attacks. The server increments its copy of the nc-value when it receives the same nonce value from a new request. Every authentication request must bear a new nonce value. Otherwise, it is a replay.

Flask-HTTPAuth's digest authentication scheme will only work if our API application provides the following callback implementations:

```
server_nonce = "H9OVSzjcB57StMQFPInmX22uZ0Kwu_4JptsWrj0oPpU"
server_opaque = "XJIXDX615CMGXXL0COHQQ0IJRG33OFTNGNFYT72VJ8XF5U3RYZ"


@auth.generate_nonce
def gen_nonce():
    return server_nonce


@auth.generate_opaque
def gen_opaque():
    return server_opaque
```

The application must generate the nonce and opaque values using the likes of the gen_nonce() and gen_opaque() callbacks. These are just trivial implementations of the methods in our ch09-auth-digest application and need better solutions that use a UUID generator or a secrets module to generate the values. The nonce generator callback must have a generate_nonce() decorator, while the opaque generator must be decorated by the generate_opaque() annotation.

Aside from these generators, the authentication scheme also needs the following validators of nonce and opaque values that the server needs from the client request:

```
@auth.verify_nonce
def verify_once(nonce):
    if nonce == server_nonce:
        return True
    else:
        return False


@auth.verify_opaque
def verify_opaque(opaque):
    if opaque == server_opaque:
        return True
    else:
        return False
```

The validators check whether the values from the request are correct based on the server's corresponding values. Now, the last requirement for the digest authentication to work is the `get_password()` callback that retrieves the password from the client for database validation of the user's existence, as shown in the following snippet:

```
@auth.get_password
def get_passwd(username):
    print(username)
    task = get_user_task_wrapper.apply_async(args=[username])
    login:Login = task.get()
    … … … … … …
    if login == None:
        return None
    else:
        return password
```

The application continuously calls the `get_password()` method every access to an API resource and provides the valid user's password as a token.

Bearer token authentication

Aside from Basic and Digest, the `Flask-HTTPAuth` module also supports the Bearer token authentication scheme by utilizing the `HTTPTokenAuth` class. The following `create_app()` snippet of the `ch09-auth-token` project sets up the Bearer token authentication:

```
from flask_httpauth import HTTPTokenAuth


def create_app(config_file):
    app = Flask(__name__,template_folder= '../modules/pages', static_
folder=  '../modules/resources')
    app.config.from_file(config_file, toml.load)
    … … … … … …
    global auth
    auth = HTTPTokenAuth('Bearer')
```

This scheme requires an additional `token` column field of the `string` type in the `Login` model to persist the token associated with the user. The application generates the token after the user signs up for login credentials. Our application uses the `PyJWT` module for token generation, as depicted in the following endpoint function:

```
from jwt import encode


@current_app.post('/login/signup')
async def add_signup():
    login_json = request.get_json()
```

```
        password = login_json["password"]
        passphrase = generate_password_hash(password)
        token = encode({'username': login_json["username"],
'exp': int(time()) + 3600}, current_app.config['SECRET_KEY'],
algorithm='HS256')
        async with db_session() as sess:
            async with sess.begin():
                repo = LoginRepository(sess)
                login = Login(username=login_json["username"],
password=passphrase, token=token,   role=login_json["role"])
                result = await repo.insert_login(login)
                … … … … … …
                return jsonify(record=login_json), 200
```

The token's **payload** contains the `username` and the token's supposed expiration time in seconds. The encoding indicated in the given `add_signup()` API method is the **HS256** symmetric algorithm, and the secret key is the application's `SECRET_KEY`. The token is always part of the request's `Authorization` header with the `Bearer` value. Now, a callback function retrieves the bearer token from the request and checks whether it is the saved token of the user. The following is `ch09-auth-token`'s callback function implementation:

```
from jwt import decode


@auth.verify_token
def verify_token(token):
    try:
        data = decode(token, app.config['SECRET_KEY'],
                          algorithms=['HS256'])
    except:
        return False
    if 'username' in data:
        return data['username']
```

The Bearer token's callback function must have the `verify_token()` method decorator. It has the `token` parameter and it returns either a boolean value or the username. It must use the same **JSON Web Token** (**JWT**) library that generated the user's token during signup. In our project, the `PyJWT` module encodes and decodes the token.

Like basic, the digest and bearer token authentication schemes use the `login_required()` decorator to impose restrictions on API endpoints. Also, both can implement role-based authorization with the `get_user_roles()` callback.

The next Flask module, `Authlib`, has core classes and methods for implementing OAuth2, OpenID Connect, and JWT Token-based authentication schemes. Let us now showcase it.

Utilizing the Authlib module

Authlib is a replacement for the **Flask-OAuthlib** module. It provides customizable classes that make the module flexible to solve different security requirements. Also, it is a big module because it supports both **OAuth 1.0** and **OAuth 2.0** specifications. Moreover, it can provide utilities for implementing **Resource Owner Password**, **Implicit**, **Authorization Code**, and **Client Credential** authorization server types. To utilize the authlib module, install its module using the following pip command:

```
pip install authlib
```

If the application to secure is not running on an HTTPS protocol, set the AUTHLIB_INSECURE_ TRANSPORT environment variable to 1 or True for Authlib to work because it is for a secured environment.

Unlike the HTTP Basic, Digest, and Bearer Token authentication schemes, the OAuth2.0 scheme uses an authorization server that provides several endpoints for authorization procedures, as well as issuing tokens, refreshing tokens, and revoking tokens. The authorization server is always part of an application that protects its resources from malicious access and attacks. Our ch09-oauth2- password project implements the Vaccine and Reports applications with the OAuth2 Resource Owner Password authorization scheme using Authlib. The following create_app() factory method shows how to set up this scheme:

```
from authlib.integrations.flask_oauth2 import AuthorizationServer
from authlib.integrations.flask_oauth2 import ResourceProtector
from modules.security.oauth2_config import PasswordGrant, query_
client, save_token


require_oauth = ResourceProtector()
oauth_server = AuthorizationServer()


def create_app(config_file):
    app = Flask(__name__,template_folder= '../modules/pages', static_
folder= '../modules/resources')

    … … … … … …

    oauth_server.init_app(app, query_client=query_client, save_
token=save_token)
    oauth_server.register_grant(PasswordGrant)
```

The AuthorizationServer class manages the authentication requests and responses of the application. It provides different types of endpoints that are suited to the authentication grant enforced by the application. Now, instantiating the class is the first step in building the OAuth2 authorization server for clients or other applications. It needs query_client() and save_token() for its token generation and the grant type of the authorization mechanism.

Authlib provides the `ResourceOwnerPasswordCredentialsGrant` class to implement the **Resource Owner Password authorization standard (RFC 6749)**. The class needs subclassing to override at least its `authenticate_user()` to perform validation before performing the `query_client()` and `save_token()` methods. The following snippet shows the `ResourceOwnerPasswordCredentialsGrant` subclass of our `ch09-oauth2-password` project:

```python
from authlib.oauth2.rfc6749.grants import
ResourceOwnerPasswordCredentialsGrant

class PasswordGrant(ResourceOwnerPasswordCredentialsGrant):
    TOKEN_ENDPOINT_AUTH_METHODS = [
      'client_secret_basic', 'client_secret_post' ]

    def authenticate_user(self, username, password):
        task = get_user_task_wrapper.apply_async(args=[username])
        login:Login = task.get()
        if login is not None and check_password_hash( login.password,
password) == True:
            return login
```

The `PasswordGrant` custom class is in the `/modules/security/oauth2_config.py` module with the `query_client()` and `save_token()` authorization server methods. The first component of `PasswordGrant` to configure is its `TOKEN_ENDPOINT_AUTH_METHODS`, which, from its default `public` value, needs to be set to `client_secret_basic`, `client_secret_post`, or both. The `client_secret_basic` is a client authentication that passes client secrets through a basic authentication scheme, while `client_secret_post` utilizes form parameters to pass client secrets to the authorization server. On the other hand, the overridden `authenticate_user()` retrieves the `username` and `password` from the token generator endpoint through basic authentication or form submission. It also retrieves the `Login` record object from the database through a `get_user_task_wrapper()` Celery task and validates the `Login`'s hashed password with the retrieved password from the client. The method returns the `Login` object that will signal the execution of the `query_client()` method. The following snippet shows our `query_client()` implementation:

```python
def query_client(client_id):
    task = get_client_task_wrapper.apply_async( args=[client_id])
    client:Client = task.get()
    return client
```

The `query_client()` is a necessary method of the `AuthorizationServer` instance. Its goal is to find the client who requested the authentication and return the `Client` object. It retrieves the `client_id` from the `AuthorizationServer` endpoint and uses it to search for the `Client` object from the database. The following snippet shows how to build the `Client` blueprint with Authlib's `OAuth2ClientMixin`:

```
from authlib.integrations.sqla_oauth2 import OAuth2ClientMixin

class Client(Base, OAuth2ClientMixin):
    __tablename__ = 'oauth2_client'
    id = Column(Integer, Sequence('oauth2_client_id_seq',
increment=1), primary_key = True)
    user_id = Column(String(20), ForeignKey('login.username'),
unique=True)

    login = relationship('Login', back_populates="client")
    ... ... ... ... ... ...
```

Authlib's `OAuth2ClientMixin` will pad all the necessary column fields to the model class, including those that are optional. The required pre-tokenization fields, such as `id`, `user_id` or `username`, `client_id`, `client_id_issued_at`, and `client_secret`, must be submitted to the database during client signup before the authentication starts. Now, if the client is valid, the `save_token()` will execute to retrieve the `access_token` from the authorization server and save it to the database. The following snippet is our implementation for `save_token()`:

```
from authlib.integrations.flask_oauth2.requests import
FlaskOAuth2Request
def save_token(token_data, request:FlaskOAuth2Request):
    if request.user:
        user_id = request.user.user_id
    else:
        user_id = request.client.user_id
    token_dict = dict()
    token_dict['client_id'] = request.client.client_id
    token_dict['user_id'] = user_id
    token_dict['issued_at'] = request.client.client_id_issued_at
    token_dict['access_token_revoked_at'] = 0
    token_dict['refresh_token_revoked_at'] = 0
    token_dict['scope'] = request.client.client_metadata["scope"]
    token_dict.update(token_data)
    token_str = dumps(token_dict)
    task = add_token_task_wrapper.apply_async( args=[token_str])
    task.get()
```

The `token_data` contains the `access_token`, and the request has the `Client` data retrieved from the `query_client()`. The method merges all these details into one `token_dict`, instantiates the `Token` class with `token_dict` as parameter, and stores the object record in the database. The following is the blueprint of the `Token` model:

```python
from authlib.integrations.sqla_oauth2 import OAuth2TokenMixin

class Token(Base, OAuth2TokenMixin):
    __tablename__ = 'oauth2_token'
    id = Column(Integer, Sequence('oauth2_token_id_seq', increment=1),
primary_key=True)
    user_id = Column(String(40), ForeignKey('login.username'),
nullable=False)

    login = relationship('Login', back_populates="token")
    ... ... ... ... ... ...
```

The `OAuth2TokenMixin` pads the `Token` class with the attributes related to `access_token`, such as `id`, `user_id`, `client_id`, `token_type`, `refresh_token`, and `scope`. By the way, `scope` is a mandatory field in Authlib that restricts access to the API resources based on some access level or role.

To trigger the authorization server, the client must access the `/oauth/token` endpoint through basic authentication or form-based transactions. The following code shows the endpoint implementation of our application:

```python
from flask import current_app, request
from modules import oauth_server

@current_app.route('/oauth/token', methods=['POST'])
async def issue_token():
    return oauth_server.create_token_response( request=request)
```

Authlib's tokenization process always happens in the POST transaction mode. The `Authorization Server` object from the `create_app()` provides the `create_token_response()` with details that the method needs to return for the user to capture the `access_token()`. Given the `client_id` of `Xd3LH9mveF524LOscPq4MzLY` and `client_secret` `t8w56Y9OBRsxdVV9vrNwdtMzQ8gY4hkKLKf4b6F6RQZ1T2zI` with the `sjctrags` username, the following `curl` command shows how to run the `/oauth/token` endpoint:

```
curl -u Xd3LH9mveF524LOscPq4MzLY:t8w56Y9OBRsxdVV9vrNwdtMzQ
8gY4hkKLKf4b6F6RQZ1T2zI -XPOST http://localhost:5000/oauth/token -F
grant_type=password -F username=sjctrags -F password=sjctrags -F
scope=user_admin -F token_endpoint_auth_method=client_secret_basic
```

A sample result of executing the preceding command will contain the following details aside from the `access_token`:

```
{"access_token": "fVFyaS06ECKIKFVtIfVj3ykgjhQjtc80JwCKyTMlZ2",
"expires_in": 864000, "scope": "user_admin", "token_type": "Bearer"}
```

As indicated in the result, the `token_type` is `Bearer`, so we can use the `access_token` to access or run an API endpoint through a bearer Token authentication, like in the following `curl` command:

```
curl -H "Authorization: Bearer fVFyaS06ECKIKFVtIfVj3y
kgjhQjtc80JwCKyTMlZ2" http://localhost:5000/doctor/profile/add
```

A secured API endpoint must have the `require_oauth("user_admin")` method decorator, wherein `require_oath` is the `ResourceProtector` instance from the `create_app()`. A sample secured endpoint is the following `add_doctor_profile()` API function:

```
from modules import require_oauth

@current_app.route('/doctor/profile/add', methods = ['GET', 'POST'])
@require_oauth("user_admin")
async def add_doctor_profile():
    … … … … … …
        async with db_session() as sess:
          async with sess.begin():
             repo = DoctorRepository(sess)
             doc = Doctor(**doctor_json)
             result = await repo.insert_doctor(doc)
             … … … … … …
             return jsonify(record=doctor_json), 200
```

Aside from the Resource Owner Password grant, Authlib has an `AuthorizationCodeGrant` class to implement an **OAuth2 authorization code grant**, as well as `JWTBearerGrant` for implementing the **OAuth2 WT Token-based** authorization scheme using its own JWT library. Our `ch09-oauth-code` project will showcase the full implementation of the OAuth2 authorization code flow, while `ch09-oauth2-jwt` will implement the JWT authorization scheme (**RFC 7519**) using the `pyjwt` module.

If Flask supports popular and ultimate authentication and authorization modules, like Authlib, it also supports unpopular but reliable extension modules that can secure web-based and API-based Flask applications. One of these modules is *Flask-Limiter*, which can prevent **Denial of Service (DoS)** attacks. Let us now apply this module to our `ch09-web-passphrase` project.

Controlling the view or API access

DoS attacks happen when a user maliciously accesses a web page or API multiple times to disrupt the traffic and make the resources inaccessible to others. `Flask-Limiter` can provide an immediate solution by managing the number of access of a user to an API endpoint. First, install the `Flask-Limiter` module using the following `pip` command:

```
pip install flask-limiter
```

Also, install the module dependency for caching its configuration details to the Redis server:

```
pip install flask-limiter[redis]
```

Now, we can set up the module's `Limiter` class in the `create_app()` factory method, like in the following snippet:

```
from flask_limiter import Limiter
from flask_limiter.util import get_remote_address

def create_app(config_file):
    app = Flask(__name__,template_folder= '../modules/pages', static_
folder=  '../modules/resources')
    … … … … … …
    global limiter
    limiter = Limiter(
      app=app, key_func=get_remote_address,
      default_limits=["30 per day", "5 per hour"],
      storage_uri="memory://", )
```

Instantiating the `Limiter` class requires at least the app instance, the host of the application through the `get_remote_address()`, the `default_limits` (e.g., `10 per hour`, `10 per 2 hours`, or `10/hour`), and the storage URI for the Redis server. The `Limiter` instance will provide each protected API with the `limit()` decorator that specifies the number of accesses not lower than the set default limit. The following API is restricted not to be accessed by a user more than a *maximum count of 5 times per minute*:

```
from modules import limiter

@current_app.route('/login/auth', methods=['GET', 'POST'])
@limiter.limit("5 per minute")
async def login_user():
    if request.method == 'GET':
        return render_template(  'login/authenticate.html'), 200
    username = request.form['username'].strip()
    password = request.form['password'].strip()
```

```
    async with db_session() as sess:
        async with sess.begin():
            repo = LoginRepository(sess)
            ... ... ... ... ... ...
                return render_template( 'login/authenticate.html'),
  200
```

Running `login_user()` more than the limit will give us the message shown in *Figure 9.5*.

Too Many Requests

5 per 1 minute

Figure 9.5 – Accessing /login/auth more than the limit

Violating the number of access rules set by Talisman will lead users to its built-in error handling mechanism: the application rendering an error page with its error message.

Summary

In this chapter, we learned that compared to FastAPI and Tornado, there is quite a list of extension modules that provide solutions to secure a Flask application against various attacks. For instance, Flask-Seasurf and Flask-WTF can help minimize CSRF attacks. When pursuing web authentication, Flask-Login can provide a reliable authentication mechanism with added password hashing and encryption mechanisms, as we learned in this chapter.

On the other hand, Flask-HTTPAuth can provide API-based applications with HTTP basic, digest, and bearer token authentication schemes. We learned that OAuth2 Authorization server grants and OAuth2 JWT Token-based types can also protect Flask applications from other applications' access.

The Flask-Talisman ensures security rules on response headers to filter the outgoing response of every API endpoint. Meanwhile, the Flask-Session module saves Flask sessions in the filesystem to avoid browser-based attacks. Escaping, stripping of whitespaces, and form validation of incoming inputs using modules like Gladiator and Flask-WTF helps prevent injection attacks by eliminating suspicious text or alphanumerics in the inputs.

This chapter proved that several updated and version-compatible modules can help protect our applications from malicious and unwanted attacks. These modules can save time and effort compared to ground-up solutions in securing our applications.

The next chapter will be about testing Flask components before running and deploying them to production servers.

Part 3:
Testing, Deploying, and Building Enterprise-Grade Applications

In this last part, you will learn some options and workarounds to test, deploy, and run our Flask 3 applications. Moreover, you will also understand the process of integrating Flask applications into GraphQL, React forms, Flutter mobile applications, and other applications built with FastAPI, Django, Tornado, and Flask using the interoperability feature.

This part includes the following chapters:

- *Chapter 10, Creating Test Cases for Flask*
- *Chapter 11, Deploying Flask Applications*
- *Chapter 12, Integrating Flask with Other Tools and Frameworks*

10

Creating Test Cases for Flask

After building the components of Flask, it is essential to create test cases to ensure their correctness and to fix their bugs. Among the types of testing, **unit testing** focuses on testing the effectiveness and performance of components independent of other modules or tasks. On the other hand, **integration testing** ensures the correctness of a Flask component's functionality and reliability with all its dependencies together at runtime.

To implement these test cases, Python has a built-in module called `unittest` that can provide a `TestCase` superclass and the `setUp()` and `tearDown()` methods that build variations of test cases and test suites. There is also a third-party module called `pytest`, which is simple, easy to use, and non-boilerplate and can provide reusable fixtures for setting up a test environment. In this chapter, we will highlight how to implement test cases using `pytest` for some selected functionalities from our projects in *Chapters 1* to *9*.

The main goal of this chapter is to provide Flask projects with the necessary test environments where we can run, study, scrutinize, analyze, and improve the Flask components without deploying the application. Another goal of this chapter is to gain the mindset that testing, at least unit testing, is an essential part of any enterprise-grade application development.

Here are the topics covered in this chapter:

- Creating test cases for web views, repository classes, and native services
- Creating test cases for components in application factory and Blueprints
- Creating test cases for asynchronous components
- Creating test cases for secured API and web components
- Creating test cases for MongoDB transactions
- Creating test cases for WebSockets

Technical requirements

All test cases will be from different applications created from *Chapters 1* to *9*. All these applications are in this GitHub repositpry: `https://github.com/PacktPublishing/Mastering-Flask-Web-Development`.

Creating test cases for web views, repository classes, and native services

The `pytest` module supports unit and integration or functional testing. It requires simple syntax to build test cases, which makes it very easy to use, and it has a platform that can automatically run all test files. Moreover, `pytest` is a free and open-source module, so install it using the following `pip` command:

```
pip install pytest
```

However, `pytest` will only work with Flask projects with directory structures managed by Blueprints and *application factories*. Our *Online Personal Counselling System* in *Chapter 1* does not follow the Flask standards on directory structure. All view modules import the app instance through `__main__`, which becomes the `pytest` module and not the `main.py` module during testing. Thus, testing our `ch01` project gives us the following runtime error message:

```
ImportError cannot import name 'app' from '__main__'
```

The error means there is no app object to import in the `pytest` module. So, a testable and new version of the *Online Personal Counselling System* is in the `ch01-testing` project, which places all the view functions inside Python functions that the `main.py` module will access to pass the app instance. The following `main.py` snippet shows these function calls replacing the view's import statements:

```
app = Flask(__name__, template_folder='pages')
… … … … … …
create_index_routes(app)
create_signup_routes(app)
create_examination_routes(app)
create_reports_routes(app)
create_admin_routes(app)
create_login_routes(app)
create_profile_routes(app)
create_certificates_routes(app)
```

The view functions enclosed in each function will utilize the app instance to implement the GET and POST routes. Moreover, to provide a testing environment from Flask, set the Testing environment to `true` in the configuration file.

Now, create a `tests` folder in the main folder of the accurately structured and circular-import-free Flask project directory adjacent to `main.py`. In this folder, implement the test cases in module files with filenames prefixed with the `test_` keyword. If the number of test cases increases, sub-folders can further organize these files according to functionality (e.g., views, repository, services, API, etc.) or type of testing (e.g., unit, integration). *Figure 10.1* shows the final directory structure of the `ch01-testing` project with the `tests` folder.

Figure 10.1 – The tests folder

Now, run the `pytest` command as a module (`python -m pytest`) to execute all the test methods, and run each test file through the following command:

```
python -m pytest tests/xxxx/test_xxxxxx.py
```

Or, run an individual test function using the following command:

```
python -m pytest tests/xxxx/test_xxxxxx.py::test_xxxxxxx
```

Let us now explore `pytest` by creating test cases for `ch01-testing`'s model classes, repository transactions, native services, and view functions.

Testing the model classes

One of the test files that showcases unit testing is `test_models.py`, which contains the following implementation:

```python
import pytest
from model.candidates import AdminUser


@pytest.fixture(scope='module', autouse=True)
def admin_details(scope="module"):
    data = {"id": 101, "position": "Supervisor","age": 45, "emp_
date": "1980-02-16", "emp_status": "regular", "username": "pedro",
"password": "pedro", "utype": 0, "firstname": "Pedro", "lastname"
:"Cruz"}
    yield data
    data = None


def test_admin_user_model(admin_details):
    admin = AdminUser(**admin_details)
    assert admin.firstname == "Pedro"
    assert admin.lastname == "Cruz"
    assert admin.age == 45
```

Unlike in `unittest`, test cases in `pytest` are in the form of *test functions*. The test function's name is unique, descriptive of its purpose, and must start with the `_test` keyword like its test file. Its code structure follows the **Given-When-Then** (**GWT**) format, where **Given** establishes the initial setup of the testing environment, **When** runs the components that need testing, and **Then** scrutinizes the expected response for each test execution. In the given `test_models.py`, the *Given* part is the creating of `admin_details` fixture, the *When* is the instantiation of the `AdminUser` class, and the *Then* depicts the series of asserts that validates if the extracted `firstname`, `lastname`, and `age` response details are precisely the same as the inputs. Unlike the `unittest`, `pytest` only uses the `assert` statement and the needed conditional expression to perform assertion.

The input to the `test_admin_user_model()` test case is an injectable and reusable admin record created through `pytest`'s `fixture()`. The `pytest` module has a decorator function called `fixture()` that defines functions as injectable resources. Like in `unittest`, `pytest`'s fixture performs `setUp()` before the call to `yield` and `tearDown()` after the yielding of the resource. In the given `test_models.py`, the fixture sets up the admin details in JSON format, and garbage collects the JSON object data after the `yield` statement. But how do test methods utilize these fixtures?

A fixture function has four scopes:

- `function`: This fixture runs only once exclusively on some selected test methods in a test file.
- `class`: This fixture runs only once on a test class containing test methods that require the resource.
- `module`: This fixture runs only once on a test file containing the test methods that require the resource.
- `package`: This fixture runs only once on a package level containing the test methods that require the resource.
- `session`: This fixture runs only once to be distributed across all test methods that require the resource in a session.

To utilize the fixture during its scoped execution, inject the resource function to test the method's parameter list. Our `admin_details()` fixture executes at the module level and is injected into `test_admin_user_model()` through the parameter list. On the other hand, `fixture()`'s `autouse` forces all test methods to request the resource during testing.

To run our test file, execute the `python -m pytest tests/repository/test_models.py` command. If the testing is successful, the console output will be similar to *Figure 10.2*:

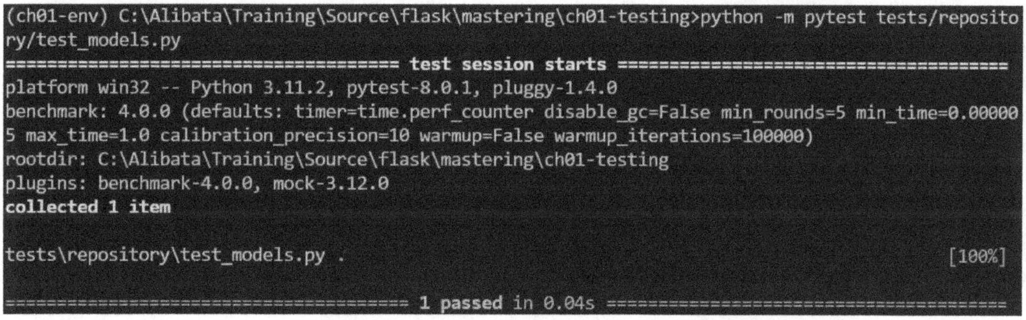

```
(ch01-env) C:\Alibata\Training\Source\flask\mastering\ch01-testing>python -m pytest tests/reposito
ry/test_models.py
================================= test session starts =================================
platform win32 -- Python 3.11.2, pytest-8.0.1, pluggy-1.4.0
benchmark: 4.0.0 (defaults: timer=time.perf_counter disable_gc=False min_rounds=5 min_time=0.00000
5 max_time=1.0 calibration_precision=10 warmup=False warmup_iterations=100000)
rootdir: C:\Alibata\Training\Source\flask\mastering\ch01-testing
plugins: benchmark-4.0.0, mock-3.12.0
collected 1 item

tests\repository\test_models.py .                                                [100%]

================================= 1 passed in 0.04s =================================
```

Figure 10.2 – The pytest result when a test succeeded

The `pytest` result includes the `pytest` plugin installed and its configuration details, a testing directory, and a horizontal green marker indicating the number of successful tests executed. On the other hand, the console output will be similar to *Figure 10.2* if a test case fails:

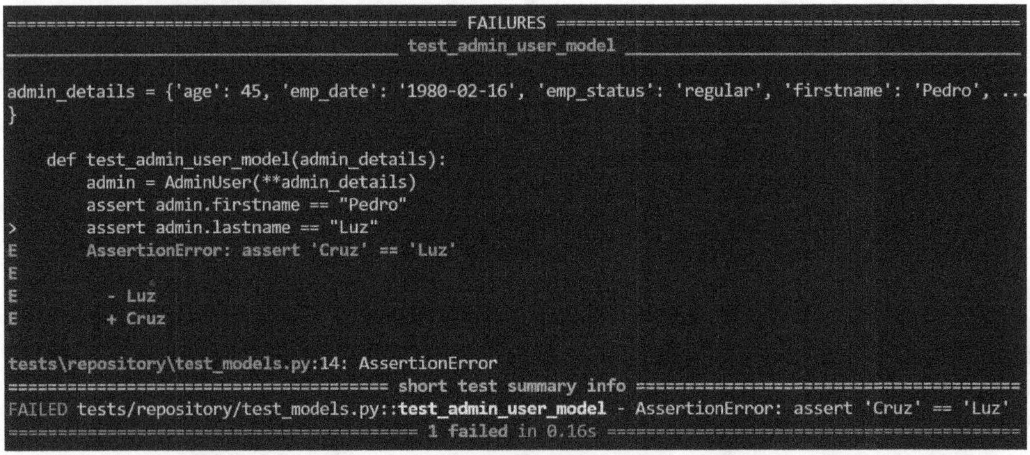

```
================================= FAILURES =================================
_____ test_admin_user_model _____

admin_details = {'age': 45, 'emp_date': '1980-02-16', 'emp_status': 'regular', 'firstname': 'Pedro', ...
}

    def test_admin_user_model(admin_details):
        admin = AdminUser(**admin_details)
        assert admin.firstname == "Pedro"
>       assert admin.lastname == "Luz"
E       AssertionError: assert 'Cruz' == 'Luz'
E
E
E         - Luz
E         + Cruz

tests\repository\test_models.py:14: AssertionError
============================ short test summary info ============================
FAILED tests/repository/test_models.py::test_admin_user_model - AssertionError: assert 'Cruz' == 'Luz'
============================== 1 failed in 0.16s ==============================
```

Figure 10.3 – The pytest result when a test failed

The console will show the assertion statement that fails and a short description of `AssertionError`. Now, test cases must only catch `AssertionError` due to failed assertions and nothing else because it is understood that codes under testing have already handled all `RuntimeError` internally using `try-except` before testing.

A few components in `ch01-testing` need unit testing. Almost all components are connected to build functionality crucial to the application, such as database connection and repository transactions.

Testing the repository classes

At this point, we will start highlighting functional or integration test cases for our application. Our `ch01-testing` project uses `psycopgy2`'s cursor methods to implement the database transactions. To impose a clean approach, a custom decorator `connect_db()` decorates all repository transactions to provide the connection object for the `execute()` and `fetchall()` cursor methods. But first, it is always a standard practice to check whether all database connection details, such as `DB_USER`, `DB_PASSWORD`, `DB_PORT`, `DB_HOST`, and `DB_NAME`, are all registered as environment variables in the configuration file. The following test case implementation showcases how to test custom decorators that provide database connection to repository transactions:

```
from config.db import connect_db

def test_connection():
    @connect_db
    def create_connection(conn):
```

```
        assert conn is not None
    create_connection()
```

The local `create_connection()` method will capture the `conn` object from the `db_connect()` decorator. Its purpose as a dummy transaction is to assert whether the `conn` object created by `psycopgy2` with the database details is valid and ready for CRUD operations. This approach will also apply to other test cases implemented to check the validity and correctness of custom decorator functions, database-oriented or not. Now, run the `python -m pytest tests/repository/test_db_connect.py` command to check whether the database configurations work.

Let us now concentrate on testing repository transactions with database connection and test data generated by `pytest`.

Passing test data to test functions

If testing the database connection is successful, the next test cases must check and refine the repository classes and their CRUD transactions. The following test function of `test_repo_admin.py` showcases how to test an `INSERT` admin detail transaction using `cursor()` from `psycopg2`:

```
import pytest
from repository.admin import insert_admin

@pytest.mark.parametrize(("id", "fname", "lname", "age", "position",
"date_employed", "status"),
  (("8999", "Juan", "Luna", 76, "Manager", "2010-10-10", "active"),
   ("9999", "Maria", "Clara", 45, "Developer", "2015-08-15",
"inactive")
))
def test_insert_admin(id, fname, lname, age, position, date_employed,
status):
    result = insert_admin(id, fname, lname, age, position, date_
employed, status)
    assert result is True
```

Pytest markers or the `pytest.mark` attribute provides additional metadata to test functions by adding built-in markers, such as the `userfixtures()`, `skip()`, `xfail()`, `filterwarnings()`, and `parametrize()` decorators. With **parameterized testing**, the `parametrize()` marker generates and provides a set of test data to test functions using the local parameter list. The test functions will utilize these multiple inputs to produce varying assert results.

`test_insert_admin()` has local parameters corresponding to the parameter names indicated in the `parametrize()` marker. The marker will pass all these inputs to their respective local parameters in the test function to make the testing happen. It will also seem to iterate all the tuples of inputs in the decorator until the test function consumes all the inputs. Running `test_insert_admin()` gave me *Figure 10.4*, proof that `@pytest.mark.parametrize()` iterates all its test inputs.

```
================================ short test summary info =================================
FAILED tests/repository/test_repo_admin.py::test_insert_admin[8999-Juan-Luna-76-Manager-2010-10-10-activ
e] - assert False is True
FAILED tests/repository/test_repo_admin.py::test_insert_admin[9999-Maria-Clara-45-Developer-2015-08-15-i
nactive] - assert False is True
=============================== 2 failed in 0.43s ===============================
```

Figure 10.4 – Result of parameterized testing

But how about if there is a need to control the behaviors of some external components connected to the functionality under testing? Let us now discuss **mocking**.

Mocking other functionality during testing

Now, there are times in integration or functionality testing when applying control to other dependencies or systems connected to a feature is necessary to test and analyze that specific feature. Controlling other connected parts requires the process of **mocking**, which is another type of unit testing that focuses on one functionality while controlling the behavior of its external dependencies using their equivalent mock objects or fake functions. To employ mocking in `pytest`, install `pytest-mock` first using the following `pip` command:

```
pip install pytest-mock
```

The `pytest-mock` plugin derives its mocking capability from the `unittest.mock` but provides a cleaner and simpler approach. Because of that, using some helper classes and methods, such as the `patch()` decorator, from `unittest.mock` will work with the `pytest-mock` module. Another option is to install and use the `mock` extension module, which is an acceptable replacement for `unittest.mock`.

The following `test_mock_insert_admin()` mocks the `psycopg2` connection to focus the testing solely on the correctness and performance of the `INSERT` admin profile details process:

```python
import pytest
from unittest.mock import patch
from repository.admin import insert_admin

@pytest.mark.parametrize(("id", "fname", "lname", "age", "position",
"date_employed", "status"),
  (("8999", "Juan", "Luna", 76, "Manager", "2010-10-10", "active"),
   ("9999", "Maria", "Clara", 45, "Developer", "2015-08-15",
"inactive")
```

```
))
@patch("psycopg2.connect")
def test_mock_insert_admin(mock_connect, id, fname, lname, age,
position, date_employed, status):
    mocked_conn = mock_connect.return_value
    mock_cur = mocked_conn.cursor.return_value
    result = insert_admin(id, fname, lname, age, position, date_
employed, status)
    mock_cur.execute.assert_called_once()
    mocked_conn.commit.assert_called_once()
    assert result is True
```

Instead of using the database connection, `test_mock_insert_admin()` mocks `psycopgy2.connect()` and replaces it with a `mock_connect` mock object through the `patch()` decorator of `unittest.mock`. The `patch()` decorator or context manager makes mocking easier by decorating the test functions in a test class or module. The first decorator passes the mock object to the first parameter of the test function, followed by other mock objects, if there are any, in the same order as their corresponding `@patch()` decorator in the layer of decorators. The `pytest` module will restore to their original state all mocked objects after testing.

A mocked object emits a `return_value` attribute to set its value when invoked or to call the mocked object's properties or methods. In the given `test_mock_insert_admin()`, the `mocked_conn` and `mock_curr` objects were derived from calling `return_value` of the mocked database connection (`mock_connect`) and the mocked `cursor()` method.

Moreover, mocked objects also emit assert methods such as `assert_called()`, `assert_not_called()`, `assert_called_once()`, `assert_called_once_with()`, and `assert_called_with()` to verify the invocation of these mocked objects during testing. The `assert_called_once_with()` and `assert_called_with()` methods verify the call of the mocked objects based on specific constraints or arguments. Our example verifies the execution of the mocked `cursor()` and `commit()` methods in the INSERT transaction under testing.

Another use of `return_value` is to mock the result of the function under test to focus on testing the performance or algorithm of the transaction. The following test case implementation shows mocking the return value of the `select_all_user()` transaction:

```
@patch("psycopg2.connect")
def test_mock_select_users(mock_connect):
    expected_rec = [(222, "sjctrags", "sjctrags", "2023-02-26"), (
567, "owen", "owen", "2023-10-22")]
    mocked_conn = mock_connect.return_value
    mock_cur = mocked_conn.cursor.return_value
    mock_cur.fetchall.return_value = expected_rec
    result = select_all_user()
    assert result is expect_rec
```

The purpose of setting `expected_rec` to `return_value` of the mocked `fetchall()` method of `mock_cur` is to establish an assertion that will complete the GWT process of the test case. The goal is to run and scrutinize the performance and correctness of the algorithms in `select_all_user()` with the mocked `cursor()` and `fetchall()` methods.

Aside from repository methods, native services also need thorough testing to examine their impact on the application.

Testing the native services

Native services or transactions in the service layer build the business processes and logic of the Flask application. The following test case implementation performs testing on `record_patient_exam()`, which stores the patient's counseling exams in the database and computes the average score given the data:

```python
import pytest
from services.patient_monitoring import record_patient_exam


@pytest.fixture
def exam_details():
    params = dict()
    params['pid'] = 1111
    params['qid'] = 568
    params['score'] = 87
    params['total'] = 100
    yield params


def test_record_patient_exam(exam_details):
    result = record_patient_exam(exam_details)
    assert result is True
```

The function-scoped fixture generated the test data for the test function. The result of testing `record_patient_exam()` will depend on the `insert_patient_score()` repository transaction with the actual database connection.

The next things to test are the view functions. What are the aspects of a view function that require testing? Is it feasible to test views without using browsers?

Testing the view functions

The Flask app instance has a `test_client()` utility to handle GET and POST routes. This method generates an object of the `Client` type, a built-in class to Werkzeug. A test file should have a fixture to set up the `test_client()` context and yield the `Client` instance to each test function for views. The following test case implementation focuses on testing GET routes with a fixture that yields the `Client` instance:

```python
import pytest
from main import app as flask_app

@pytest.fixture(autouse=True)
def client():
    with flask_app.test_client() as client:
        yield client

def test_default_page(client):
    res = client.get("/")
    assert "OPCS" in res.data.decode()
```

This `test_default_page()` runs the root page using the `Client` instance and checks whether the rendered Jinja template contains the `"OPCS"` substring. `res.data` is always in bytes, so decoding it will give us the string equivalent:

```python
def test_home_page(client):
    res = client.get("/home")
    assert "Welcome" in res.data.decode()
    assert res.request.path == "/home"
```

On the other hand, `test_home_page()` runs the /home GET route and verifies whether there is a `"Welcome"` word on its template page. Also, it checks whether the path of the rendered page is still the /home URL path:

```python
def test_exam_page(client):
    res = client.get("/exam/assign")
    assert res.status_code == 200
```

It is also ideal to verify the status code of the `client.get()`'s response. The given `test_exam_page()` checks whether running the /exam/assign URL will result in an HTTP Status Code 200.

On the other hand, the `Client` instance has a `post()` method to test and run `POST` routes. The following implementation shows how to simulate form-handling transactions:

```python
import pytest
from main import app as flask_app

@pytest.fixture(autouse=True)
def client():
    with flask_app.test_client() as client:
        yield client

@pytest.fixture(autouse=True)
def form_data():
    params = dict()
    params["username"] = "jean"
    params["password"] = "jean"
    ... ... ... ... ... ...
    yield params
    params = None

def test_signup_post(client, form_data):
    response = client.post("/signup/submit", data=form_data)
    assert response.status_code == 200
```

Since form parameters are ideally in a hashtable format, fixtures must yield these form parameters with their corresponding values inside a dictionary collection, like in our `form_data()` fixture. Then, we pass this yielded form data to the data parameter of the `client.post()` method. Afterward, we perform the necessary assertions to verify the correctness of the view procedure and its response.

Aside from checking the rendered URL path, content, and status code of a GET route, it is also feasible to test redirections in a view using `pytest`. The following implementation showcases how to test whether a POST transaction redirected a user to another view page:

```python
def test_assign_exam_redirect(client, form_data):
    res = client.post('/exam/assign', data=form_data, follow_
redirects=True)
    assert res.status == '200 OK'
    assert res.request.path == url_for('redirect_success_exam')
```

The goal of `test_assign_exam_redirect()` is to test the `/exam/assign` POST transaction and see whether it can successfully persist the score details (`form_data`) from the counseling exam and compute the rating based on the total number of exam items. The `client.post()` method has a `follow_redirects` parameter that can enforce redirection during testing when set to `True`. In our case, `client.post()` will run the `/exam/assign` POST transaction with redirection. If the view performs redirection during testing, its resulting `status` must be `"200 OK"` or its `status_code` is `200` and not `"302 FOUND"` or `302` because `follow_redirects` ensures that redirection or the HTTP Status Code 302 will happen. So, the assertion will be there is redirection (`200`) or none.

Another option to verify redirection is to set `follow_redirects` to `False` and then assert whether the `status_code` is `302`. The following test method shows this kind of testing approach:

```python
def test_assign_exam_redirect_302(client, form_data):
    res = client.post('/exam/assign', data=form_data)
    assert res.status_code == 302
    assert res.location.split('?')[0] == url_for('redirect_success_
exam')
```

In this approach, the expected status code is `302` because there is no `follow_redirects` parameter set in `client.post()`. Also, `res.location` is the appropriate attribute to extract the URL path because `res.request.path` will give the URL path of the POST transaction instead.

Aside from asserting `status_code`, verifying the correctness of the redirection includes checking the correct redirected path and the content type. Mocking can also be an additional strategy to closely examine the internals of the POST transactions and their redirections if there are any. **Monkey patching** can help refine the view processes through testing.

Let's now learn how to use monkey patching in testing view functions.

Applying the monkey patching

Monkey patching is a `pytest` feature that involves intercepting a function in a view transaction and replacing it with a custom-implemented mock function that returns our desired result. The mock function must have the same parameter list and return type as the original one. Otherwise, monkey patching will not work. The following is a test case for redirection that uses monkey patching:

```python
@connect_db
def insert_question_details(conn, id:int, cid:str, pid:int, exam_
date:date, duration:int):
        return True


@pytest.fixture
def insert_question_patched(monkeypatch):
```

```
    monkeypatch.setattr( "views.examination.insert_question_details",
insert_question_details)

def test_assign_mock_exam(insert_question_patched, client, form_data):
    res = client.post('/exam/assign', data=form_data, follow_
redirects=True)
    assert res.status == '200 OK'
    assert res.request.path == url_for('redirect_success_exam')
```

monkeypatch is an object injected into the fixture function. It can emit a variety of methods for faking attributes and functions of other objects in a package and modules. In the given example, the objective is to test the /exam/assign form transaction with a mocked insert_question_details(). Instead of using the patch() decorator, the monkeypatch object of the fixture replaces the original function with a dummy insert_question_details() using its setattr() method. The dummy method needs to return a True value because the test needs to examine the behavior of the view function whenever the INSERT transaction is successful. Now, to enable monkey patching, you must inject the fixture containing the monkeypatch objects into the test functions like a typical fixture that yields resources.

Monkey patching does not replace the actual code of the mocked function. In the given setattr(), the views.examination.insert_question_details expression indicates the repository method in the /exam/assign route and not in its actual repository class. So, this is just replacing the state of the method call in the view function and not modifying the method's actual implementation.

Testing the repository, service, and view layers requires integration testing with or without mocking and parametrize() markers to find all the bugs and inconsistencies in the algorithms. Regardless, it is easier to set up test classes and files in organized applications that utilize application factories and Blueprints because those projects do not need directory restructuring, such as the one imposed on the ch01 application.

Let us discuss now the benefit of using create_app() and Blueprints in testing Flask components.

Creating test cases for components in application factory and Blueprints

Application factory functions and Blueprints help solve circular import problems by managing the context loading and allowing the Flask app instance to be accessible across the application without tapping the __main__ top-level module. Since every component and layer is in its proper place, it is easier to set up the testing environment.

Our applications in *Chapters 2* and *3* have essential Flask components that need testing, such as the repository transaction built by the SQLAlchemy, exceptions, and standard API functions. All these components are built by the create_app() factory and Blueprints.

Let us start formulating test cases for SQLAlchemy repository transactions.

Testing ORM transactions

The *Online Shipping* app in *Chapter 2* uses the standard SQLAlchemy ORM to implement the CRUD transactions. Integration testing can help test the repository layer of our application. Let us examine the following test case implementation that runs the `insert()` transaction of `ProductRepository`:

```python
import pytest
from mock import patch
from main import app as flask_app
from modules.product.repository.product import ProductRepository
from modules.model.db import Products


@pytest.fixture(autouse=True)
def form_data():
    params = dict()
    params["name"] = "eraser"
    params["code"] = "SCH-8977"
    params["price"] = "125.00"
    yield params
    params = None


@patch("modules.model.config.db_session")
def test_mock_add_products(mocked_sess, form_data):
    db_sess = mocked_sess.return_value
    with flask_app.app_context() as context:
        repo = ProductRepository(db_sess)
        prod = Products(price=form_data["price"], code=form_data["code"], name=form_data["name"])
        res = repo.insert(prod)
        db_sess.add.assert_called_once()
        db_sess.commit.assert_called_once()
        assert res is True
```

`test_mock_add_products()` focuses on examining the flow of the INSERT transaction in adding a new product line to the database. It mocks the `db_session` from SQLAlchemy's scoped_session because the test is on the lines of codes and not with the `db_session`'s add() method. `assert_called_once()` of mocked add() and commit() will verify the execution of these methods during the test.

Now, the `ch02-blueprint` project uses the `before_request()` and `after_request()` events to track down the requests of every view and the user who accesses the views. These two application-level events become the core implementation of the application's custom authentication mechanism. All view pages in the project happen to be secured. So, running and testing the `/ch02/products/add` view, for instance, without logging in as a valid user, will lead to a redirection to the login page, as verified by the following test case:

```python
def test_add_product_no_login(form_data, client):
    res = client.post("/ch02/products/add", data=form_data)
    assert res.status_code == 302
    assert res.status == "302 FOUND"
    assert res.location.split('?')[0] == "/ch02/login/auth"
```

Running the `add_product()` view directly will redirect us to `login_db_auth()` from the `login_bp` Blueprint, thus the HTTP Status Code 302. To prove that login authentication is required for the user to access the `add_product()` view, create a test case that will include the `/ch02/login/auth` access, like in the following test case:

```python
def test_add_product_with_login(form_data, login_data, client):
    res_login = client.post("/ch02/login/auth", data=login_data)
    with client.session_transaction() as session:
        assert 'admin' == session["username"]
        assert res_login.location.split('?')[0] == url_for('home_
bp.menu')
        res = client.post("/ch02/products/add", data=form_data)
        assert res.status_code == 200
```

Testing and running `/ch02/login/auth` must be the initial goal before running `/ch02/products/add`. The `login_data()` fixture must provide a valid user detail for authentication. Since Flask's built-in `session` is responsible for storing `username`, you can open a session in the test using `session_transaction()` of the test `Client` and the `with` context manager. Within the context of the simulated session, check and confirm whether `/ch02/login/auth` saved `username` in the `session` object. Also, assert whether the aftermath of a successful authentication will redirect the user to the `home_bp.menu` page. If all these verifications are `True`, run and test now the `add_product()` view and perform the proper verifications afterward.

Next, we will test Flask API functions with `pytest`.

Testing API functions

Chapter 3 introduced and used Flask API endpoint functions in building our *Online Pizza Ordering System*. Testing these API functions is not the same as consuming them. A test `Client` provides the utility methods, such as `get()`, `post()`, `put()`, `delete()`, and `patch()`, to run the API and consume its resources, while extension modules such as `requests` build client applications to access and consume the APIs.

Like in testing view pages, it is still the test `Client` class that can run, test, and mock our Flask API functions. The following test cases show how to examine, scrutinize, and analyze the performance and responses of the APIs in the `ch03` project using `pytest`:

```
import pytest
from mock import patch, MagicMock
from main import app as flask_app
import json


@pytest.fixture
def client():
    with flask_app.test_client() as client:
        yield client


def test_index(client):
    res = client.get('/index')
    data = json.loads(res.get_data(as_text=True))
    assert data["message"] == "This is an Online Pizza Ordering
System."
```

The given `test_index()` is part of the `test_http_get_api.py` test file and has the task of scrutinizing if calling `/index` will have a response of `{"message": "This is an Online Pizza Ordering System."}`. Like in the views, the response will always give data in bytes. However, by using `get_data(as_text=True)` with the `json.loads()` utility, the data response will become a JSON object.

Now, the following is a test case that performs adding new order details to the existing database:

```
@pytest.fixture
def order_data():
    order_details = {"date_ordered": "2020-12-10", "empid": "EMP-101"
, "cid": "CUST-101", "oid": "ORD-910"}
    yield order_details
    order_details = None


def test_add_order(client, order_data):
    res = client.post("/order/add", json=order_data)
```

```
    assert res.status_code == 201
    assert res.content_type == 'application/json'
```

Like in views, the `client.post()` method consumes the POST API transactions, but with input details passed to its `json` parameter and not `data`. The given `test_add_order()` performs and asserts the `/order/add` API to verify whether SQLAlchemy's `add()` function works successfully given the configured `db_session`. The test expects `content_type` of `application/json` in its response.

Aside from `get()` and `post()`, the test `Client` has a `delete()` method to run `HTTP DELETE` API transactions. The following test function runs `/order/delete` with a path variable of `ORD-910`:

```
def test_delete_order(client):
    res = client.delete("/order/delete/ORD-910")
    assert res.status_code == 201
```

The given test class studies the deletion of an order with `ORD-910` as the order ID will not throw runtime errors even if the order does not exist.

Now, the test `Client` also has a `patch()` method to run `PATCH API` transactions and a `put()` method for `PUT API`.

Determining what exception a repository, service, API, or view under test will throw is a testing mechanism called **exception testing**. This type of testing is essential in verifying whether the Flask application can handle the major, if not all, possibilities that might cause runtime problems. But how does `pytest` implement exception testing?

Implementing exception testing

There are many variations of implementing exception testing in `pytest`, but the most common is to use the `raises()` function and the `xfail()` marker of `pytest`.

The `raises()` utility applies to testing features that explicitly call `abort()` or `raise()` methods to throw specific built-in or custom `HTTPException` classes. Its goal is to verify whether the functionality under test is throwing the exact exception class. For instance, the following test case checks whether the `/ch03/employee/add` API raises `DuplicateRecordException` during duplicate record insert:

```
import pytest
from main import app as flask_app
from app.exceptions.db import DuplicateRecordException


@pytest.fixture
def client():
    with flask_app.test_client() as client:
```

```
       yield client

@pytest.fixture
def employee_data():
    order_details = {"empid": "EMP-101", "fname": "Sherwin John" ,
"mname": "Calleja", "lname": "Tragura", "age": 45 , "role": "clerk",
"date_employed": "2011-08-11", "status": "active", "salary": 60000.99}
    yield order_details
    order_details = None


def test_add_employee(client, employee_data):
    with pytest.raises(DuplicateRecordException) as ex:
        res = client.post('/employee/add', json=employee_data)
        assert res.status_code == 200
    assert str(ex.value) == "insert employee record encountered a
problem"
```

The add_employee() endpoint function raises DuplicateRecordException if the return value of EmployeeRepository's insert() is False. test_add_employee() checks whether the endpoint raises the exception given an order_details() fixture that yields an existing employee record. If status_code is not 200, then there is a glitch in the code.

On the other hand, the xfail() marker applies to testing components with overlooked and unhandled risky lines of code that have a considerable chance of messing up the application anytime at runtime. But xFail() can also apply to test classes that verify known custom exceptions, like in the following snippet:

```
@pytest.mark.xfail(strict=True, raises=NoRecordException, reason="No
existing record.")
def test_update_employee(client, employee_data):
    res = client.patch(f'/employee/update/ {employee_data["empid"]}',
json=employee_data)
    assert res.status_code == 201
```

test_update_employee() runs PATCH API, the /ch03/employee/update, and verifies whether the employee_data() fixture provides new details for an existing employee record. If the employee, determined by empid, is not in the database record throwing NoRecordException, pytest will trigger the xFail() marker and render the marker's *"No existing record."* reason.

Because of the organized directory structures that applications in *Chapter 2* and *Chapter 3* follow, testing becomes clear-cut, isolated, reproducible, and categorized based on functionality. Using the application factories and Blueprints will not only give benefits to the development side but also to the testing environment.

Let us try using `pytest` to test asynchronous transactions in Flask. Do we need to install additional modules to work out the kind of testing?

Creating test cases for asynchronous components

Flask 3.x supports asynchronous transactions with the `asyncio` platform. *Chapter 5* introduced creating asynchronous API endpoint functions, web views, background tasks and services, and repository transactions using the `async/await` features. The test `Client` class of Flask 3.x is part of the `Flask[async]` core libraries, so there will be no problem running the `async` components with `pytest`.

The following test cases on the asynchronous repository layer, Celery tasks, and API endpoints will provide proof on `pytest` supporting Flask 3.x asynchronous platform.

Testing asynchronous views and API endpoint function

The test `Client` can run and test this `async` route function similar to running standard Flask routes using its `get()`, `post()`, `delete()`, `patch()`, and `put()` methods. In other words, the same testing rules apply to testing asynchronous view functions, as shown in the following test case:

```python
import pytest
from main import app as flask_app


@pytest.fixture(scope="module", autouse=True)
def client():
    with flask_app.test_client() as app:
        yield app


def test_add_vote_ws_client(client):
    res = client.get('/ch05/votecount/add')
    assert res.status_code == 200
```

In our *Online Voting* application, the `ch05-web` project has an `async` WebSocket client, the `add_vote_count_client()` view function. The given `test_add_vote_ws_client()` runs and tests the `HTTP GET` request transaction using `client.get()`. So, this is the same when running a standard view function using the test `Client` class and also with an asynchronous API endpoint function, as shown by the following test case implementation:

```python
@pytest.fixture(autouse=True, scope="module")
def login_details():
    data = {"username": "sjctrags", "password":"sjctrags"}
    yield data
    data = None
```

```
@pytest.xfail(reason="An exception is encountered")
def test_add_login(client, login_details):
    res = client.post("/ch05/login/add", json=login_details)
    assert res.status_code == 201
```

The add_login() API with the /ch05/login/add URL pattern is an async API endpoint function that adds new login details to the database. test_add_login() performs exception testing on the API to check whether adding existing records will throw an error. So, the process and formulation of the test cases are the same as testing their standard counterparts. But what if the transactions under testing are asynchronous such that test functions need to await to execute them? How can pytest directly call an async method? Let us take a look at testing asynchronous SQLAlchemy transactions.

Testing the asynchronous repository layer

The ORM used in ch05-web and ch05-api projects is the asynchronous SQLAlchemy. In an async ORM, all CRUD operations run as coroutines that need the await keyword to execute them. Likewise, test functions need to await these asynchronous components under test to execute them as coroutines. However, pytest requires an extension module called pytest-asyncio to add support for implementing asynchronous test functions. So, install the pytest-asyncio module using the following pip command before implementing the test case:

```
pip install pytest-asyncio
```

The implementation is the same as the previous ones except for the pytest_plugins component, which imports the necessary pytest extension, such as pytest-asyncio. The pytest_plugins component imports the installed pytest extensions and adds features to the testing environment that pytest alone cannot perform. With pytest-asyncio, testing transactions run by coroutines, like in the following snippet, is now feasible:

```
import pytest
from app.model.config import db_session
from app.model.db import Login
from app.repository.login import LoginRepository


pytest_plugins = ('pytest_asyncio',)

@pytest.fixture
def login_details():
    login_details = {"username": "user-1908", "password": "pass9087" }
    login_model = Login(**login_details)
    return login_model

@pytest.mark.asyncio
```

```
async def test_add_login(login_details):
    async with db_session() as sess:
        async with sess.begin():
            repo = LoginRepository(sess)
            res = await repo.insert_login(login_details)
            assert res is True
```

Calling an asynchronous method for testing always requires a test function to be `async` because it needs to await the function under test. The given `test_add_login()` is an `async` method because it needs to call and await an asynchronous `insert_login()` transaction. However, for `pytest` to run an `async` test function, it will require the test functions to be decorated by `@pytest.mark.asyncio()` provided by the `pytest-asyncio` library. But what will be the case when Celery background tasks will undergo testing?

Testing Celery tasks

`Pytest` needs the `pytest-celery` extension module to run Celery tasks under testing. Thus, the test file needs to include `pytest_celery` in its `pytest_plugins`. The following test function runs the `add_vote_task_wrapper()` task to add a candidate's votes to the database:

```
import pytest
from app.services.vote_tasks import add_vote_task_wrapper
import json
from main import app as flask_app

pytest_plugins = ('pytest_celery',)

@pytest.fixture(scope='session')
def celery_config():
    yield {
        'broker_url': 'redis://localhost:6379/1',
        'result_backend': 'redis://localhost:6379/1'
    }

@pytest.fixture
def vote():
    login_details = {"voter_id": "BCH-111-789", "election_id": 1,
"cand_id": "PHL-102" , "vote_time": "09:11:19" }
    login_str = json.dumps(login_details)
    return login_str

def test_add_votes(vote):
    with flask_app.app_context() as context:
        login_task = add_vote_task_wrapper.apply_async( args=[vote])
```

```
    result = login_task.get()
    assert bool(result) is True
```

Before writing the test cases for Celery tasks, import the `python_celery` library in `pytest_plugins`. Then, create a test function, such as `test_add_votes()`, run the Celery task the usual way with the app's context, and perform the needed verifications. By the way, running the Celery task with the application's context means that testing will utilize the configured Redis broker. However, if testing decides not to use the configured Redis configurations (e.g., `broker_url`, `result_backend`) of `app`, `pytest_celery` can allow `pytest` to inject dummy Redis configurations into its test functions through the fixture, like the given `celery_config()`, or override the built-in configuration through `@pytest.mark.celery(result_backend='xxxxx')`. Running the task without the default Redis details will lead to a `kombu.connection:connection.py:669 no hostname was supplied` error.

Can `pytest` create a test case for asynchronous file upload for some web-based applications?

Testing asynchronous file upload

Chapter 6 showcases an *Online Housing Pricing Prediction and Analysis* application, which highlights creating views that capture data from uploaded *XLSX* files for data analysis and graphical plotting. The project also has `pytest` test files that analyze the file uploading process of some form views and verify the rendition types of their Flask responses.

`Pytest` supports running and testing web views that involve uploading files of any mime type and converting them to `FileStorage` objects for content processing. Uploading a *multipart* file requires the `client.post()` function to have the `content_type` parameter set to `multipart/form-data`, the `buffered` parameter to `True`, and its `data` parameter to a *dictionary* consisting of the form parameter name of the file-type form component as the key, and the file object opened as a binary file for reading as its value. The following test case verifies whether `/ch06/upload/xlsx/analysis` can upload an XLSX file, extract some columns, and render them on HTML tables:

```
import os

def test_upload_file(client):
    test_file = os.getcwd() + "/tests/files/2011Q2.xlsx"
    data = {
        'data_file': (open(test_file, 'rb'), test_file)
    }
    response = client.post("/ch06/upload/xlsx/analysis",
 buffered=True, content_type='multipart/form-data', data=data)
    assert response.status_code == 200
    assert response.mimetype == "text/html"
```

`test_upload_file()` fetches some XLSX sample files within the project and opens these as binary files for reading. The object extracted from the `open()` file becomes the value of the `data_file` form parameter of the Jinja template. `client.post()` will run `/ch06/upload/xlsx/analysis` and use the file object as input. If the `pytest` execution has no uploading-related exceptions, the response should emit `status_code` of `200` with the `content-type` header of `text/html`.

After testing unsecured components, let us now deal with test cases that run views or APIs that require authentication and authorization.

Creating test cases for secured API and web components

All applications in *Chapter 9* implement the authentication methods essential to small-, middle-, or large-scale Flask applications. `pytest` can test secured components, both standard and asynchronous ones. This chapter will cover testing Cross-Site Request Forgery- or CSRF-protected views running on an HTTPS with `flask-session` managing the user session, HTTP basic authenticated views, and web views secured by the `flask-login` extension module.

Testing secured API functions

Chapter 9 showcases a *Vaccine Reporting and Management* system with web-based and API-based versions. The `ch09-web-passphrase` project is a web version of the prototype with views protected by a custom authentication mechanism using the `flask-session` module, web forms that are CSRF-protected, and all components running on an HTTPS protocol.

The `/ch09/login/auth` route is the entry point to the application, where users must log in using their `username` and `password` credentials. To test the secured view routes, the `/ch09/login/auth` route must have the first execution in the test function to allow access to other views. The following test case runs the `/ch09/patient/profile/add` view without user authentication:

```python
import pytest
from flask import url_for
from main import app as flask_app

@pytest.fixture
def client():
    flask_app.config["WTF_CSRF_ENABLED"] = False
    with flask_app.test_client() as app:
        yield app

@pytest.fixture(scope="module")
def user_credentials():
    params = dict()
    params["username"] = "sjctrags"
    params["password"] = "sjctrags"
```

```
        return params

def test_patient_profile_add_invalid_access(client):
    res = client.get("/ch09/patient/profile/add", base_url='https://
localhost')
    assert res.status_code == 302
    assert res.location.split('?')[0] == url_for('login_user')
```

For pytest to access and run /ch09/patient/profile/add, which is a form view, you must first disable the CSRF protection of the flask_wtf.csrf module using its WTF_CSRF_ENABLED built-in environment variable before extracting the test Client instance in the client() fixture. Running the given test function with the correct user credentials will show a successful result since the access is unauthenticated, causing a redirection to the /ch09/login/auth view. So far, the application uses custom authentication but with a database-encrypted username managed by the flask-session extension module.

This test proves that accessing any view from our *Online Vaccine Registration* application requires user authentication from its /ch09/login/auth view page. Any unauthenticated attempt to its view will redirect the user to the login page. The following snippet builds the proper access flow to our application's views with user authentication:

```
def test_patient_profile_add_valid_access(client, user_credentials):
    res_login = client.post('/ch09/login/auth', data=user_
credentials, base_url='https://localhost')
    assert res_login.status_code == 302
    assert res_login.location.split('?')[0] == url_for('view_
signup')
    res = client.get("/patient/profile/add", base_url='https://
localhost:5000')
    assert res.status_code == 200
    with client.session_transaction() as session:
        assert session["user"] == "sjctrags"
```

test_patient_profile_add_valid_access() has the same test flow as in the ch02-blueprint project. The only difference is the presence of the base_url parameter in client.post() since the view runs on an HTTPS platform. The goal of the test is to run /ch09/profile/add successfully after logging into /ch09/login/auth with the correct login details. Also, this test function verifies whether the flask-session module is working on saving the user data in the session object.

How about testing APIs secured by HTTP-based authentication mechanisms that use the Authorization header? How does pytest run these types of secured API endpoint functions?

Testing HTTP Basic authentication

The ch09-api-auth-basic project, an API-based version of the *Online Vaccine Registration* application, uses the HTTP basic authentication scheme to secure all API endpoint access. An Authorization header with the base64-encoded username:password credential must be part of the request headers to access an API. Moreover, the access is also restricted by the flask-cors extension module. The following test case accesses /ch09/vaccine/add without authentication:

```python
import pytest
from main import app as flask_app
import base64


@pytest.fixture
def client():
    with flask_app.test_client() as app:
        yield app

@pytest.fixture
def vaccine():
    vacc = {"vacid": "VAC-899", "vacname": "Narvas", "vacdesc": "For
Hypertension", "qty": 5000, "price": 1200.5, "status": True}
    return vacc

def test_add_vaccine_unauth(client, vaccine):
    res = client.post("/ch09/vaccine/add", json=vaccine,
headers={'Access-Control-Allow-Origin': "http://localhost:5000"})
    assert res.status_code == 201
```

The test Client methods have header parameters that can contain a dictionary of request headers, such as Access-Control-Allow-Headers, Access-Control-Allow-Methods, Access-Control-Allow-Credentials, and Access-Control-Allow-Origin for managing the application's **cross-origin resource sharing (CORS)** header details. However, the given test will still return an HTTP Status Code 403 even with the CORS headers because no Authorization header is present with the Basic credential. The following snippet is the correct test case for successful access to the API endpoint secured by the basic scheme:

```python
@pytest.fixture
def auth_header():
    credentials = base64.b64encode(b'sjctrags:sjctrags')
.decode('utf-8')
    return credentials


def test_add_vaccine_auth(client, vaccine, auth_header):
```

```
    res = client.post("/vaccine/add", json=vaccine,
headers={'Authorization': 'Basic ' + auth_header, 'Access-Control-
Allow-Origin': "http://localhost:5000"})
    assert res.status_code == 201
```

The preceding test case will show a successful result given the correct base64-encoded credentials. The inclusion of the Authorization header with the Basic and base64-encoded credentials from the auth_header() fixture in the header parameter of client.post() will fix the HTTP Status Code 403 error.

The Authorization header must be in the header parameter of any Client method when testing and running the API endpoint secured by HTTP basic, digest, and bearer-token authentication schemes. In the Authorization Digest header, the *nonce*, *opaque*, and *nonce count* must be part of the header details. On the other hand, the token-based scheme needs a secure **JSON Web Token** (**JWT**) token in the Authorization Bearer header or with the token_auth parameter of the test Client methods.

But how does pytest scrutinize the view routes secured by the flask-login extension? Is there an added behavior that pytest should adopt when testing views secured by flask-login?

Testing web logins

The ch09-web-login application is another version of the *Online Vaccine Registration* application that uses the flask-login module as its source for security. It uses the flask-session module to store the user session in the file system instead of the browser. Like in the ch09-web-passphrase and ch02-blueprint projects, users must log into the application before accessing any views. Otherwise, the application will redirect them to the login page. The following test case is similar to the previous test files where the test accesses /ch09/login/auth first before accessing any views or APIs:

```
import pytest
from flask_login import current_user
from main import app as flask_app


def test_add_admin_profile(client, admin_details, user_credentials):
    res_login = client.post('/ch09/login/auth', data=user_credentials)
    assert res_login.status_code == 200

    with client.session_transaction() as session:
        assert current_user.username == "sjctrags"
        res = client.post("/ch09/admin/profile/add", data=admin_
details)
        assert res.status_code == 200
```

The given `test_admin_admin_profile()` will run the `/ch09/admin/profile/add` route with a successful result given the valid `user_credentials()` fixture. One advantage of using the `flask-login` module compared to custom session-handling is the `current_user` object it has that can give proof if the user login transaction created a session, if a user depicted in the `user_credentials()` fixture is the one stored in its session, or if the authentication was done using the *remember me* feature. The given test function verifies whether `username` indicated in the `user_credentials()` fixture is the one saved in the `flask-login` session.

Another feature of `flask-login` that is beneficial to `pytest` is its capability to turn off all the authentication mechanisms during testing. The following test class runs the same `/ch09/admin/profile/add` route successfully without logging in:

```
@pytest.fixture
def client():
    flask_app.config["LOGIN_DISABLED"] = True
    with flask_app.test_client() as app:
        yield app


def test_add_admin_profile(client, admin_details):
    res = client.post("/admin/profile/add", data=admin_details)
    assert res.status_code == 200
```

Disabling authentication in `flask-login` requires setting its built-in `LOGIN_DISABLED` environment variable to `True` at the configuration level. The setup should be part of the `client()` fixture before extracting the test `Client` object from the Flask's app instance.

Pytest and its add-ons can test all authentication schemes and authorization rules applied to Flask 3.x apps. Using the same GWT unit testing strategy and behavioral testing mechanisms, such as mocking and monkey patching, `pytest` is a complete and adequate testing library to run and verify secured APIs and view routes.

How can `pytest` mock a MongoDB connection when running and testing routes? Let's learn how.

Creating test cases for MongoDB transactions

The formulation of the test files, classes, and functions is the same when testing components with the Flask application running on MongoDB. The only difference is how `pytest` will mock the MongoDB connection to pursue testing.

This chapter showcases the `mongomock` module and its `MongoClient` mock object that can replace a configured MongoDB connection. So, install the `mongomock` module using the following `pip` command before creating the test file:

```
pip install mongomock
```

Chapter 7 has a *Tutor Finder* application with components running on NoSQL databases such as MongoDB. The application uses the connect() method of the mongoengine module to establish a MongoDB connection for a few of the APIs. Instead of using the configured connection in the Flask's app context, a MongoClient object from mongomock can replace the mongoengine's connect() method with a fake one. The following snippet of the test_tutor_login.py file mocks the MongoDB connection to run insert_login() of LoginRepository:

```
import pytest
import mongomock
from mongoengine import connect, get_connection, disconnect
from main import app as flask_app
from modules.repository.mongo.tutor_login import TutorLoginRepository
from bcrypt import hashpw, gensalt


@pytest.fixture
def login_details():
    login = dict()
    login["username"] = "sjctrags"
    login["password"] = "sjctrags"
    login["encpass"] = hashpw(str(login['username']) .encode(),
gensalt())
    return login



@pytest.fixture
def client():
    disconnect()
    with flask_app.test_client() as client:
        yield client

@pytest.fixture
def connect_db():
    connect(host='localhost', port=27017, db='tfs_test',
uuidRepresentation='standard', mongo_client_class=mongomock.
MongoClient)
    conn = get_connection()
    return conn
```

The given connect_db() recreates the MongoClient object using the same mongoengine's connect() method but now with the fake MongoClient. However, the parameter values of connect(), like the values of db, host, and port, must be part of the testing environment setup. Also, the uuidRepresentation parameter must be present in the mocking.

After the mocked `connect()` setup, it needs to call the mongoengine's `get_connection()` and yield it to the test function. So, the connection created from a mocked `MongoClient` is fake but with the existing database configuration details.

Now, before injecting the `connect_db()` fixture to the test functions, call the `disconnect()` method to kill an existing connection in the Flask `app` context and avoid multiple connections running in the background, which will cause an error. The following test function has the injected mocked MongoDB connection for testing the `insert_login()` MongoDB repository transaction:

```
def test_add_login(client, connect_db, login_details):
    repo = TutorLoginRepository()
    res = repo.insert_login(login_details)
    assert res is True
```

Aside from `mongomock`, the `pytest-mongo` and `pytest-mongodb` modules allow mocking `mongoengine` models and collections by using the actual MongoDB database configuration.

Can `pytest` run and test WebSocket endpoints created by `flask-sock`? Let us implement a test case that will analyze WebSockets.

Creating test cases for WebSockets

WebSockets are components of our `ch05-web` and `ch05-api` projects. The applications use `flask-sock` to implement the WebSocket endpoints. So far, `pytest` can only provide the testing environment for WebSockets. However, it needs the `websockets` module to run, test, and assert the response of our WebSocket endpoints. So, install this module using the following `pip` command:

```
pip install websockets
```

There are three components that the `websockets` module can provide to `pytest`:

- The simulated route that will receive the message from the client
- The mock server
- The test function that will serve as the client

All these components must be `async` because running WebSockets requires the `asyncio` platform. So, also install the `pytest-asyncio` module to give asynchronous support to these three components:

```
Pip install pytest-asyncio
```

Then, start implementing the simulated or mocked view similar to the following implementation to receive and process the messages sent by a WebSocket:

```
import websockets
import pytest
import json
import pytest_asyncio


pytest_plugins = ('pytest_asyncio',)


async def simulated_add_votecount_view(websocket):
    async for message in websocket:
        print("received: ",message)
        # Place here the VoteCount repo transactions
        await websocket.send("data added")
```

simulated_add_votecount_view() will serve as the mocked WebSocket endpoint function, which receives and saves the tallied votes into the database.

Next, create a mock server using the websockets.serve() method to run the simulated route in an **event loop**. This method needs host, port, and the simulated view name, such as simulated_add_votecount_view, to operate. The following is our WebSocket server, which will run at the ws://localhost:5001 address:

```
@pytest_asyncio.fixture
async def create_ws_server():
    async with websockets.serve( simulated_add_votecount_
view,  "localhost", 5001) as server:
        yield server
```

Since create_ws_server() must be async, decorating it with @pytest.fixture will cause an error. So, use @pytest_asyncio.fixture to declare the asynchronous fixtures for pytest.

Finally, we start our test function implementation with the context manager that opens the websockets client object for WebSocket endpoint execution and closes it afterward. The following implementation shows a test function for the add_vote_count_server() WebSocket with the ws://localhost:5001/ch05/vote/save/ws URL address:

```
@pytest.mark.asyncio
async def test_votecount_ws(create_ws_server, vote_tally_details):
        async with websockets.connect( "ws://localhost:5001/ch05/vote/
save/ws") as websocket:
            await websocket.send(json.dumps( vote_tally_details))
            response = await websocket.recv()
            assert response == "data added"
```

A successful connection to the mock server will create a client object through the `websockets.connect()` method with the URI of the WebSocket as its parameter argument. The client object can send a string or numeric message to the simulated route and receive a string or numeric response from that server. This send-and-receive process will only happen once per execution of the test function. Since the `with-context` manager, `send()`, and `recv()` are all awaited, the test function must be `async`. Now, use the `assert` statement to verify whether our client receives the proper message from the server.

Another way to test the WebSocket endpoint is to use the actual development environment, for instance, running our `ch05-web` project with the PostgreSQL database, Redis, and the **Web Server Gateway Interface (WSGI)** server operating altogether. The following `test_websocket_actual()` method runs the same WebSocket server without monkey patching or a mocked server:

```python
import pytest
import websockets
import json


pytest_plugins = ('pytest_asyncio',)

@pytest.fixture(scope="module", autouse=True)
def vote_tally_details():
    tally = {"election_id":"1", "precinct": "111-C", "final_tally":
"6000", "approved_date": "2024-10-10"}
    yield tally
    tally = None


@pytest.mark.asyncio
async def test_websocket_actual(vote_tally_details):
    async with websockets.connect("ws://localhost:5001/ch05/ vote/
save/ws") as websocket:
            await websocket.send(json.dumps( vote_tally_details))
            response = await websocket.recv()
            assert response == "data not added"
```

The test method adds a new vote tally to the database. If the voting precinct number of the record is not yet in the table, then the WebSocket will return the `"data added"` message to the client. Otherwise, it will return the `"data not added"` message. This approach also tests the correct configuration details of the Redis and PostgreSQL servers used by the WebSocket endpoint. Others may mock the Redis connectivity and PostgreSQL database connection to focus on the WebSocket implementation and refine its client response.

Testing Flask components should focus on different perspectives to refine the application's performance and quality. Unit testing components using monkey patching or mocking is an effective way of refining, streamlining, and scrutinizing the inputs and results. However, most often, the integration testing with the servers, internal modules, and external dependencies included can help identify and resolve major technical issues, such as compatibility and versioning problems, bandwidth and connection overhead, and performance issues.

Summary

There are many strategies and approaches in testing Flask applications, but this chapter focuses on the components found in our applications from *Chapters 1* to *9*. Also, the goal is to build test cases using the straightforward syntax of the `pytest` module.

This chapter started with testing the standard Flask components with the web views, API functions, repository transactions, and native services. Aside from simply running the components and verifying their response details using the `assert` statement, mocking becomes an essential ingredient in many test cases of this chapter. The `patch()` decorator from the `unittest` module mocks the `psycopg2` connections, repository transactions in views and services, and the SQLAlchemy utility methods. This chapter also discussed monkey patching, which replaces a function with a mock one, and exception testing, which determines raised exceptions and undetected bugs.

This chapter also established proof that it is easier to test asynchronous `Flask[async]` components, such as asynchronous SQLAlchemy transactions, services, views, and API endpoints, using `pytest` and its `pytest-asyncio` module. On the other hand, another module called `pytest-celery` helps `pytest` examine and verify the Celery tasks.

However, the most challenging part is how this chapter uses `pytest` to examine components from secured applications, run repository transactions that connect to MongoDB, and analyze and build WebSockets.

It is always recommended to apply testing on Flask components during development to study the process flows, runtime performance, and the feasibility of the implementations.

The next chapter will discuss the different deployment strategies of our Flask applications.

11

Deploying Flask Applications

When the development of a Flask application is over, you can always decide to deploy it somewhere outside Werkzeug's HTTP server. The final application needs a production server that is fast and reliable, with minimal or no potential security risks, configurable, and easy to manage. Instead of utilizing the built-in Werkzeug server, the product needs a non-development server not for development, debugging, or testing but for running the software product. Flask deployment requires a stable and independent Python server or a hosting platform.

This chapter will focus on different approaches, options, and procedures for deploying Flask applications to production servers suited for the product's scope, environment, and objectives.

The following topics will be covered in this chapter:

- Running the application on Gunicorn and uWSGI
- Running the application on Uvicorn
- Deploying the application to the Apache HTTP Server
- Deploying the application to Docker
- Deploying the application to Kubernetes
- Creating an API gateway using NGINX

Technical requirements

Our application will be using PostgreSQL to manage its data. The projects will also be applying the `Blueprint` approach of managing Flask components. The project prototype will focus on simple e-commerce, inventory, and stocking transactions for a small-scale grocery store, and it will be called an *Online Grocery* application. All these applications can be found at `https://github.com/PacktPublishing/Mastering-Flask-Web-Development/tree/main/ch11`.

Getting ready for deployment

In this chapter, we'll create an *Online Grocery* application that can be deployed to different platforms. The application is an API-based type with administration, login, inventory, stocking, order, and purchase modules designed for small business transactions of a small shopping or grocery store.

The Peewee ORM builds the application's model and repository layer. To utilize the standard `Peewee` module, install it and the `psycopg2` driver using the following `pip` command:

```
pip install psycopg2 peewee
```

The Peewee ORM provides the standard *INSERT*, *UPDATE*, *DELETE*, and *SELECT* transactions, thus including the `psycopg2` driver as a dependency library. Let's begin structuring the model layer of the Peewee ORM.

Classes and methods for the standard Peewee ORM

Our *Online Grocery* application is deployed to a **Gunicorn** server and uses the standard Peewee helper classes and methods to establish the model layer and the repository classes. Here is a typical Peewee configuration for the PostgreSQL database connection:

```
(app/models/config.py)
from peewee import PostgresqlDatabase
database = PostgresqlDatabase(
     'ogs', user='postgres', password='admin2255', autocommit=False,
host='localhost', port=5432)
```

Peewee has `PostgresqlDatabase`, `MySQLDatabase`, and `SqliteDatabase` driver classes that will create a connection object for the application. Our option is `PostgresqlDatabase`, as shown in the preceding code, since our application uses the **PostgreSQL** database platform. Note that you should always set the `autocommit` constructor parameter to `False` to enable transaction management for CRUD operations.

The `database` connection object will map Peewee's model classes to their actual table schemas. The following are some model classes of our applications:

```
(app/models/db.py)
from app.models.config import database
from peewee import Model, CharField, IntegerField, BigIntegerField,
ForeignKeyField, DateField

class Product(Model):
    id = BigIntegerField(primary_key=True, null=False,
sequence="product_id_seq")
    code = CharField(max_length="20", unique="True", null=False)
    name = CharField(max_length="100", null=False)
```

```
    btype = ForeignKeyField(model=Brand, null=False, to_field="code",
backref="brand")
    ctype = ForeignKeyField(model=Category, null=False, to_
field="code", backref="category")
    ... ... ... ... ... ...
    discount = ForeignKeyField(model=Discount, null=False, to_
field="code", backref="discount")

    class Meta:
        db_table = "product"
        database = database
```

The given `Product` model class represents the record details of a product sold by the grocery store, while the following `Stock` model creates stock information about a product:

```
class Stock(Model):
    id = BigIntegerField(primary_key=True, null=False,
sequence="stock_id_seq")
    sid = ForeignKeyField(model=Supplier, null=False, to_field="sid",
backref="supplier")
    invcode = ForeignKeyField(model=InvoiceRequest, null=False, to_
field="code", backref="invoice")
    qty = IntegerField(null=False)
    payment_date = DateField(null=True)
    received_date = DateField(null=False)
    recieved_by = CharField(max_length="100")

    class Meta:
        db_table = "stock"
        database = database
    ... ... ... ... ... ...
```

All model classes must subclass Peewee's `Model` class to become the logical representations of the database tables. The Peewee model classes, like the given `Product` and `Stock`, have the `Meta` class, which holds the `database` and `db_table` attributes responsible for mapping them to the physical tables of our database. Peewee's column helper classes build the column attributes of the model classes. Now, the `main.py` module must enable the `before_request()` glocal event of Flask to handle the database connection. The following snippet shows the implementation of the `before_request()` global event:

```
from app import create_app
from app.models.config import database

app = create_app('../config_dev.toml')
```

```
@app.before_request
def db_connect():
    database.connect()


@app.teardown_request
def db_close(exc):
    if not database.is_closed():
        database.close()
```

Here, `teardown_request()` closes the connection during server shutdown.

Like in SQLAlchemy, the Peewee ORM needs the model classes to create the transaction layer to perform the CRUD operations. The following is a `ProductRepository` class that manages and executes SQL statements using the standard Peewee transactions:

```
from app.models.db import Product
from app.models.db import database
from typing import Dict, Any


class ProductRepository:
    def insert_product(self, details:Dict[str, Any]) -> bool:
        try:
            with database.atomic() as tx:
                Product.create(**details)
                tx.commit()
                return True
        except Exception as e:
            print(e)
        return False
```

The Peewee repository class derives its transaction management from the `database` connection object. Its emitted `atomic()` method provides a transaction object that performs `commit()` and `rollback()` during SQL execution. The given `insert_product()` function performs an INSERT operation of a `Product` record by calling the model's `create()` class method with the `kwargs` variable of details and returns `True` if the operation is successful. Otherwise, it returns `False`.

On the other hand, an UPDATE operation in standard Peewee requires a transaction layer to retrieve the record object that needs an update, access its concerned field(s), and replace them with new values. The following `update_product()` function shows the implementation of a `Product` update:

```
def update_product(self, details:Dict[str,Any]) -> bool:
    try:
        with database.atomic() as tx:
            prod = Product.get( Product.code==details["code"])
            prod.rate = details["name"]
```

```
            prod.code = details["btype"]
            prod.rate = details["ctype"]
            prod.code = details["unit_type"]
            prod.rate = details["sell_price"]
            prod.code = details["purchase_price"]
            prod.rate = details["discount"]
            prod.save()
            tx.commit()
            return True
    except Exception as e:
        print(e)
    return False
```

The get() method of the model class retrieves a single instance matching the given query constraint. The goal is to update only one record, so be sure that the constraint parameters in the record object retrieval only involve the unique or primary key column fields.

Now, the save() method of the record object will eventually merge the new record object with the old one linked to the database. This commit() will finally persist and flush the updated record to the table.

When it comes to deletion, the initial step is similar to updating a record, which involves retrieving the record object for deletion. The following delete_product_code() repository method depicts this initial process:

```
def delete_product_code(self, code:str) -> bool:
    try:
        with database.atomic() as tx:
            prod = Product.get(Product.code==code)
            prod.delete_instance()
            tx.commit()
            return True
    except Exception as e:
        print(e)
    return False
```

The record object has a delete_instance() function that removes the record from the schema. In the case of delete_product_code(), it deletes a Product record through the record object retrieved by its product code.

When retrieving records, the Peewee ORM has a `select()` method that builds variations of query implementations. The following `select_product_code()` and `select_product_id()` functions show how to retrieve single records based on unique or primary key constraints:

```
def select_product_code(self, code:str):
    prod = Product.select(Product.code==code)
    return prod.to_json()

def select_product_id(self, id:int):
    prod = Product.select(Product.id==id)
    return prod.to_json()
```

On the other hand, the following `select_all_product()` function retrieves all records in the product table:

```
def select_all_product(self):
    prods = Product.select()
    records = [log.to_json() for log in prods]
    return records
```

All model classes retrieved by the `select()` method are non-serializable or non-JSONable. So, in the implementation, be sure to include the conversion of all model objects into JSON records using any accepted method. In the given sample, all our model classes have a `to_json()` method that returns a JSON object containing all the `Product` fields and values. The query transactions include a list comprehension in its procedure to generate a list of JSONable records of `Product` details using the `to_json()` method.

Classes and methods for the Async Peewee ORM

Some parts of our deployed *Online Grocery* application runs on the **asyncio** platform with async API endpoints and async repository transactions. Peewee has an async version that supports asynchronous request transactions in Flask. To utilize the **Async Peewee ORM**, install the peewee-async module using the following `pip` command:

```
pip install aiopg peewee-async
```

Also, include the `aiopg` module, which provides PostgreSQL asynchronous database access through the *DB API* specification.

Async Peewee has `PooledPostgresqlDatabase`, `AsyncPostgresqlConnection`, and `AsyncMySQLConnection` driver classes that create database connection objects in `async` mode. Our configuration uses the `PooledPostgresqlDatabase` driver class to include the creation of a connection pool:

```python
from peewee_async import PooledPostgresqlDatabase
database = PooledPostgresqlDatabase(
        'ogs', user='postgres', password='admin2255',
        host='localhost', port='5432', max_connections = 3,
        connect_timeout = 3, autocommit=False)
```

The given configuration has a maximum pool size of 3 with `autocommit` set to `False`.

The Async Peewee ORM handles database connectivity differently: it does not use the `before_request()` and `teardown_request()` events but rather uses configuration with the `create_app()` factory method. The following snippet shows how to establish a PostgreSQL database connection using the `peewee-async` module:

```python
from app.models.config import database
from peewee_async import Manager

def create_app(config_file):
    app = Flask(__name__)
    app.config.from_file(config_file, toml.load)

    global conn_mgr
    conn_mgr = Manager(database)
    database.set_allow_sync(False)
    … … … … … …
```

Here, `Manager` establishes an `asyncio` database connection pattern without using `before_request()` to connect to and `teardown_request()` to disconnect from the database. However, it can emit the `connect()` and `close()` methods to manage the database connection during query execution explicitly. Instantiating the `Manager` class requires the database connection object and an optional `asyncio` event loop. Through the `Manager` object, you can invoke its `set_allow_sync()` method and set it to `False` to restrict the usage of non-async utility Peewee methods.

The `conn_mgr` and `database` objects are equally essential for building the repository layer, as depicted in the following `DiscountRepository` implementation:

```python
from app.models.db import Discount
from app.models.db import database
from app import conn_mgr
from typing import Dict, Any
```

```
class DiscountRepository:

    async def insert_discount(self, details:Dict[str, Any]) -> bool:
        try:
            async with database.atomic_async() as tx:
                await conn_mgr.create(Discount, **details)
                await tx.commit()
                return True
        except Exception as e:
            print(e)
        return False
```

Although the implementation of the model layer is similar to the standard Peewee, its repository layer is not the same because of the `asyncio` platform used by the ORM to perform the CRUD transactions. For instance, the following `insert_discount()` function emits `atomic_async()` from the `conn_mgr` instance to generate an async transaction layer, which will commit the inserted `Discount` record performed by the `create()` method of `conn_mgr`, not by `Discount`. The use of the `async`/`await` keywords is present in the implementations.

On the UPDATE operation, the `get()` method of `conn_mgr` retrieves the record object that needs updating, and its `update()` method flushes the newly updated fields to the table. Again, the async `Manager` methods operate the transaction, not the model class. The following `update_discount()` function showcases Peewee's async approach to updating table records:

```
    async def update_discount(self, details:Dict[str,Any]) -> bool:
        try:
            async with database.atomic_async():
                discount = await conn_mgr.get(Discount,
code=details["code"])
                discount.rate = details["rate"]
                await conn_mgr.update(discount, only=("rate", ))
                return True
        except Exception as e:
            print(e)
        return False
```

The local parameters of the `update()` method of `conn_mgr` include the record object with the updated fields and the `only` parameter for controlling a tuple of field names that need updating in the table.

On the other hand, the `DELETE` operation uses the same async `get()` method of `conn_mgr` in `update_discount()` to retrieve the record object for deletion. As shown in the following `delete_discount_code()` function, the async `delete()` method of `conn_mgr` deletes the record from the table using the record object:

```
async def delete_discount_code(self, code:str) -> bool:
    try:
        async with database.atomic_async():
            discount = await conn_mgr.get(Discount, code=code)
            await conn_mgr.delete(discount)
            return True
    except Exception as e:
        print(e)
    return False
```

When implementing async query transactions, the Async Peewee ORM uses the `Manager` class's async `get()` method to retrieve a single record and the `execute()` method to wrap and run the `select()` statement for retrieving a single or all the records asynchronously. The following snippets show the query implementation for `DiscountRepository`:

```
async def select_discount_code(self, code:str):
    discount = await conn_mgr.get(Discount, code=code)
    return discount.to_json()

async def select_discount_id(self, id:int):
    discount = await conn_mgr.get(Discount, id=id)
    return discount.to_json()

async def select_all_discount(self):
    discounts = await conn_mgr.execute( Discount.select())
    records = [log.to_json() for log in discounts]
    return records
```

So, all these bundled methods in the `Manager` class's instance provide the operations for implementing the CRUD transactions in the asynchronous transaction layer.

Peewee is a simple and flexible ORM for small to middle-scale Flask applications. Although SQLAlchemy offers more powerful utilities, it is not suited for a small application like our *Online Grocery* application, which has less scope and complexity.

Next, we'll deploy our applications that utilize both the standard and asynchronous Peewee ORM for their repository layers.

Running the application on Gunicorn and uWSGI

The main reason why Flask applications start by running the `flask run` command or by calling `app.run()` in `main.py` during development is because of the built-in WSGI server that the `werkzeug` module has. However, there are limitations that this server possesses, such as its inability to respond to more requests from clients without slowing down and its incapability to maximize the resources of the production server. Moreover, the built-in server has several vulnerabilities, which pose security risks. For standard Flask applications, it is best to use another WSGI server for production, such as **Gunicorn** or **uWSGI**.

Let's start by deploying our application to the *Gunicorn* server.

Using the Gunicorn server

Gunicorn is the most common WSGI-based HTTP server that runs on a POSIX environment. If Windows is the environment that's used for development, it will be a requirement to deploy our application to a UNIX-based server with the installed Gunicorn server. In our case, we will use WSL on Windows PowerShell to access the Ubuntu running on Windows and deploy our `ch11-guni` application. But first, we must install the `gunicorn` module in the application's virtual environment using the following `pip` command:

```
pip install gunicorn
```

Then, run the `gunicorn` command with the module name and the app instance in `{module}`:`{flask_app}` format, the binding host address, and the port. The following is the complete command to run a standard Flask application on the Gunicorn server with a single worker:

```
gunicorn --bind 127.0.0.1:8000 main:app
```

Figure 11.1 shows the server log after successfully running the given command with the default single worker:

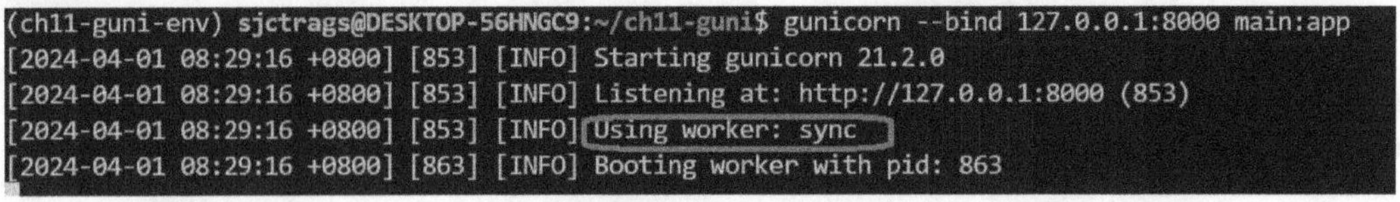

```
(ch11-guni-env) sjctrags@DESKTOP-56HNGC9:~/ch11-guni$ gunicorn --bind 127.0.0.1:8000 main:app
[2024-04-01 08:29:16 +0800] [853] [INFO] Starting gunicorn 21.2.0
[2024-04-01 08:29:16 +0800] [853] [INFO] Listening at: http://127.0.0.1:8000 (853)
[2024-04-01 08:29:16 +0800] [853] [INFO] Using worker: sync
[2024-04-01 08:29:16 +0800] [863] [INFO] Booting worker with pid: 863
```

Figure 11.1 – Server log after starting the Gunicorn server

A *Gunicorn* worker is a Python process that manages one HTTP request-response transaction at a time. A default Gunicorn server has one worker process running in the background. Logically, the more workers that are spawned to manage the requests and responses, the better the application's performance. However, for Gunicorn, the number of workers depends on the count of CPU processors on the server machine and is derived using the $(2*CPU)+1$ formula. These child processes will manage HTTP requests simultaneously, utilizing the maximum level of resources that the hardware can provide. One of the advantages of Gunicorn is its capability to leverage the resources efficiently to manage the runtime performance:

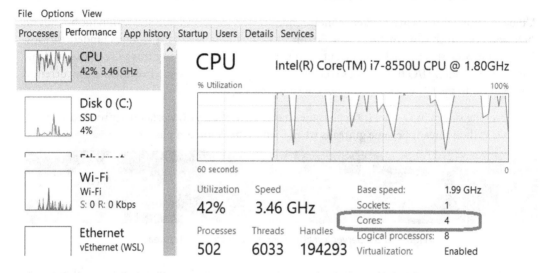

Figure 11.2 – The CPU utilization dashboard of a Windows system

Figure 11.2 shows that our production server machine has 4 CPU cores, which means that the acceptable number of workers that our Gunicorn server can utilize is 9. Thus, the following command runs a Gunicorn server with 9 workers:

```
gunicorn --bind 127.0.0.1:8000 main:app --workers 9
```

Adding the --workers setting in the command statement allows us to include the appropriate worker count in the HTTP request processing.

Adding workers to the Gunicorn server that does not improve the total CPU-bound performance of the application is a waste of resources. A remedy is to add more threads to a worker rather than add unhelpful workers.

Workers or processes consume more memory space. Additionally, no two workers can share memory space, unlike with threads. A *thread* consumes less memory space since it is more lightweight than a worker. To experience the best server performance, each worker must spawn at least 2 threads that will work concurrently on HTTP requests and responses. So, running the following Gunicorn command can start a server with 1 worker with 2 threads:

```
gunicorn --bind 127.0.0.1:8000 main:app --workers 1 --threads 2
```

The `--threads` setting allows us to add at least 2 threads per worker.

Although setting threads in a worker connotes concurrency, the threads are still within the bounds of their workers, which run synchronously. So, the blocking limitation of the workers hinders threads from performing their actual concurrent performance. However, having threads can manage the overhead of handling I/O transactions compared to the pure worker setup because the concurrency that's applied to the I/O blockings will not consume more space.

The server log shown in *Figure 11.3* depicts the change from the `sync` worker to `gthread` since all spawned Python threads become gthreads when used in the Gunicorn platform:

```
(ch11-guni-env) sjctrags@DESKTOP-56HNGC9:~/ch11-guni$ gunicorn --bind 127.0.0.1:8000 main:app --workers
1 --threads 2
[2024-04-01 16:05:13 +0800] [19828] [INFO] Starting gunicorn 21.2.0
[2024-04-01 16:05:13 +0800] [19828] [INFO] Listening at: http://127.0.0.1:8000 (19828)
[2024-04-01 16:05:13 +0800] [19828] [INFO] Using worker: gthread
[2024-04-01 16:05:13 +0800] [19829] [INFO] Booting worker with pid: 19829
```

Figure 11.3 – Server log after running Gunicorn with threads

Now, when the number of features that require I/O transactions increases, Gunicorn, along with workers and servers, will not help speed up the processing of HTTP requests and responses. Another solution is to add *pseudo-threads* or *green-threads*, through the `eventlet` and `gevent` libraries, to the Gunicorn server as worker classes. Both libraries use asynchronous utilities and `greenlet` threads to interface and execute the standard Flask components, especially I/O transactions, for more efficiency. They use the *monkey-patching* mechanism to replace the standard or blocking components with their asynchronous counterparts.

To deploy our application to Gunicorn with the `eventlet` library, install the `greenlet` module first using the following `pip` command, followed by `eventlet`:

```
pip install greenlet eventlet
```

For `psycopg2` or database-related monkey-patching, install the `psycogreen` module with the following `pip` command:

```
pip install psycogreen
```

Then, apply monkey-patching for Peewee and `psycopg2` transactions by calling the `patch_psycopg()` function of the `psycogreen.eventlet` module in the uppermost portion of the `main.py` file before calling the `create_app()` method. The following snippet shows the portion of the `main.py` file with the `psycogreen` setup:

```
import psycogreen.eventlet
psycogreen.eventlet.patch_psycopg()

from app import create_app
from app.models.config import database

app = create_app('../config_dev.toml')
... ... ... ... ... ...
```

The `psycogreen` module provides a blocking interface or wrapper for `psycopg2` transactions to interact with coroutines or asynchronous components of the `eventlet` worker without altering the standard Peewee codes.

To deploy our *Online Grocery* application (`ch11-guni-eventlet`) to the Gunicorn server that uses 1 `eventlet` worker with 2 threads, run the following command:

```
gunicorn --bind 127.0.0.1:8000 main:app --workers 1 --worker-
class  eventlet --threads 2
```

Figure 11.4 shows the server log after running the Gunicorn server:

```
(ch11-guni-eventlet-env) sjctrags@DESKTOP-56HNGC9:~/ch11-guni-eventlet$ gunicorn --bind 127.0.0.1:8000 main:app
--workers 1 --worker-class  eventlet --threads 2
[2024-04-01 18:08:55 +0800] [26588] [INFO] Starting gunicorn 21.2.0
[2024-04-01 18:08:55 +0800] [26588] [INFO] Listening at: http://127.0.0.1:8000 (26588)
[2024-04-01 18:08:55 +0800] [26588] [INFO] Using worker: eventlet
[2024-04-01 18:08:55 +0800] [26589] [INFO] Booting worker with pid: 26589
```

Figure 11.4 – Server log after starting the Gunicorn server using the eventlet worker

The log depicts that the worker that was used by the server is an `eventlet` worker type.

The `eventlet` library provides concurrent utilities that run standard or non-async Flask components asynchronously using task switching, a shift from sync to async tasks internally without explicitly programming it.

Aside from `eventlet`, `gevent` can also manage concurrent requests from I/O-bound tasks of the applications. Like `eventlet`, `gevent` is a coroutine-based library but relies more on its stack of `greenlet` objects and their event loops. The `gevent` library's `greenlet` is a lightweight and powerful thread that executes in a cooperative scheduling fashion. To operate a `gevent` worker in the Gunicorn server, install the `greenlet`, `eventlet`, and `gevent` modules using the following `pip` command:

```
pip install greenlet eventlet gevent
```

Also, install `psycogreen` to monkey-patch the database-related transactions of the application using its `gevent patch_psycopg()`. The following snippet shows a portion of the `main.py` file of the `ch11-guni-gevent` project, a version of our *Online Grocery* application that needs to run on Gunicorn with `gevent` workers:

```
import gevent.monkey
gevent.monkey.patch_all()

import psycogreen.gevent
psycogreen.gevent.patch_psycopg()
import gevent

from app import create_app
... ... ... ... ... ...
app = create_app('../config_dev.toml')
... ... ... ... ... ...
```

In `gevent`, the main module must call its `patch_all()` method from the `gevent.monkey` module, above anything else, to explicitly interface all the events at runtime to run asynchronously like coroutines. Afterward, it needs to call the `psycogreen` module's `patch_psycopg()`, but this time under the `gevent` sub-module.

To start the Gunicorn server using the 2 `gevent` workers with 2 thread utilization each, run the following command:

```
gunicorn --bind 127.0.0.1:8000 main:app --workers 2 --worker-class
gevent --threads 2
```

Figure 11.5 shows the server log after starting up the Gunicorn server:

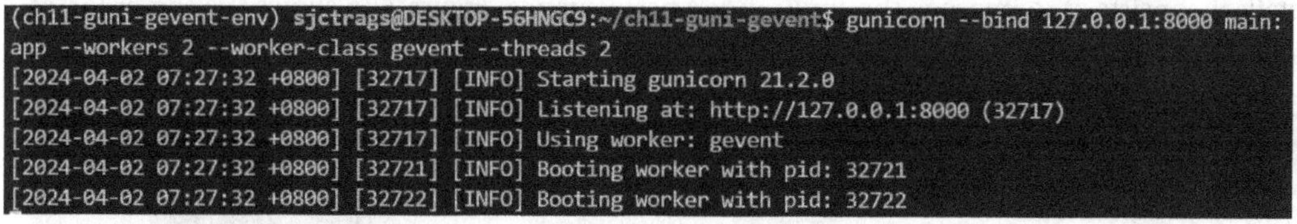

```
(ch11-guni-gevent-env) sjctrags@DESKTOP-56HNGC9:~/ch11-guni-gevent$ gunicorn --bind 127.0.0.1:8000 main:
app --workers 2 --worker-class gevent --threads 2
[2024-04-02 07:27:32 +0800] [32717] [INFO] Starting gunicorn 21.2.0
[2024-04-02 07:27:32 +0800] [32717] [INFO] Listening at: http://127.0.0.1:8000 (32717)
[2024-04-02 07:27:32 +0800] [32717] [INFO] Using worker: gevent
[2024-04-02 07:27:32 +0800] [32721] [INFO] Booting worker with pid: 32721
[2024-04-02 07:27:32 +0800] [32722] [INFO] Booting worker with pid: 32722
```

Figure 11.5 – Server log after starting the Gunicorn server using the gevent workers

The worker used by the Gunicorn is now a `gevent` worker, as depicted in the preceding server log.

Now, let's use uWSGI as our production application server.

Using uWSGI

uWSGI is a highly configurable, fast, and flexible server that can run any WSGI-based application. This server container can only operate in a POSIX-based operating system such as Gunicorn. To utilize uWSGI, install the `pyuwsgi` module using the following `pip` command:

```
pip install pyuwsgi
```

uWSGI has several required and optional setting options. One is the `-w` setting, which requires the WSGI module that the server needs to run. The `-p` setting indicates the number of workers or processes that can manage HTTP requests. The `--http` setting denotes the address and the port the server will be listening to. The `--enable-threads` setting allows the server to utilize Python threads for background processes.

To deploy our *Online Grocery* application (`ch11-uwsgi`) to a uWSGI server with 4 workers and background Python threads, run the following command:

```
uwsgi --http 127.0.0.1:8000 --master -p 4 -w main:app --enable-threads
```

Here, `--master` is an optional setting that allows the master process and its workers to shut down and restart gracefully.

Unlike Gunicorn, uWSGI generates a long server log mentioning the several manageable configuration details it consists of to improve the application's performance. *Figure 11.6* shows the server log of uWSGI after its startup:

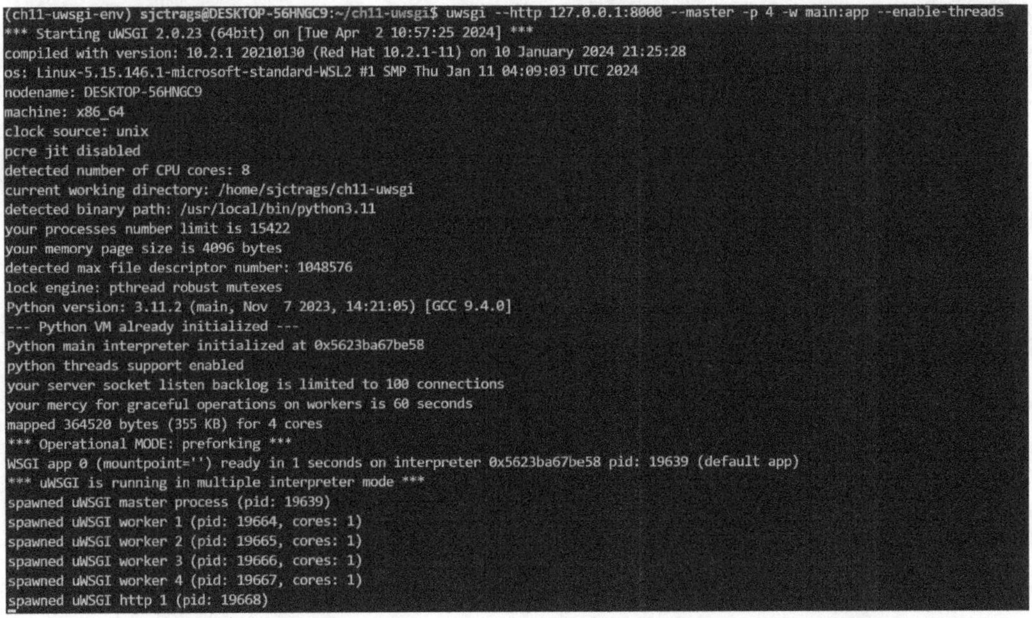

```
(ch11-uwsgi-env) sjctrags@DESKTOP-56HNGC9:~/ch11-uwsgi$ uwsgi --http 127.0.0.1:8000 --master -p 4 -w main:app --enable-threads
*** Starting uWSGI 2.0.23 (64bit) on [Tue Apr  2 10:57:25 2024] ***
compiled with version: 10.2.1 20210130 (Red Hat 10.2.1-11) on 10 January 2024 21:25:28
os: Linux-5.15.146.1-microsoft-standard-WSL2 #1 SMP Thu Jan 11 04:09:03 UTC 2024
nodename: DESKTOP-56HNGC9
machine: x86_64
clock source: unix
pcre jit disabled
detected number of CPU cores: 8
current working directory: /home/sjctrags/ch11-uwsgi
detected binary path: /usr/local/bin/python3.11
your processes number limit is 15422
your memory page size is 4096 bytes
detected max file descriptor number: 1048576
lock engine: pthread robust mutexes
Python version: 3.11.2 (main, Nov  7 2023, 14:21:05) [GCC 9.4.0]
--- Python VM already initialized ---
Python main interpreter initialized at 0x5623ba67be58
python threads support enabled
your server socket listen backlog is limited to 100 connections
your mercy for graceful operations on workers is 60 seconds
mapped 364520 bytes (355 KB) for 4 cores
*** Operational MODE: preforking ***
WSGI app 0 (mountpoint='') ready in 1 seconds on interpreter 0x5623ba67be58 pid: 19639 (default app)
*** uWSGI is running in multiple interpreter mode ***
spawned uWSGI master process (pid: 19639)
spawned uWSGI worker 1 (pid: 19664, cores: 1)
spawned uWSGI worker 2 (pid: 19665, cores: 1)
spawned uWSGI worker 3 (pid: 19666, cores: 1)
spawned uWSGI worker 4 (pid: 19667, cores: 1)
spawned uWSGI http 1 (pid: 19668)
```

Figure 11.6 – Server log after starting the uWSGI server with 4 workers

Shutting down the uWSGI server with the `--master` setting allows us to send the master process and its workers the `SIGTERM` signal to impose graceful shutdown, restart, or reload, which is better than the abrupt kill process. *Figure 11.7* shows the advantage of having the `--master` setting in the command:

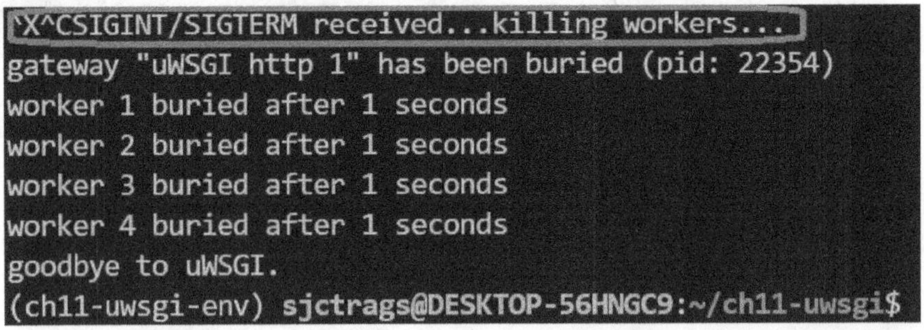

```
^X^CSIGINT/SIGTERM received...killing workers...
gateway "uWSGI http 1" has been buried (pid: 22354)
worker 1 buried after 1 seconds
worker 2 buried after 1 seconds
worker 3 buried after 1 seconds
worker 4 buried after 1 seconds
goodbye to uWSGI.
(ch11-uwsgi-env) sjctrags@DESKTOP-56HNGC9:~/ch11-uwsgi$
```

Figure 11.7 – Server log after shutting down the uWSGI server with the --master setting

Managing uWSGI is complex compared to the easy-to-configure Gunicorn. So far, Gunicorn is still the recommended server to use when deploying standard Flask applications.

Now, let's deploy *Flask[async]* to an ASGI server called *Uvicorn*.

Deploying the application to Uvicorn

Uvicorn is a popular ASGI-based HTTP server that's used by the Starlette and FastAPI frameworks. But Uvicorn remains an easy-to-use ASGI development server. It is still ideal to deploy **Flask[async]** or **FastAPI** applications to the production server using Gunicorn with `uvicorn.workers.UvicornWorker` as its HTTP server.

Even though Gunicorn is a WSGI-based server, it can support running Flask applications in standard and async mode through its `--worker-class` setting. For Flask[async] applications, Gunicorn can utilize the `aiohttp` or `uvicorn` worker class types.

Our async *Online Grocery* application (`ch11-async`) uses Gunicorn with a `uvicorn` worker as its deployment platform. Before applying the worker type, install the `uvicorn` module first by running the following `pip` command:

```
pip install uvicorn
```

Then, import `WsgiToAsgi` from the `uvicorn` module's `asgiref.wsgi` module to wrap the Flask app instance. The following snippet shows how to transform a WSGI application into an ASGI type:

```
from asgiref.wsgi import WsgiToAsgi
from app import create_app

app = create_app('../config_dev.toml')

asgi_app = WsgiToAsgi(app)
```

The Gunicorn server will run `asgi_app` instead of the original Flask app. To start Gunicorn using two Uvicorn workers with two threads each, run the following command:

```
gunicorn main:asgi_app --bind 0.0.0.0:8000 --workers 2 --worker-class
uvicorn.workers.UvicornWorker --threads 2
```

Here, `UvicornWorker`, a Gunicorn-compatible worker class from the `uvicorn` library, provides an interface to an ASGI-based application so that Gunicorn can communicate with all the HTTP requests from the coroutines of the applications and eventually handle those requests.

Figure 11.8 shows the server log after running the Gunicorn server:

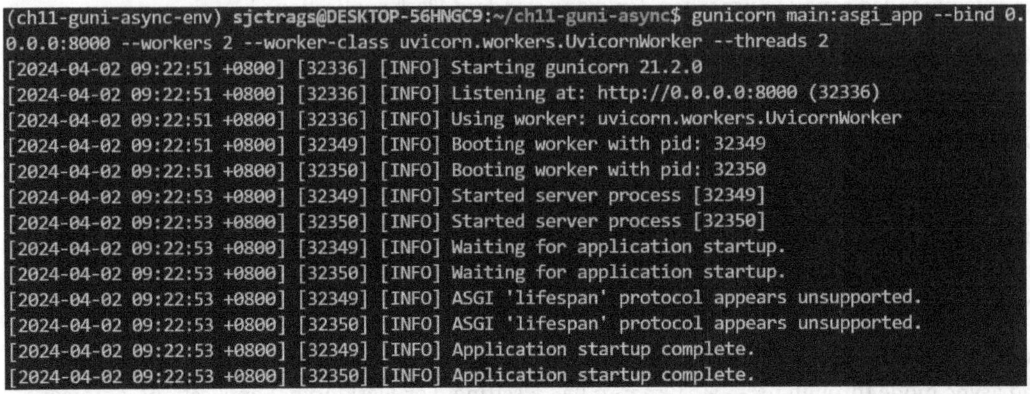

```
(ch11-guni-async-env) sjctrags@DESKTOP-56HNGC9:~/ch11-guni-async$ gunicorn main:asgi_app --bind 0.
0.0.0:8000 --workers 2 --worker-class uvicorn.workers.UvicornWorker --threads 2
[2024-04-02 09:22:51 +0800] [32336] [INFO] Starting gunicorn 21.2.0
[2024-04-02 09:22:51 +0800] [32336] [INFO] Listening at: http://0.0.0.0:8000 (32336)
[2024-04-02 09:22:51 +0800] [32336] [INFO] Using worker: uvicorn.workers.UvicornWorker
[2024-04-02 09:22:51 +0800] [32349] [INFO] Booting worker with pid: 32349
[2024-04-02 09:22:51 +0800] [32350] [INFO] Booting worker with pid: 32350
[2024-04-02 09:22:53 +0800] [32349] [INFO] Started server process [32349]
[2024-04-02 09:22:53 +0800] [32350] [INFO] Started server process [32350]
[2024-04-02 09:22:53 +0800] [32349] [INFO] Waiting for application startup.
[2024-04-02 09:22:53 +0800] [32350] [INFO] Waiting for application startup.
[2024-04-02 09:22:53 +0800] [32349] [INFO] ASGI 'lifespan' protocol appears unsupported.
[2024-04-02 09:22:53 +0800] [32350] [INFO] ASGI 'lifespan' protocol appears unsupported.
[2024-04-02 09:22:53 +0800] [32349] [INFO] Application startup complete.
[2024-04-02 09:22:53 +0800] [32350] [INFO] Application startup complete.
```

Figure 11.8 – Server log after starting the Gunicorn server using UvicornWorker

The server log depicts the use of `uvicorn.workers.UvicornWorker` as the Gunicorn worker, and it also shows the *"ASGI 'lifespan' protocol appears unsupported."* log message, which means Flask does not yet support ASGI with the lifespan protocol used to manage server startup and shutdown.

The Apache HTTP Server, a popular production server for most PHP applications, can also host and run standard Flask applications. So, let's explore the process of migrating our applications to the *Apache HTTP Server*.

Deploying the application on the Apache HTTP Server

Apache HTTP Server is an open source server under the Apache projects that can run on Windows and UNIX-based platforms to provide an efficient, simple, and flexible HTTP server for various applications.

Before anything else, download the latest server from `https://httpd.apache.org/download.cgi` and unzip the file to the production server's installation directory. Then, download the latest *Microsoft Visual C++ Redistributable* from `https://learn.microsoft.com/en-us/cpp/windows/latest-supported-vc-redist`, install it, and run the server through the `httpd.exe` file of its `/bin` folder.

After the installation, follow these steps to deploy our application to the Apache HTTP Server:

1. Build your Flask application, as we did with our *Online Grocery* application, run it using the built-in WSGI server, and refine the components using `pytest` testing.

2. Next, install the `mod_wsgi` module, which enables the Apache HTTP Server's support to run WSGI applications. Install the module using the following `pip` command:

    ```
    pip install mod_wsgi
    ```

3. If the installation encounters an error similar to what's shown in the error log in *Figure 11.9*, run the `set` command to assign the **Apache HTTP Server's installation directory** to the `MOD_WSGI_APACHE_ROOTDIR` environment variable:

    ```
    set "MOD_WSGI_APACHE_ROOTDIR= C:/.../Server/Apache24"
    ```

4. Apply *forward slashes* (/) to create the directory path. Afterward, re-install the `mod_wsgi` module:

```
Traceback (most recent call last):
  File "<string>", line 2, in <module>
  File "<pip-setuptools-caller>", line 34, in <module>
  File "C:\Users\alibatasys\AppData\Local\Temp\pip-install-u13ke1b1\mod-wsgi_0befb6e70dbc4f6
1bca16219734935fa\setup.py", line 81, in <module>
    raise RuntimeError('No Apache installation can be found. Set the '
  RuntimeError: No Apache installation can be found. Set the MOD_WSGI_APACHE_ROOTDIR environme
nt to its location.
  [end of output]
```

Figure 11.9 – No MOD_WSGI_APACHE_ROOTDIR error

5. Again, if the re-installation of `mod_wsgi` gives another error stating the required **Microsoft Visual C++ tool**, do the following:

    I.   Download `VisualStudioSetup.exe` from `https://visualstudio.microsoft.com/downloads`.

    II. Run the `VisualStudioSetup.exe` file; a menu dashboard will appear, as shown in *Figure 11.10*.

III.   Click the **Desktop Development with C++** menu option to show the installation details on the right-hand side of the dashboard:

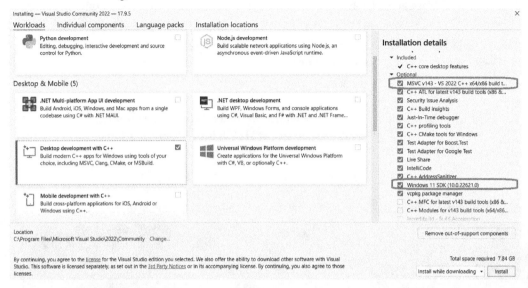

Figure 11.10 – Microsoft Visual Studio Library dashboard

This installation is different from the previous Microsoft Visual C++ Redistributable installation procedure.

6. Now, select **C++ build tools** from the left-hand side menu and choose the **Windows 10 SDK** or **Windows 11 SDK** and the **MSVC 142 ***** options from the list of checkboxes. It is optional to select the other libraries if they are crucial to the mod_wsgi installation.

7. After choosing the necessary components, click the **Install** button at the bottom right of the dashboard.

8. After installing **Microsoft Visual C++ tool**, run pip install mod_wsgi once more. This time, the mod_wsgi installation must proceed successfully.

9. The mod_wsgi module needs a configuration file inside the project that the Apache HTTP Server needs to load during startup. This file should be in a separate folder, say wsgi, and must be in the main project folder. In our ch11-apache project, the configuration file is conf. wsgi and has been placed in the wsgi folder. Be sure to add the __init__.py file to this folder too. The following is the content of conf.wsgi:

```
import sys
sys.path.insert(0, 'C:/Alibata/Training/ Source/flask/mastering/
ch11-apache')
from main import app as application
```

The conf.wsgi configuration file provides the Apache HTTP Server a channel to access the Flask app instance for deployment and execution through the mod_wsgi module.

10. Run the mod_wsgi-express module-config command to generate the LoadModule configuration statements that the Apache HTTP Server needs to integrate with the project directory. The following are the LoadModule snippets that have been generated for our *Online Grocery* application:

```
LoadFile "C:/Alibata/Development/Language/ Python/Python311/
python311.dll"
LoadModule wsgi_module "C:/Alibata/Training/Source/ flask/
mastering/ch11-apache-env/Lib/site-packages/mod_wsgi/server/mod_
wsgi.cp311-win_amd64.pyd"
WSGIPythonHome "C:/Alibata/Training/Source/ flask/mastering/
ch11-apache-env"
```

11. Place these LoadModule configuration statements in the Apache HTTP Server's /conf/ http.conf file, specifically anywhere in the LoadModule area under the **Dynamic Shared Object (DSO) Support** segment.

12. At the end of the /conf/http.conf file, import the custom VirtualHost configuration file of the project. The following is a sample import statement for our *Online Grocery* application:

```
Include conf/ch11_apache.conf
```

13. Now, create the VirtualHost configuration file referenced in *Step 10*. The following is a sample configuration setup in our ch11_apache.conf file:

```
<VirtualHost *:8080>
    ServerName localhost
    WSGIScriptAlias / C:/Alibata/Training/Source/ flask/
mastering/ch11-apache/wsgi/conf.wsgi
        <Directory C:/Alibata/Training/Source/ flask/mastering/ch11-
apache>
            Require all granted
        </Directory>
</VirtualHost>
```

The VirtualHost configuration defines the host address and port that the server will listen to so that it can run our application. Its WSGIScriptAlias directive gives reference to the mod_wsgi configuration file of the application. Moreover, the configuration permits the server to access all files in the ch11-apache project.

14. Now, open a terminal and run or restart the server through httpd.exe. Access all the APIs using pytest or API clients.

Choosing the Apache HTTP Server as the production server is a common approach in many deployment plans for Flask projects involving the standalone server platform. Although the deployment process is tricky and lengthy, the server's fast and stable performance, once configured and managed well, makes it a better choice for setting up a significantly effective production environment for Flask applications.

There is another way of deploying Flask applications that involves fewer tweaks and configurations but provides an enterprise-grade production setup: **the containerized deployment approach**. Let's discuss how to deploy the application to *Docker* containers.

Deploying the application on Docker

Docker is a powerful tool for deploying and running applications using software units instead of hardware setups. Each independent, lightweight, standalone, and executable unit, called a **container**, must contain all the files of the applications that it needs to run. Docker is the core container engine that manages all the containers and packages applications in their appropriate containers. To download Docker, download the **Docker Desktop** installer that's appropriate for your system from `https://docs.docker.com/engine/install/`. Be sure to enable the Window's **Hyper-V service** before installing Docker. Use your Docker credentials to log in to the application. *Figure 11.11* shows a sample account dashboard of the Docker Desktop application:

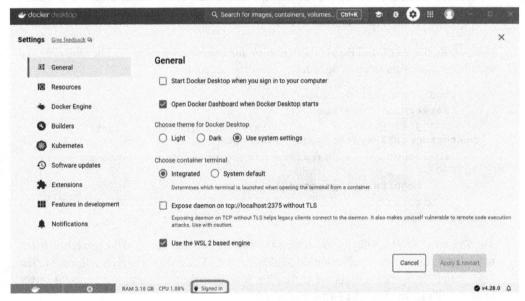

Figure 11.11 – A Desktop Docker profile

Docker requires some rules when deploying applications to its containers. The first requirement is to create a Dockerfile inside the project's *main* directory, on the same level as the `main.py` and `.toml` configuration files. The following is the content of the `ch11-asgi` file's Dockerfile:

```
FROM python:3.11
WORKDIR /usr/src/ch11-asgi
RUN pip install --upgrade pip
COPY ./requirements.txt /usr/src/ch11-asgi/requirements.txt
RUN pip install -r requirements.txt
COPY . /usr/src/ch11-asgi


EXPOSE 8000
CMD ["gunicorn", "main:asgi_app", "--bind", "0.0.0.0:8000", "--worker-
class", "uvicorn.workers.UvicornWorker", "--threads", "2"]
```

A **Dockerfile** contains a series of instructions made by Docker commands that the engine will use to assemble an image. A **Docker image** is a software template containing the needed project files, folders, Python modules, server details, and commands to start the Flask server. Docker will run the image to generate a running image instance called a container.

The first line of our Dockerfile is the FROM instruction, which creates a stage or a copy of the base image from the Docker repository. Here are the guidelines to follow when choosing the base image:

- Ensure it is complete with libraries, tools, filesystem structure, and network structures so that the container will be stable.

- Ensure it can be updated in terms of operating system plugins and libraries.

- Ensure it's equipped with up-to-date and stable Python compilers and core libraries.

- Ensure it's loaded with extensions and additional plugins for additional complex integrations.

- Ensure it has a smaller file size.

Choosing the right base image is crucial for the application to avoid problems during production phases.

The next instruction is the WORKDIR command, which creates and sets the new application's working directory. The first RUN command updates the container's `pip` command, which will install all the libraries from the `requirements.txt` file copied by the COPY command from our local project folder. After installing the modules in the container, the next instruction is to COPY all the project files from the local folder to the container.

The EXPOSE command defines the port the application will listen on. The CMD command, on the other hand, tells Docker how to start the Gunicorn server with `UvicornWorker` when the container starts.

After composing the Dockerfile, open a terminal to run the `docker login` CLI command and input your credentials. The `docker login` command enables access to your Docker repository using other Docker's CLI commands, such as `docker run` to execute the instructions from the Dockerfile. By the way, aside from our Flask[async] application, there is a need to pull an image to generate a container for the PostgreSQL database of our application. Conventionally, to connect these containers, such as our PostgreSQL and Redis containers, to the Python container with the Flask application, Docker networking, through running the `docker network` command, creates the network connections that will link these containers to establish the needed connectivity. But this becomes complex if there are more containers to attach. As a replacement to save time and effort, *Docker Compose* can establish all these step-by-step networking procedures by only running the `docker-compose` command. There is no need to install Docker Compose since it is part of the bundle that's installed by the Docker Desktop installer. Docker Compose uses Docker Engine, so installing the engine also includes Compose. To start Docker Compose, just run `docker login` and enter a valid Docker account.

Using Docker Compose

Docker Compose is an open source orchestration tool that manages and deploys multiple containers to one server host using a single service implemented in a series of rules in its configuration file, `docker-compose.yaml`. The following is the configuration file that's used by our `ch11-asgi-deployment` project:

```
version: '3.0'
services:
  api:
    build: ./ch11-asgi
    volumes:
      - ./ch11-asgi/:/usr/src/ch11-asgi/
    ports:
      - 8000:8000
    depends_on:
      - postgres
  postgres:
    image: «bitnami/postgresql:latest»
    ports:
      - 5432:5432
    env_file:
      - db.env # configure postgres
    volumes:
      - database-data:/var/lib/postgresql/data/
  volumes:
    database-data:
```

The `version` directive indicates the Compose syntax version the configuration will use in the Compose instructions. Our Compose configuration file uses version `3.0`, which is the latest at the time of writing this book. Lower versions mean deprecated keywords and commands.

Now, the `services` directive defines all the containers that Compose will create and run. Ours include the *Online Grocery* application (`api`) and the PostgreSQL database platform (`postgres`). Here, `api` is the name of the service for our application. It contains the following required sub-directives:

- `build`: Points to the location of the local project folder containing the Dockerfile.

- `ports`: Maps the container's ports to the host's ports, either TCP or UDP.

- `volumes`: Attaches the local project files to the specified directory of the container, which spares the image from rebuilding if there are changes in the project files.

- `depends_on`: Mentions the service name considered as one of the container's dependencies.

Another service is `postgres`, which provides the database platform for the `api` service, thus the dependency between the two services. Instead of using the `build` directive, its `image` directive will pull the latest `bitnami/postgresql` image to create a container for the PostgreSQL platform with an empty database schema. Its `ports` directive indicates that the container will use port `5432` to listen for database connectivity. The database credentials are in the `db.env` file indicated by the `env_file` directive. The following snippet shows the content of the `db.env` file:

```
POSTGRES_USER=postgres
POSTGRES_PASSWORD=admin2255
POSTGRES_DB=ogs
```

The `volumes` directive for the `postgres` service is essential for data persistence because its absence in the configuration means data cleanup after the container restarts.

After finalizing the `docker-compose.yaml` file, run the `docker-compose --build` command to build or rebuild the services, then once again after the `docker-compose up` command to create and run the containers. *Figure 11.12* shows the command logs after running the `docker-compose up --build` commands:

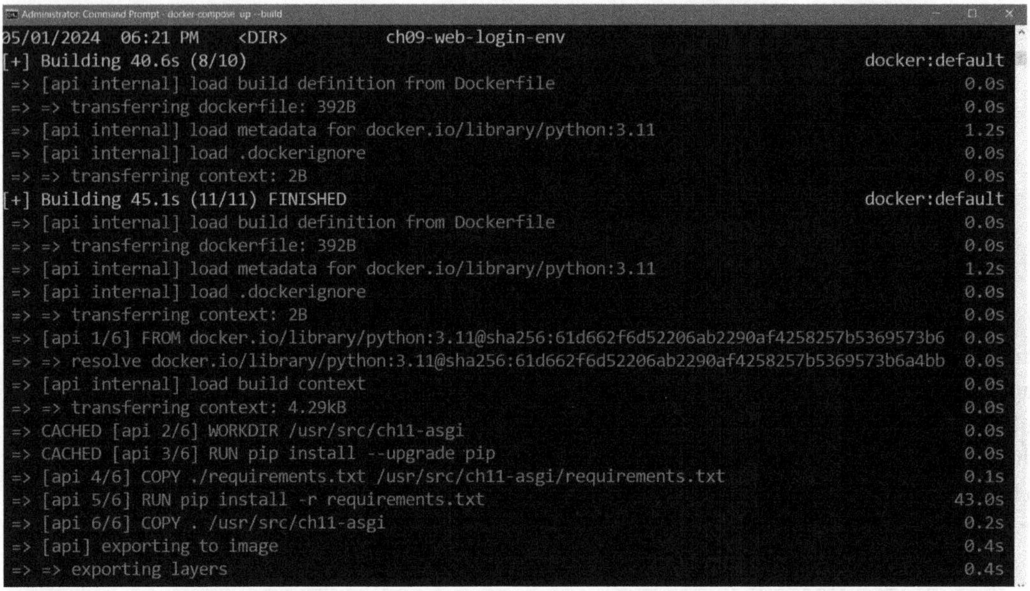

```
Administrator: Command Prompt - docker-compose up --build                                                    — □ ×
05/01/2024  06:21 PM    <DIR>              ch09-web-login-env
[+] Building 40.6s (8/10)                                                           docker:default
=> [api internal] load build definition from Dockerfile                                     0.0s
=> => transferring dockerfile: 392B                                                         0.0s
=> [api internal] load metadata for docker.io/library/python:3.11                           1.2s
=> [api internal] load .dockerignore                                                        0.0s
=> => transferring context: 2B                                                              0.0s
[+] Building 45.1s (11/11) FINISHED                                                  docker:default
=> [api internal] load build definition from Dockerfile                                     0.0s
=> => transferring dockerfile: 392B                                                         0.0s
=> [api internal] load metadata for docker.io/library/python:3.11                           1.2s
=> [api internal] load .dockerignore                                                        0.0s
=> => transferring context: 2B                                                              0.0s
=> [api 1/6] FROM docker.io/library/python:3.11@sha256:61d662f6d52206ab2290af4258257b5369573b6  0.0s
=> => resolve docker.io/library/python:3.11@sha256:61d662f6d52206ab2290af4258257b5369573b6a4bb  0.0s
=> [api internal] load build context                                                        0.0s
=> => transferring context: 4.29kB                                                          0.0s
=> CACHED [api 2/6] WORKDIR /usr/src/ch11-asgi                                               0.0s
=> CACHED [api 3/6] RUN pip install --upgrade pip                                            0.0s
=> [api 4/6] COPY ./requirements.txt /usr/src/ch11-asgi/requirements.txt                     0.1s
=> [api 5/6] RUN pip install -r requirements.txt                                            43.0s
=> [api 6/6] COPY . /usr/src/ch11-asgi                                                       0.2s
=> [api] exporting to image                                                                  0.4s
=> => exporting layers                                                                       0.4s
```

Figure 11.12 – Logs when running the docker-compose up --build command

The Docker Desktop dashboard, on the other hand, will display the following container structure in *Figure 11.13* after successfully running the generated containers:

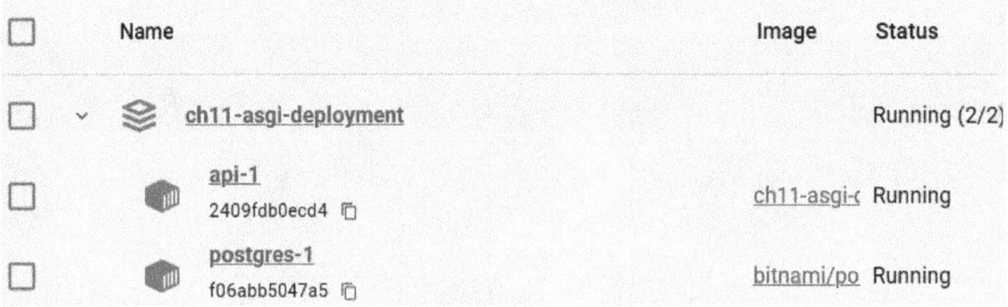

	Name	Image	Status
☐	⌄ ⧉ ch11-asgi-deployment		Running (2/2)
☐	🗔 api-1 2409fdb0ecd4 ⧉	ch11-asgi-c	Running
☐	🗔 postgres-1 f06abb5047a5 ⧉	bitnami/po	Running

Figure 11.13 – Docker Desktop showing ch11-asgi and the PostgreSQL containers

Here, `ch11-asgi-deployment` in the given container structure is the name of the deployment folder containing the `db.env` and `docker-compose.yaml` files, and the directory where the terminal invocation of the `docker-compose` commands happened. Inside the Compose container structure are the two containers that were generated by the services. Clicking the `api-1` container will provide us with the Gunicorn server logs presented in *Figure 11.14*:

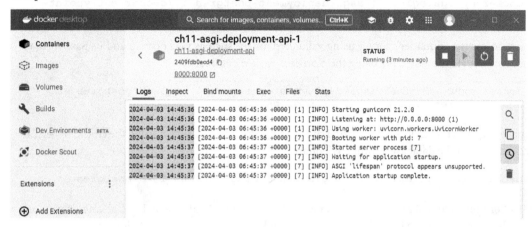

Figure 11.14 – The Gunicorn server log from ch11-asgi app in the api-1 container

On the other hand, clicking the `postgres-1` container will show the logs shown in *Figure 11.15*:

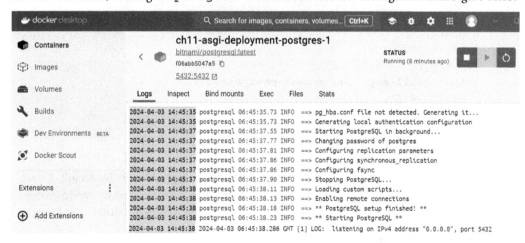

Figure 11.15 – The PostgreSQL server log in the postgres-1 container

Now, the database schema in the postgres-1 container is empty. To populate the database with the tables and data from the local PostgreSQL server, run pg_dump to create a .sql dump file. Then, in the directory location of the .sql backup file, run the following docker copy command to copy the backup file, say ogs.sql, to the entrypoint directory of the container:

```
docker cp ogs.sql ch11-asgi-deployment-postgres-1:/docker-entrypoint-
initdb.d/ogs.sql
```

Then, access the container's server using valid credentials, such as postgres and its password, to spool or execute the .sql file using the docker exec command:

```
docker exec -it ch11-asgi-deployment-postgres-1 psql -U postgres -d
ogs -f docker-entrypoint-initdb.d/ogs.sql
```

Finally, log in to the ch11-asgi-deployment-postgres-1 server using the docker exec command with the database admin credentials:

```
docker exec -it ch11-asgi-deployment-postgres-1 psql -U postgres
```

Also, don't forget to replace the host parameter of the PooledPostgresqlDatabase driver class with the container's name instead of localhost and its port to 5432. The following snippet shows the changes in the driver class configuration that can be found in the app/models/config module:

```
from peewee_async import PooledPostgresqlDatabase

database = PooledPostgresqlDatabase(
        'ogs',
        user='postgres',
        password='admin2255',
        host='ch11-asgi-deployment-postgres-1',
        port='5432',
        max_connections = 3,
        connect_timeout = 3
    )
```

Now, problems arise when one or some of the containers fail during production. By default, it does support automatic container restart when there are runtime errors in the application or some memory-related issues. Moreover, Compose cannot perform container orchestration in a distributed setup.

Another powerful approach to deploying applications to different hosts rather than to a single server is through *Kubernetes*. In the next section, we'll use Kubernetes to deploy our ch11-asgi application with Gunicorn as the server.

Deploying the application on Kubernetes

Like Compose, **Kubernetes** or **K8** manages multiple containers with or without dependencies on each other. Kubernetes can utilize volume storage for data persistence and has CLI commands to manage the life cycle of the containers. The only difference is that Kubernetes can run containers in a distributed setup and uses Pods to manage its containers.

Among the many ways to install Kubernetes, this chapter utilizes the **Kubernetes** feature in Docker Desktop's **Settings**, as shown in *Figure 11.16*:

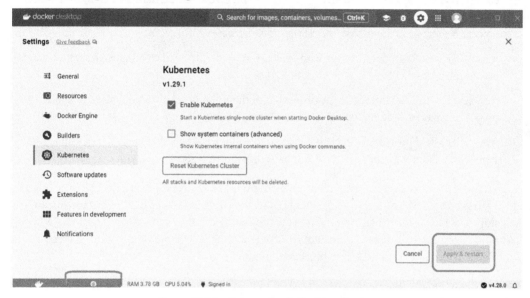

Figure 11.16 – Kubernetes in Desktop Docker

Check the **Enable Kubernetes** checkbox from the **Settings** area and click the **Apply & restart** button in the lower right portion of the dashboard. It will take a while for Kubernetes to appear running or *green* in the lower left corner of the dashboard, depending on the number of containers running on Docker Engine.

When the Kubernetes engine fails, click the **Reset Kubernetes Cluster** button to remove all containers and files of the Kubernetes stack. Additionally, for Windows users, delete the Docker fragment files in the C:\Users\alibatasys\AppData\Local\Temp folder before restarting Docker Desktop.

Kubernetes uses YAML files to define and create Kubernetes objects, such as **Deployment**, **Pods**, **Services**, and **PersistentVolume**, all of which are required to establish some container rules, manage the host resources, and build containerized applications. An object definition in YAML format always consists of the following manifest fields:

- `apiVersion`: The field that indicates the appropriate and stable Kubernetes API for a Kubernetes object creation. This field must always appear first in the file. Kubernetes has several APIs, such as `batch/v1`, `apps/v1`, `v1`, and `rbac.authorization.k8s.io/v1`, but the more common is `v1` for `PersistentVolume`, `PersistentVolumeClaims`, `Service`, `Secret`, and `Pod` object creation and `apps/v1` for `Deployment` and `ReplicaSets` objects. So far, `v1` is the first stable release of Kubernetes API.

- `kind`: The field that identifies the Kubernetes object the file needs to create. Here, `kind` can be `Secret`, `Service`, `Deployment`, `Role`, or `Pod`.

- `metadata`: This field specifies the properties of the Kubernetes object defined in the file. The properties may include the *name*, *labels*, and *namespace*.

- `spec`: This field provides the specification of the object in key-value format. The same object type with a different `apiVersion` can have different specification details.

In this chapter, the Kubernetes deployment involves pulling our `ch11-asgi` file's Docker image and the latest `bitnami/postgresql` image from the Docker registry hub. But before creating the deployment file, our first manifest focuses on containing the `Secret` object definition, which aims to store and secure the database PostgreSQL credentials. The following is our `kub-secrets.yaml` file, which contains our `Secret` object definition:

```
apiVersion: v1
kind: Secret
metadata:
  name: postgres-credentials
data:
  # replace this with your base4-encoded username
  user: cG9zdGdyZXM=
  # replace this with your base4-encoded password
  password: YWRtaW4yMjU1
```

A `Secret` object contains protected data such as a password, user token, or access key. Instead of hardcoding these confidential data in the applications, it is safe to store them in Pods so that they can be accessed by other Pods in the cluster.

Our second YAML file, kub-postgresql-pv.yaml, defines the object that will create persistent storage resources for our PostgreSQL, the PersistentVolume object. Since our Kubernetes runs on a single-node server, the default storage class is hostpath. This storage will hold the data of the PostgreSQL permanently, even after the removal of our containerized application. The following kub-postgresql-pv.yaml file defines the PersistentVolume object that will manage our application's data storage:

```yaml
apiVersion: v1
kind: PersistentVolume
metadata:
    name: postgres-pv-volume
    labels:
        type: local
spec:
    storageClassName: manual
    capacity:
        storage: 5Gi
    accessModes:
        - ReadWriteOnce
    hostPath:
        path: "/mnt/data"
```

In Kubernetes, utilizing storage from the PersistentVolume object requires a PersistentVolumeClaims object. This object requests a portion of the cluster storage that Kubernetes *Pods* will use for the application's read and write. The following kub-postgresql-pvc.yaml file creates an PersistentVolumeClaims object for the deployment's storage:

```yaml
kind: PersistentVolumeClaim
apiVersion: v1
metadata:
  name: postgresql-db-claim
spec:
  accessModes:
    - ReadWriteOnce
  resources:
    requests:
        storage: 5Gi
```

The PersistentVolumeClaims and PersistentVolume objects work together to dynamically claim a new volume storage for the bitnami/postgresql container. The *manual* StorageClass type indicates that there is a binding from PersistentVolumeClaims to PersistentVolume for the request of the storage.

After creating the configuration files for the `Secret`, `PersistentVolume`, and `PersistentVolumeClaims` objects, the next crucial step is to create the deployment configuration files that will connect the `ch11-asgi` and `bitnami/postgresql` Docker images with database configuration details from the `Secret` object, utilize the volume claims for PostgreSQL data persistency, and deploy and run them all together with Kubernetes Services and Pods. Here, `Deployment` manages a set of Pods to run an application workload. A Pod, as Kubernetes' fundamental building block, represents a single running process within the Kubernetes cluster. The following `kub-postgresql-deployment.yaml` file tells Kubernetes to manage an instance that will hold the PostgreSQL container:

```
apiVersion: apps/v1
kind: Deployment
metadata:
  name: ch11-postgresql
```

For this deployment configuration, `v1` or `apps/v1` is the proper choice for the `apiVersion` metadata. The `kub-postgresql-deployment.yaml` file is a `Deployment` type of Kubernetes document, as indicated in the `kind` metadata, which will generate a container named `ch11-postgresql`:

```
spec:
  replicas: 1
  selector:
    matchLabels:
      app: ch11-postgresql
  template:
    metadata:
      labels:
        app: ch11-postgresql
    spec:
      terminationGracePeriodSeconds: 180
      containers:
        - name: ch11-postgresql
          image: bitnami/postgresql:latest
          imagePullPolicy: IfNotPresent
          ports:
            - name: tcp-5432
              containerPort: 5432
```

From the overall state indicated in the `spec` metadata, the deployment will create *1 replica* in a Kubernetes pod, with `ch11-postgresql` as its label, to run the PostgreSQL server. Moreover, the deployment will pull the `bitnami/postgresql:latest` image to create the PostgreSQL container, bearing the `ch11-postgresql` label also. The configuration also includes a `terminationGracePeriodSeconds` value of `180` to shut down the database server safely:

```
env:
- name: POSTGRES_USER
  valueFrom:
    secretKeyRef:
      name: postgres-credentials
      key: user
- name: POSTGRES_PASSWORD
  valueFrom:
    secretKeyRef:
      name: postgres-credentials
      key: password
- name: POSTGRES_DB
  value: ogs
- name: PGDATA
  value: /var/lib/postgresql/data/pgdata
```

The `env` or environment variables portion provides the database credentials, `POSTGRES_USER` and `POSTGRES_DB`, to the database, which are base64-encoded values from the previously created `Secret` object, `postgres-credentials`. Note that this deployment will also auto-generate the database with the name `ogs`:

```
volumeMounts:
  - name: data-storage-volume
    mountPath: /var/lib/postgresql/data
resources:
  requests:
    cpu: "50m"
    memory: "256Mi"
  limits:
    cpu: "500m"
    memory: "256Mi"
volumes:
  - name: data-storage-volume
    persistentVolumeClaim:
      claimName: postgresql-db-claim
```

The deployment will also allow us to save all data files in the `/var/lib/postgresql/data` file of the generated container in the `ch11-postgresql` pod, as indicated in the `volumeMounts` metadata. Specifying the `volumeMounts` metadata avoids data loss when the database shuts down and makes the database and tables accessible across the network. The pod will access the volume storage created by the `postgres-pv-volume` and `postgresql-db-claim` objects.

Aside from the `Deployment` object, this document defines a `Service` type that will expose our PostgreSQL container to other Pods within the cluster at port `5432` through a *ClusterIP*:

```
---
apiVersion: v1
kind: Service
metadata:
  name: ch11-postgresql-service
  labels:
    name: ch11-postgresql
spec:
  ports:
    - port: 5432
  selector:
    app: ch11-postgresql
```

The `---` symbol is a valid separator syntax separating the `Deployment` and `Service` definitions.

Our last deployment file, `kub-app-deployment.yaml`, pulls the `ch11-asgi` Docker image and assigns the generated container to the Pods:

```
apiVersion: apps/v1
kind: Deployment
metadata:
  name: ch11-app
  labels:
    name: ch11-app
```

The `apiVersion` field of our deployment configuration file is `v1`, an appropriate Kubernetes version for deployment. In this case, our container will be labeled `ch11-app`, as indicated in the `metadata/name` configuration:

```
spec:
  replicas: 1
  selector:
    matchLabels:
      app: ch11-app
```

The spec field describes the overall state of the deployment, starting with the number of `replicas` the deployment will create, how many `containers` the Pods will run, the environment variables – namely `username`, `password`, and `SERVICE_POSTGRES_SERVICE_HOST` – that `ch11-app` will use to connect to the PostgreSQL container, and the `containerPort` variable the container will listen to:

```yaml
  template:
    metadata:
      labels:
        app: ch11-app
    spec:
      containers:
      - name: ch11-app
        image: sjctrags/ch11-app:latest
        env:
            - name: SERVICE_POSTGRES_SERVICE_HOST
              value: ch11-postgresql-service. default.svc.cluster.
local
            - name: POSTGRES_DB_USER
              valueFrom:
                secretKeyRef:
                  name: postgres-credentials
                  key: user
            - name: POSTGRES_DB_PSW
              valueFrom:
                secretKeyRef:
                  name: postgres-credentials
                  key: password
        ports:
        - containerPort: 8000
```

Also included in the YAML file is the `Service` type that will make the application to the users:

```yaml
---
apiVersion: v1
kind: Service
metadata:
  name: ch11-app-service
spec:
  type: LoadBalancer
  selector:
    app: ch11-app
```

```
ports:
  - protocol: TCP
    port: 8000
    targetPort: 8000
```

The definition links the `postgres-credentials` object to the pod's environment variables that refer to the database credentials. It also defines a *LoadBalancer* `Service` to expose our containerized Flask[async] to the HTTP client at port `8000`.

To apply these configuration files, Kubernetes has a `kubectl` client command to communicate with Kubernetes and run its APIs defined in the manifest files. Here is the order of applying the given YAML files:

1. `kubectl apply -f kub-secrets.yaml`.

2. `kubectl apply -f kub-postgresql-pv.yaml`.

3. `kubectl apply -f kub-postgresql-pvc.yaml`.

4. `kubectl apply -f kub-postgresql-deployment.yaml`.

5. `kubectl apply -f kub-app-deployment`.

To learn about the status and instances that run the applications, run `kubectl get pods`. To view the Services that have been created, run `kubectl get services`. *Figure 11.17* shows the list of Services after applying all our deployment files:

```
C:\Alibata\Training\Source\flask\mastering\ch11-asgi-dep-kub>kubectl get services
NAME                      TYPE           CLUSTER-IP      EXTERNAL-IP   PORT(S)          AGE
ch11-app-service          LoadBalancer   10.97.233.40    localhost     8000:32193/TCP   8s
ch11-postgresql-service   ClusterIP      10.109.65.234   <none>        5432/TCP         36s
kubernetes                ClusterIP      10.96.0.1       <none>        443/TCP          19d
```

Figure 11.17 – Listing all Kubernetes Services with their details

To learn all the details about the Services and Pods that have been deployed and the status of each pod, run `kubectl get all`. The result will be similar to what's shown in *Figure 11.18*:

```
C:\Alibata\Training\Source\flask\mastering\ch11-asgi-dep-kub>kubectl get all
NAME                                    READY    STATUS     RESTARTS    AGE
pod/ch11-app-76546b4555-6m22b           1/1      Running    0           84s
pod/ch11-postgresql-b7fc578f4-5d6kn     1/1      Running    0           112s


NAME                                TYPE           CLUSTER-IP      EXTERNAL-IP    PORT(S)           AGE
service/ch11-app-service            LoadBalancer   10.97.233.40    localhost      8000:32193/TCP    84s
service/ch11-postgresql-service     ClusterIP      10.109.65.234   <none>         5432/TCP          112s
service/kubernetes                  ClusterIP      10.96.0.1       <none>         443/TCP           19d


NAME                                READY    UP-TO-DATE    AVAILABLE    AGE
deployment.apps/ch11-app            1/1      1             1            84s
deployment.apps/ch11-postgresql     1/1      1             1            112s


NAME                                          DESIRED    CURRENT    READY    AGE
replicaset.apps/ch11-app-76546b4555           1          1          1        84s
replicaset.apps/ch11-postgresql-b7fc578f4     1          1          1        112s
```

Figure 11.18 – Listing all the Kubernetes cluster details

All the Pods and the containerized applications can be viewed on Docker Desktop, as shown in *Figure 11.19*:

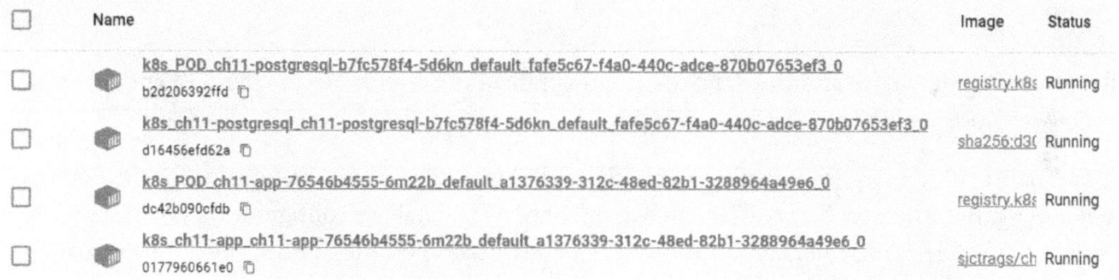

	Name		Image	Status
☐		k8s_POD_ch11-postgresql-b7fc578f4-5d6kn_default_fafe5c67-f4a0-440c-adce-870b07653ef3_0		
☐		b2d206392ffd	registry.k8s	Running
☐		k8s_ch11-postgresql_ch11-postgresql-b7fc578f4-5d6kn_default_fafe5c67-f4a0-440c-adce-870b07653ef3_0		
		d16456efd62a	sha256:d3(	Running
☐		k8s_POD_ch11-app-76546b4555-6m22b_default_a1376339-312c-48ed-82b1-3288964a49e6_0		
		dc42b090cfdb	registry.k8s	Running
☐		k8s_ch11-app_ch11-app-76546b4555-6m22b_default_a1376339-312c-48ed-82b1-3288964a49e6_0		
		0177960661e0	sjctrags/ch	Running

Figure 11.19 – Docker Desktop view of all Pods and applications

Before accessing the `ch11-asgi` container, populate the empty PostgreSQL database with the `.sql` dump file from the local database. Use the `Pod` name (for example, `ch11-postgresql-b7fc578f4-6g4nc`) of the deployed PostgreSQL container and copy the `.sql` file to the `/temp` directory of the container (for example, `ch11-postgresql-b7fc578f4-6g4nc:/temp/ogs.sql`) using the `kubectl cp` command and the pod. Be sure to run the command in the location of the `.sql` file:

```
kubectl cp ogs.sql ch11-postgresql-b7fc578f4-6g4nc:/tmp/ogs.sql
```

Run the `.sql` file in the `/temp` folder of the container using the `kubectl exec` command and the pod:

```
kubectl exec -it ch11-postgresql-b7fc578f4-6g4nc -- psql -U postgres
-d ogs -f /tmp/ogs.sql
```

Also, replace the `user`, `password`, `port`, and `host` parameters of Peewee's `Pooled PostgresqlDatabase` with the environment variables declared in the `kub-app-deployment.yaml` file. The following snippet shows the changes in the driver class configuration found in the `app/models/config` module:

```python
from peewee_async import PooledPostgresqlDatabase
import os

database = PooledPostgresqlDatabase(
        'ogs',
        user=os.environ.get('POSTGRES_DB_USER'),
        password=os.environ.get('POSTGRES_DB_PSW'),
        host=os.environ.get( 'SERVICE_POSTGRES_SERVICE_HOST'),
        port='5432',
        max_connections = 3,
        connect_timeout = 3
    )
```

After migrating the tables and the data, the client application can now access the API endpoints of our *Online Grocery* application (`ch11-asgi`).

A Kubernetes pod undergoes **Running**, **Waiting**, and **Terminated** states. The goal is for the Pods to stay *Running*. But when problems arise, such as encountering database configuration errors, binding to existing ports, lack of Kubernetes objects, lack of permission on files, and applications throwing memory and runtime errors, Pods emit `CrashLoopBackOff` and stay in **Awaiting** mode. To avoid Pods crashing, always carefully review the definitions files before applying them and monitor the logs of running Pods from time to time.

Sometimes, a Docker or Kubernetes deployment requires adding a reverse proxy server to manage all the incoming requests of the deployed applications. In the next section, we'll add the *NGINX* gateway server to our containerized `ch11-asgi` application.

Creating an API gateway using NGINX

Our deployment needs **NGINX** to manage the high traffic of incoming requests from clients, load balance the requests across the server groups, add some HTTP caches, or add security to filter suspicious access. NGINX is a stable HTTP server that can be installed on Linux-based operating systems. In this chapter, NGINX is part of our Docker deployment, which consists of our `ch11-asgi` app and PostgreSQL database platform. It will serve as the facade of the Gunicorn server running our application.

Here, `ch11-asgi-dep-nginx` is a Docker Compose folder consisting of the `ch11-asgi` project directory, which contains a Dockerfile, the `docker-compose.yaml` file, and the `nginx` folder containing a Dockerfile and our NGINX configuration settings. The following is the `nginx.conf` file that's used by Compose to set up our NGINX server:

```
server {
    listen 80;
    server_name localhost;

    location / {
        proxy_pass http://ch11-asgi-dep-nginx-api-1:8000/;
        proxy_set_header X-Forwarded-For $proxy_add_x_forwarded_for;
        proxy_set_header X-Forwarded-Proto $scheme;
        proxy_set_header X-Forwarded-Host $host;
        proxy_set_header X-Forwarded-Prefix /;
    }
}
```

The NGINX configuration depends on its installation setup, the applications that have been deployed to the servers, and the server architecture. Ours is for a reverse proxy NGINX server of our application deployed on a single server. NGINX will allow access to our application through `localhost` and port `80` instead of `http://ch11-asgi-dep-nginx-api-1:8000`, as indicated in `proxy_pass`. Since we don't have a new domain name, `localhost` will be the proxy's hostname. The de facto request headers, such as `X-Forwarded-Host`, `X-Forwarded-Proto`, `X-Forwarded-Host`, and `X-Forwarded-Prefix`, will collectively help the load balancing mechanism during NGINX's interference on a request.

When the `docker-compose` command runs the YAML file, NGINX's Dockerfile will pull the latest `nginx` image and copy the given `nginx.conf` settings to the `/etc/nginx/conf.d/` directory of its container. Then, it will instruct the container to run the NGINX server using the `nginx -g daemon off` command.

Adding NGINX makes the deployed application manageable, scalable, and maintainable. It can also centralize user request traffic in a microservice architecture, ensuring that the access reaches the expected API endpoints, containers, or sub-modules.

Summary

There are several solutions and approaches to migrating a Flask application from the development to the production stage. The most common server that's used to run Flask's WSGI applications in production is Gunicorn. uWSGI, on the other hand, can run WSGI applications in more complex and refined settings. Flask[async] applications can run on Uvicorn workers with a Gunicorn server.

For external server-based deployment, the Apache HTTP Server with Python provides a stable and reliable container for running Flask applications with the support of Python's `mod_wsgi` module.

Flask applications can also run on containers through Docker and Docker Compose to avoid the nitty gritty configuration and installations in the Apache HTTP Server. In Dockerization, what matters is the Dockerfile for a single deployment or the `docker-compose.yaml` file for multiple deployments and the combinations of Docker instructions that will contain these configuration files. For a more distributed, flexible, and complex orchestration, Kubernetes's Pods and Services can aid a better deployment scheme for multiple deployments.

To manage incoming requests across the servers, the Gunicorn servers running in containers can work with NGINX for reverse proxy, load balancing, and additional HTTP security protocols. A good NGINX setting can provide a better facade for the entire production setup.

Generally, the deployment procedures that were created, applied, and utilized in this chapter are translatable, workable, and reversible to other more modern and advanced approaches, such as deploying Flask applications to Google Cloud and AWS cloud services. Apart from deployment, Flask has the edge to compete with other frameworks when dealing with innovation and building enterprise-grade solutions.

In the next chapter, we will showcase the use of the Flask platform in providing middleware solutions to many popular integrations.

12

Integrating Flask with Other Tools and Frameworks

Flask's flexibility, seamlessness, and pluggability provide ease for building various applications, from simple form-based counseling systems to Docker-based applications. *Chapters 1* to *10* have showcased its minimalistic but powerful framework with several extension modules and libraries providing support, fast solutions, and clean coding for web and API applications.

Although Flask is not favored to handle large enterprise solutions like Django can, it can serve as a middleware or component to many enterprise-grade systems and can even be a good solution for building microservices. The superb flexibility of Flask makes it an ideal inclusion to the recipes of many software infrastructures of many business processes.

To give a clear picture as to where to place Flask in the list of popular Python frameworks, the goal of this last chapter is to highlight the feasibility of having a backend Flask implementation for mobile and frontend applications, a solution to a microservice architecture, and an implementation to many modern requirements such as creating and running queries and CRUD transactions through GraphQL.

These are the topics covered in this chapter:

- Implementing microservice applications involving FastAPI, Django, and Tornado
- Implementing Flask instrumentation
- Applying OpenAPI 3.x specification with Swagger
- Providing REST services to a Flutter mobile application
- Consuming REST endpoints with a React application
- Building a GraphQL application

Technical requirements

This chapter uses an *Online Library Management System* to expound on building a microservice application that incorporates FastAPI, Django, and Flask sub-applications with Tornado as the facade application and the server. The sub-applications are mounted using different URL prefixes. Here are the services offered by each of the mounted applications:

- Django sub-module – managing student book borrowers

- Flask sub-module – managing faculty book browsers

- FastAPI sub-module – managing feedback and complaints from borrowers

- Flask main application – the core transactions

- Tornado application – the facade application

Figure 12.1 shows the flow of transactions of these mounted applications.

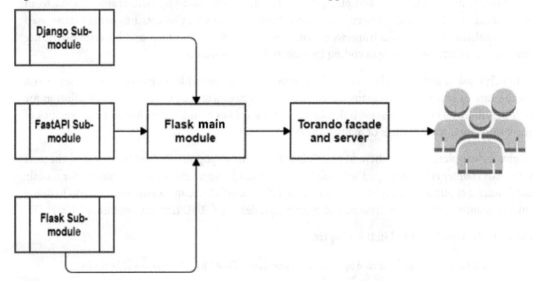

Figure 12.1 – Combined Django, Flask, FastAPI, and Tornado applications

The sub-modules use **SQLAlchemy** as the ORM, while the Flask main application uses the standard **Peewee** ORM. All the projects of this chapter are uploaded at https://github.com/PacktPublishing/ Mastering-Flask-Web-Development/tree/main/ch12.

Implementing microservice applications involving FastAPI, Django, and Tornado

Flask 3.x provides a `DispatcherMiddleware` class from *Werkzeug* that combines isolated and valid WSGI-based applications to form a complete and larger system. These combined applications can be all Flask or different WSGI-based applications such as Django, each with a unique URL prefix. *Figure 12.2* shows the directory structure of our combined projects:

Figure 12.2 – Django, FastAPI, Flask, and Tornado in one project structure

All views, repositories, services, models, and configuration files of the main Flask application are in the `modules` folder. On the other hand, all the application files of the FastAPI app are in the `modules_fastapi` folder, all components of the Django app are in the `modules_django` folder, all Tornado API handlers are in `modules_tornado`, and all GraphQL components are in the `modules_sub_flask` directory.

When it comes to their respective module scripts, the FastAPI app instance is in `main_fastapi.py`, the Flask sub-module's `app` instance is in `main_sub_flask.py`, and the Flask main module's `app` instance together with the Tornado server is in `main.py`.

Now, let us discuss how all these sub-applications can run together using one Tornado server.

Adding the Flask sub-application

`DispatcherMiddleware` requires a Flask app instance to its first parameter and a dictionary containing the mounts of sub-applications, where the *key* is the URL pattern mapped to their corresponding WSGI app instance. Mounting Flask sub-applications to a main Flask application is straightforward. The following snippets show how to mount the `flask_sub_app` instance of the Flask sub-application to the core Flask `app` instance:

```
(main_sub_flask.py)
from modules_sub_flask import create_app_sub
from flask_cors import CORS
... ... ... ... ... ...
flask_sub_app = create_app_sub("../config_dev_sub.toml")
CORS(flask_sub_app)


(main.py)
from werkzeug.middleware.dispatcher import DispatcherMiddleware
from main_sub_flask import flask_sub_app
... ... ... ... ... ...
from modules import create_app
app = create_app('../config_dev.toml')
... ... ... ... ... ...
final_app = DispatcherMiddleware(app, {
    '/fastapi': ASGIMiddleware(fast_app),
    '/django': django_app,
    '/flask': flask_sub_app
})
```

The Flask sub-application must have its dedicated module script (e.g., `main_flask_sub.py`) for its `flask_sub_app` instantiation. `main.py` must import the `flask_sub_app` instance from a dedicated module rather than creating it in `main.py` for traceability, easy debugging, and code-clean reasons. Combining Flask apps to form a larger unit does not need additional configurations, unlike adding a FastAPI application to the main context. How do we register FastAPI applications to `DispatcherMiddleware`?

Adding the FastAPI sub-application

Not all ASGI-based applications are compatible with Flask's context and can be part of the `DispatcherMiddleware` mounts. For FastAPI, the workaround is to convert the app instance to WSGI at runtime using `ASGIMiddleware` from the `a2wsgi` module. To utilize the ASGI-to-WSGI converter, first install `a2wsgi` using the following `pip` command:

```
pip install a2wsgi
```

ASGIMiddleware does not depend on many external modules, so the conversion is straightforward from its built-in mechanisms. It will not consume more memory for its conversion. But, if the FastAPI application has several background tasks to perform, the utility class has a constructor parameter, wait_time, to set an allowable time duration for every background task to finish running before finishing a request. Moreover, its constructor has a loop parameter to allow setting another event loop in case the core platform needs a different type of event loop. Now, the following main.py snippet shows how to add our FastAPI app instance to the mounted applications:

```
(main_fastapi.py)
from fastapi import FastAPI
from fastapi.middleware.cors import CORSMiddleware
from modules_fastapi.api import faculty


fast_app = FastAPI()
fast_app.include_router(faculty.router, prefix='/ch12')
fast_app.add_middleware(
    CORSMiddleware, allow_origins=['*'],
    allow_credentials=True, allow_methods=['*'], allow_headers=['*'])

(main.py)
from main_fastapi import fast_app
from a2wsgi import ASGIMiddleware
… … … … … …

final_app = DispatcherMiddleware(app, {
    '/fastapi': ASGIMiddleware(fast_app),
    '/django': django_app,
    '/flask': flask_sub_app
})
```

The a2wsgi module works perfectly with FastAPI applications. Not all ASGI-based applications can undergo seamless conversions with a2wsgi like FastAPI apps.

Let us now add our Django sub-module to our mounted applications.

Adding the Django sub-application

Django is a pure WSGI framework but can run on ASGI servers with additional configurations. Unlike in FastAPI, adding a Django application to the mounts requires several steps, which include the following procedures in the `main.py` module and Django admin's `settings.py`:

1. Since `module_django` is not the main project folder, import the `os` module and set the default value of the `DJANGO_SETTINGS_MODULE` environment variable to `modules_django.modules_django.settings`:

    ```
    os.environ.setdefault('DJANGO_SETTINGS_MODULE', 'modules_django.
    modules_django.settings')
    ```

    This setting defines the location of `settings.py` of the Django admin folder. Failure to adjust this setting will lead to the following runtime error:

    ```
    attributeerror: module 'modules django.modules django.settings'
    has no attribute 'logging_config'
    ```

2. Specifying the package name of `settings.py` with the Django directory name requires adjusting some package names in the Django project. Among the modifications is the change of `ROOT_URLCONF` in `settings.py` from `'modules_django.urls'` to `'modules_django.modules_django.urls'`.

3. Registering a Django application to the `INSTALLED_APPS` settings must include the Django project name:

    ```
    INSTALLED_APPS = [
        'django.contrib.admin',
        'django.contrib.auth',
        'django.contrib.contenttypes',
        'django.contrib.sessions',
        'django.contrib.messages',
        'django.contrib.staticfiles',
        'rest_framework',
        'corsheaders',
        'modules_django.olms'
    ]
    ```

4. Also, include the Django project folder in defining the Django application object in `settings.py`:

    ```
    WSGI_APPLICATION = 'modules_django.wsgi.application'
    ```

5. Anywhere inside the application (e.g., `modules_django.olms`), import all the custom components with the project folder included. The following snippet shows the implementation of REST services that manage student borrowers using Django RESTful services and Django ORM:

```
(views.py)
from rest_framework.response import Response
from rest_framework.decorators import api_view
from modules_django.olms.serializer import
BorrowedHistSerializer, StudentBorrowerSerializer
from modules_django.olms.models import StudentBorrower,
BorrowedHist


@api_view(['GET'])
def getData(request):
    app = StudentBorrower.objects.all()
    serializer = StudentBorrowerSerializer(app, many=True)
    return Response(serializer.data)


@api_view(['POST'])
def postData(request):
    serializer = StudentBorrowerSerializer(data=request.data)
    if serializer.is_valid():
        serializer.save()
        return Response(serializer.data)
    else:
        return Response({"message:error"})
```

6. Since we are mounting our Django application as a WSGI sub-application, set `DJANGO_ALLOW_ASYNC_UNSAFE` to `false` using the `os` module:

```
os.environ["DJANGO_ALLOW_ASYNC_UNSAFE"] = "true"
```

Failure to set this setting to `true` will cause this runtime exception:

```
django.core.exceptions.SynchronousOnlyOperation: You cannot call
this from an async context - use a thread or sync_to_async.
```

7. Lastly, import `get_wsgi_application` from the `django.core.wsgi` module and register it to `DispatcherMiddleware`. For Django web applications, import `StaticFilesHandler` from the `django.contrib.staticfiles.handlers` module and wrap the `get_wsgi_application()`'s returned object to access the static web files (e.g., CSS, JS, images):

```
from django.core.wsgi import get_wsgi_application
from django.contrib.staticfiles.handlers import
StaticFilesHandler

… … … … … …

django_app = StaticFilesHandler( get_wsgi_application())
```

```
… … … … … …
final_app = DispatcherMiddleware(app, {
    '/fastapi': ASGIMiddleware(fast_app),
    '/django': django_app,
    '/flask': flask_sub_app
})
```

Now, it is time to mount the main Flask app to the Tornado server after mounting the FastAPI, Django, and Flask sub-applications.

Putting it all together with Tornado

Although it is ideal to use Gunicorn to run Flask applications in the production server, sometimes using the non-blocking Tornado server is a perfect choice for a Flask project that focuses more on event-driven transactions, WebSocket, and **Server-Sent Events** (**SSE**). For this chapter, our design is to mount the main Flask application, which implements the core *Online Library Management System's* transactions, to a Tornado application. By the way, Tornado is a Python framework and asynchronous networking library whose strength is more on event-driven, non-blocking, and long polling transactions that require a long-live connection to a user. It has a bundled non-blocking HTTP server, unlike FastAPI.

To run a compatible WSGI application on Tornado's HTTP server, it has a `WSGIContainer` utility class that wraps a core Flask `app` instance and runs the application on the non-blocking or asynchronous server mode.

> **Important note**
>
> At the moment of writing this book, `Flask[async]` with async HTTP requests mounted on the Tornado server throws a `SynchronousOnlyOperation` error. Thus, this chapter focuses on the standard Flask request transactions only.

The following `main.py` snippet shows the integration of our core Flask app with the Tornado server:

```
from tornado.wsgi import WSGIContainer
from tornado.web import FallbackHandler, Application
from tornado.platform.asyncio import AsyncIOMainLoop
from modules_tornado.handlers.home import MainHandler
import asyncio
… … … … … …
from modules import create_app
app = create_app('../config_dev.toml')
… … … … … …
final_app = DispatcherMiddleware(app, {
    '/fastapi': ASGIMiddleware(fast_app),
```

```
    '/django': django_app,
    '/flask': flask_sub_app
})


main_flask = WSGIContainer(final_app)
application = Application([
    (r"/ch12/tornado", MainHandler),
    (r".*", FallbackHandler, dict(fallback=main_flask)),
])


if __name__ == "__main__":
    loop = asyncio.get_event_loop()
    application.listen(5000)
    loop().run_forever()
```

Tornado is not a WSGI application and is not thread-safe. It uses one thread to manage one process at a time. It is a framework designed for event-driven applications with a built-in server created for running non-blocking I/O sockets. But, it can now directly use `asyncio` to run our mounted applications asynchronously in replacement of `IOLoop` for events. In the given snippet, our Tornado and main Flask applications run on the `asyncio` platform using the main event loop retrieved by `get_event_loop()`. By the way, our complete system has a Tornado handler, `MainHandler`, in `modules_tornado` that renders a welcome message in JSON format.

At this point, our application has become complex because of the mounting of different WSGI and ASGI applications, so it needs a microservice mechanism called **instrumentation** to monitor the low-level behavior of each mounted application and capture low-level metrics such as database-related logs, memory usage, and library-specific logs and issues.

Implementing Flask instrumentation

Instrumentation is a mechanism that generates, collects, and exports data about the runtime diagnostics of an application, microservice, or distributed setup. Usually, this observable data includes traces, logs, and metrics that can provide an understanding of the system. Among the many ways to implement instrumentation, **OpenTelemetry** offers easy configuration and vendor-agnostic approaches to monitoring and observing the system's internal state. To utilize OpenTelemetry for Flask, install the following external modules using the `pip` command:

```
pip install opentelemetry-api opentelemetry-sdk opentelemetry-
instrumentation-flask opentelemetry-instrumentation-requests
```

The following snippet added to the `create_app()` factory in the `__init__.py` of the main Flask application provides the console-based instrumentation:

```
from opentelemetry import trace
from opentelemetry.sdk.resources import Resource
from opentelemetry.sdk.trace import TracerProvider
from opentelemetry.sdk.trace.export import BatchSpanProcessor
from opentelemetry.sdk.trace.export import ConsoleSpanExporter
from opentelemetry.sdk.resources import SERVICE_NAME, Resource
from opentelemetry.instrumentation.flask import FlaskInstrumentor

def create_app(config_file):
    provider = TracerProvider(resource= Resource.create({SERVICE_
NAME: "packt-flask-service"}))
    processor = BatchSpanProcessor(ConsoleSpanExporter())
    provider.add_span_processor(processor)

    trace.set_tracer_provider(provider)

    global tracer
    tracer = trace.get_tracer("packt-flask-tracer")

    app = OpenAPI(__name__, info=info)
    ... ... ... ... ... ...
    FlaskInstrumentor(app).instrument(  enable_commenter=True,
commenter_options={})
    ... ... ... ... ... ...
```

OpenTelemetry requires the Flask application to set up the Tracing API consisting of `TracerProvider`, `Tracer`, and `Span(s)`. The first component, `TracerProvider`, is the entry point of the API and the registry for creating the tracers. A tracer is responsible for creating spans. On the other hand, a span is an API that can monitor any part of the application.

After instantiating `TracerProvider`, part of the setup to create a tracer is to apply `Batch SpanProcessor`, which preprocesses the spans per batch before exporting them to another system, tool, or backend. It requires a specific exporter class in its constructor parameter, such as `ConsoleSpanExporter`, which sends the span to the console. To complete the `TracerProvider` setup, add the processor to the `TracerProvider` object using its `add_span_processor()` method.

Finally, import the trace API object from the `opentelemetry` module and invoke its `set_tracer_provider()` class method to set the created `TracerProvider` instance. To extract the `tracer` object, invoke its `get_tracer()` method and specify its name, such as `packt-flask-tracer`.

Now, import the `tracer` object anywhere inside the application. The following module script imports the `tracer` object to monitor the `add_login()` endpoint:

```
from modules import tracer


@current_app.post("/login/add)
def add_login():
    with tracer.start_as_current_span('users_span'):
        login_json = request.get_json()
        repo = LoginRepository()
        result = repo.insert_login(login_json)
        if result == False:
            return jsonify(message="error"), 500
        else:
            return jsonify(record=login_json)
```

Invoking the `start_as_current_span()` method of a `tracer` object creates a span, a single operation within a trace. For larger systems, spans can be nested to form a trace tree for detailed monitoring. A nested trace contains a root span, which typically describes the upper-level operation, and one or more child spans for its lower-level operations. *Figure 12.3* shows the console log after running the `add_login()` endpoint:

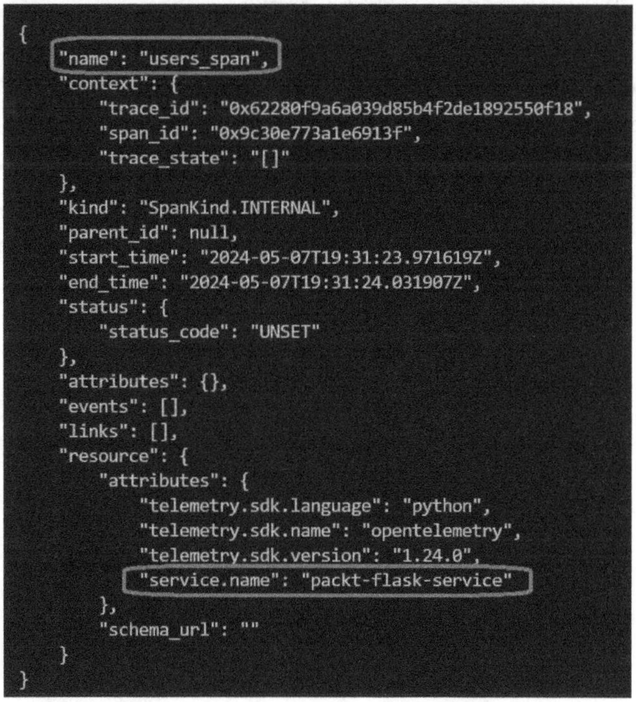

```
{
    "name": "users_span",
    "context": {
        "trace_id": "0x62280f9a6a039d85b4f2de1892550f18",
        "span_id": "0x9c30e773a1e6913f",
        "trace_state": "[]"
    },
    "kind": "SpanKind.INTERNAL",
    "parent_id": null,
    "start_time": "2024-05-07T19:31:23.971619Z",
    "end_time": "2024-05-07T19:31:24.031907Z",
    "status": {
        "status_code": "UNSET"
    },
    "attributes": {},
    "events": [],
    "links": [],
    "resource": {
        "attributes": {
            "telemetry.sdk.language": "python",
            "telemetry.sdk.name": "opentelemetry",
            "telemetry.sdk.version": "1.24.0",
            "service.name": "packt-flask-service"
        },
        "schema_url": ""
    }
}
```

Figure 12.3 – A tracer log exported to the console

Additionally, Jaeger, a distributed tracing platform, can visualize all the logs from the trace in graphical views. OpenTelemetry has an exporter class that can export the spans to the Jaeger platform after preprocessing. But first, install Jaeger either through its Docker image or binaries. In this chapter, we start the Jaeger server through the `jaeger-all-in-one.exe` command from its binaries.

Then, install the OpenTelemetry module for Jaeger support using the following `pip` command:

```
pip install opentelemetry-exporter-jaeger
```

After the installation, add the following snippet to the previous OpenTelemetry setup in the `create_app()` factory:

```
from opentelemetry.exporter.jaeger.thrift import JaegerExporter

… … … … … …
    trace.set_tracer_provider(provider)

    jaeger_exporter = JaegerExporter(agent_host_name= "localhost",
agent_port=6831,)
    trace.get_tracer_provider().add_span_processor(
BatchSpanProcessor(jaeger_exporter))

    global tracer
    tracer = trace.get_tracer("packt-flask-tracer")
… … … … … …
```

`JaegerExporter` sends the traces to a thrift server running through the HTTP protocol. The constructor parameters of the exporter class are all about the server details of the thrift server. In our case, `agent_host_name` is `localhost` and `agent_port` is `6831`.

Restart the Tornado server, run the monitored APIs, and open the Jaeger dashboard at `http://localhost:16686/` using the browser to check the traces. *Figure 12.4* shows the **Search** page of the Jaeger dashboard after searching four traces:

Figure 12.4 – Jaeger search result for span traces

The left portion of Jaeger's **Traces** view renders a list of Tracing API services that monitor our application. Ours is `packt-flask-service`, which gave four searches at the time of its search. On the right portion of the dashboard is the search result listing the traces produced by the spans that monitored the transaction performed. Clicking each row leads to trace details in a graphical format. On the other hand, the graph in the header portion summarizes all the traces at a specified duration.

Let us now explore how to add OpenAPI 3.x documentation to the API endpoints of our Flask application.

Applying OpenAPI 3.x specification with Swagger

Aside from instrumentation, some solutions provide well-formatted API documentation, as in FastAPI. One of these solutions is to use `flask_openapi3`, which applies OpenAPI 3.x specification to implement Flask components and document the API endpoints with Swagger, ReDoc, and RapiDoc.

`flask_openapi3` is not a library or dependency module of Flask but a separate framework based on the current Flask 3.x with the `pydantic` module to support OpenAPI documentation. It also supports `Flask[async]` components.

After installing `flask_openapi3` using the `pip` command, replace the `Flask` class with `OpenAPI` and `Blueprint` with `APIBlueprint`. These are still the original Flask classes but with the feature of adding API documentation. The following is the `create_app()` factory method of our main Flask application that uses `OpenAPI` to create the application object with the added documentation components:

```
from flask_openapi3 import Info
from flask_openapi3 import OpenAPI

... ... ... ... ... ...
info = Info(title="Flask Interoperability (A Microservice)",
version="1.0.0")
... ... ... ... ... ...
def create_app(config_file):

    ... ... ... ... ... ...
    app = OpenAPI(__name__, info=info)
    app.config.from_file(config_file, toml.load)
    cors = CORS(app)
    app.config['CORS_HEADERS'] = 'Content-Type'
    ... ... ... ... ... ...
```

The `Info` utility class provides the project title of the OpenAPI documentation. Its instance is part of the constructor parameter values of OpenAPI.

To document an API, add a summary of the endpoint, a complete description of the API transaction, tags, request field descriptions, and response details. The following snippet shows a simple documentation of the `list_login()` endpoint:

```
from flask_openapi3 import Tag
list_login_tag = Tag(name="list_login", description="List all user
credentials.")
... ... ... ... ... ...
@current_app.get("/login/list/all", summary="List all login records.",
tags=[list_login_tag])
def list_login():
    """
    API for retrieving all the records from the olms database.
    """
    with tracer.start_as_current_span('users_span'):
        repo = LoginRepository()
        result = repo.select_all_login()
        print(result)
        return jsonify(records=result)
```

The HTTP method decorators of OpenAPI allow additional arguments such as a summary and tags describing the API endpoint. The Tags class is a flask_openapi3 utility that enables the creation of tags for an endpoint. An endpoint can be associated with a list of tag objects. On the other hand, the documentation comment placed in the first lines of the API function becomes the complete and detailed description of the API implementation.

Now, an application using the flask_openapi3 framework has an additional endpoint, /openapi/swagger, which, when run on a web browser, will render a Swagger documentation, as illustrated in *Figure 12.5*:

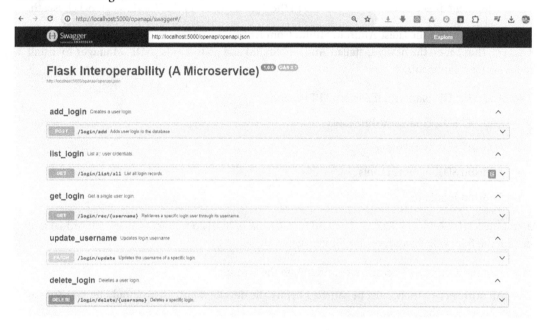

Figure 12.5 – An OpenAPI/Swagger documentation in Flask

As long as flask-openapi3 is always up to date and synchronized with the Flask releases, it is a helpful and feasible solution to build Flask applications with documentation implemented using OpenAPI/Swagger. The framework can provide acceptable and standard API documentation with less additional YAML or JSON configuration in main.py.

In the next section, let us discuss other ways our Flask application can provide services outside the Python platform, starting with mobile development.

Providing REST services to a Flutter mobile application

Flask applications can be a potential backend API service provider for many popular mobile platforms, such as Flutter. Flutter is an open source mobile toolkit created by Google to provide commercially accepted and natively compiled applications for mobile platforms. It can serve as a frontend framework for our microservice applications.

To start Flutter, download the latest Flutter SDK release – in my case, for Windows, from the `https://docs.flutter.dev/release/archive?tab=windows` download site. Unzip the file to your development directory and register `%FLUTTER_HOME%\bin` in the Windows global classpath.

Afterward, download the latest Android Studio from `https://developer.android.com/studio`. Then, open the newly installed Android Studio and go to **Tools | SDK Manager** to install the following components:

- Android SDK platform, the latest API version
- Android SDK command-line tools
- Android SDK build tools
- Android SDK platform tools
- Android Emulator

After a successful Android SDK update and installation of SDK components, open a terminal and perform the following procedures for Flutter diagnostics:

1. Run the `flutter doctor` command on the terminal.
2. All settings and tools must be satisfied before the development. There should be no issues at this point of the installation.
3. Using the VS Code editor with the Flutter extension, create a Flutter project; it should have the project structure presented in *Figure 12.6*:

Figure 12.6 – Our ch12-flutter-flask project

Inside /lib/olms/provider, providers.dart implements the transactions that consume the HTTP GET and POST of our APIs for user management. Here are the dart codes for our service providers:

```dart
import 'dart:convert';
import 'package:flutter/material.dart';
import 'package:http/http.dart' as http;
import 'package:library_app/olms/models/login.dart';

class LoginProvider with ChangeNotifier{
  List<Login> _items = [];
  List<Login> get items {
    return [..._items];
  }
}
```

```
Future<void> addLogin(String username, String password, String role
) async {
  String url = 'http://<actual IP address>:5000/login/add';
  try{
    if(username.isEmpty || password.isEmpty || role.isEmpty){
      return;
    }
    Map<String, dynamic> request = {"username": username,
"password": password, "role": int.parse(role)};
    final headers = {'Content-Type': 'application/json'};
    final response = await http.post(Uri.parse(url), headers:
headers, body: json.encode(request));
    Map<String, dynamic> responsePayload = json.decode(response.
body);
    final login = Login(
        username: responsePayload["username"],
        password: responsePayload["password"],
        role: responsePayload["role"]
    );
    print(login);
    notifyListeners();
  }catch(e){
    print(e);
  }
}
```

The given addLogin() consumes the add_login() API from our Flask microservice app:

```
Future<void> get getLogin async {
  String url = 'http://<actual IP address>:5000/login/list/all';
  var response;
  try{
    response = await http.get(Uri.parse(url));
    Map body = json.decode(response.body);

    List<Map> loginRecs = body["records"].cast<Map>();
    print(loginRecs);
    _items = loginRecs.map((e) => Login(
        id: e["id"],
        username: e["username"],
        password: e["password"],
        role: e["role"],
    )
    ).toList();
  }catch(e){
```

```
        print(e);
      }
    notifyListeners();
  }
}
```

The `Login` model class mentioned in the code is a Flutter class that the `addLogin()` and `getLogin()` provider transactions will map to the JSON records from the Flask API endpoints. The following is Flutter's `Login` model class derived from the `ch12-microservices-interop`'s SQLAlchemy model layer:

(/lib/olms/models/login.dart
```
class Login{
  int? id;
  String username;
  String password;
  int role;
  Login({ required this.username, required this.password, required
this.role, this.id=0});
}
```

Now, the given `getLogin()` retrieves the `Login` records in JSON format from our `list_login()` endpoint function. Notice that Flutter requires the *actual IP address* of the host server for its services to access the API endpoint resources.

The `/lib/olms/tasks/task.dart` file implements the form widgets and the corresponding events that invoke these two service methods. The following part of the implementation shows the `Login` form:

(/lib/olms/tasks/task.dart)
```
... ... ... ... ... ...
class _TasksWidgetState extends State<LoginViewWidget> {
  ... ... ... ... ... ...

  @override
  Widget build(BuildContext context) {
    return Padding(
    ... ... ... ... ... ...
      children: [
        Row(
          children: [
            Expanded(
              child: TextFormField(
                controller: userNameController,
                decoration: const InputDecoration(
```

```
              labelText: 'Username',
              border: OutlineInputBorder(),
            ),
    ... ... ... ... ... ...
          Expanded(
            child: TextFormField(
    ... ... ... ... ... ...
              labelText: 'Password',
              border: OutlineInputBorder(),
              ),
            ),
          ),
          Expanded(
            child: TextFormField(
    ... ... ... ... ... ...
              labelText: 'Role',
              border: OutlineInputBorder(),
              ),
    ... ... ... ... ... ...
          const SizedBox(width: 10,),
          ElevatedButton(
    ... ... ... ... ... ...
              child: const Text("Add"),
              onPressed: () {
                Provider.of<LoginProvider>(context, listen:
false).addLogin(userNameController.text, passwordController.text,
roleController.text);
    ... ... ... ... ... ...
              }
          )
        ],
      ),
```

In general, `LoginViewWidget` component returns a `Padding` widget composed of two sub-widgets. The preceding code renders a horizontal form of three `TextFormField` widgets that will accept login details from the user and an `ElevatedButton` that will trigger the `add_login()` provider transaction to invoke our Flask `/login/add` endpoint. The following code shows the next part of `LoginViewWidget` that renders the list of `Login` records from our database:

```
... ... ... ... ... ...
FutureBuilder(future: Provider.of<LoginProvider>(context, listen:
false).getLogin,
        builder: (ctx, snapshot) =>
```

```
            snapshot.connectionState == ConnectionState.waiting
            ? const Center(child: CircularProgressIndicator())
            : Consumer<LoginProvider>(
                … … … … … …
                builder: (ctx, loginProvider, child) =>
                … … … … … …
                    Container(
                    … … … … … … …
                      child: SingleChildScrollView(
                        scrollDirection: Axis.horizontal,
                          child: DataTable(
                            columns: <DataColumn>[
                              DataColumn(
                                label: Text(
                                  'Username',
                                  style:
                                    … … … … … …
                              ),
                              DataColumn(
                                label: Text(
                                  'Password',
                                  style:
                                  … … … … … …
                              ),
                              DataColumn(
                                label: Text(
                                  'Role',
                                  … … … … … …
                              ),],
                        rows: <DataRow>[
                          DataRow(cells: <DataCell>[
                            DataCell(Text( loginProvider. items[i].
username)),
                            DataCell(Text(loginProvider. items[i].
password)),
                            DataCell(Text(loginProvider. items[i].
role.toString())),
                              ],
                        … … … … … …
```

The preceding code renders a FutureWidget component composed of ListView of DataColumn and DataRow widgets to showcase the Login records from our Flask /login/list/all endpoint. It has a vertical scrollbar rendered by its SingleChildScrollView widget. *Figure 12.7* shows the resulting LoginViewWidget form after running our ch12-flask-flutter application using the flutter run command inside the /library_app directory.

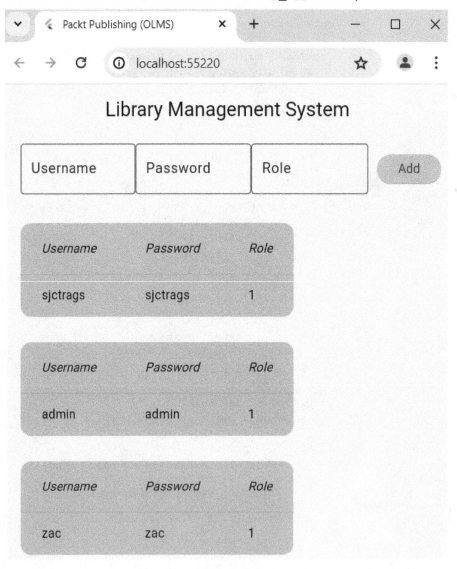

Figure 12.7 – A Flutter form for login transactions

In addition to mobile development, Flask APIs can also provide data services to popular frontend frameworks, such as React. Let us see how in the next section.

Consuming REST endpoints with a React application

React is a popular frontend server-side JavaScript library used to build scalable user interfaces for websites, mostly **single-page applications** (**SPAs**). It is a popular library in rendering pages that changes its state without reloading the page.

After creating a React project using the `create-react-app` command, our `ch12-react-flask` application has implemented the following functional component to build the form page and the table listing all the faculty borrowers of the *Online Library Management System*:

```
export const FacultyBorrowers =(props)=>{
    const [id] = React.useState(0);
    const [firstname, setFirstname] = React.useState('');
    const [lastname, setLastname] = React.useState('');
    const [empid, setEmpid] = React.useState('');
    const [records, setRecords] = React.useState([]);
```

The given `useState()` hook methods define the state variables that the form components will use to capture form data before submitting the details to the `add_faculty_borrower()` API endpoint of our FastAPI sub-application:

```
    React.useEffect(() => {
        const url_get = 'http://localhost:5000/fastapi/ ch12/faculty/
borrower/list/all';
        fetch(url_get)
        .then((response) =>  response.json() )
        .then((json) =>  { setRecords(json)})
        .catch((error) => console.log(error));
    }, []);


    const addRecord = () =>{
        const url_post = 'http://localhost:5000/fastapi/ ch12/
faculty/borrower/add';
        const options = {
            method: 'POST',
            headers:{
                'Content-Type': 'application/json'
            },
            body: JSON.stringify(
                {
                'id': id,
```

```
                    'firstname': firstname,
                    'lastname': lastname,
                    'empid': empid
                    }
            )
        }
        fetch(url_post, options)
            .then((response) => { response.json() })
            .then((json) => { console.log(json)})
            .catch((error) => console.log(error));

            const url_get = 'http://localhost:5000/fastapi/ ch12/
faculty/borrower/list/all';
            fetch(url_get)
            .then((response) =>  response.json() )
            .then((json) =>  { setRecords(json)})
            .catch((error) => console.log(error));
    }
```

The `addRecord()` event method forms the JSON data from the state variables before submitting it to the `add_faculty_borrower()` API endpoint. Likewise, it retrieves all the faculty borrowers from the microservice through the `list_all_faculty_borrowers()` endpoint of the same FastAPI sub-application:

```
    return <div>
       <form id='idForm1' onSubmit={ addRecord }>
          Employee ID: <input type='text' onChange={ (e) => {setEmpid(e.
target.value)}} /><br/>
          First Name: <input type='text' onChange={ (e) =>
{setFirstname(e.target.value) }} /><br/>
          Last Name: <input type='text' onChange={ (e) =>
{setLastname(e.target.value) }}/><br/>
          <input type='submit' value='ADD Faculty Borrower'/>
            </form>
            <br/>
            <h2>List of Faculty Borrowers</h2>
            <table >
              <thead>
                  <tr><th>Id</th>
                      <th>Employee ID</th>
                      <th>First Name</th>
                      <th>Last Name</th>
              </tr></thead>
              <tbody>
```

```
            {records.map((u) => (
              <tr>
                <td>{u.id}</td>
                <td>{u.empid}</td>
                <td>{u.firstname}</td>
                <td>{u.lastname}</td>
              </tr>
            ))}
        </tbody></table>
    </div>}
```

The `records.map()` function builds the table of records after adding a new faculty borrower record to the database. `addRecord()`, with the help of the `useEffect()` hook method, captures all the records from the `list_all_faculty_borrowers()` API and stores the list of JSON-formatted data to the state variable, `records`.

Aside from building service providers for other platforms, Flask can also build GraphQL applications for easy CRUD operations. Let's learn about it in the next discussion.

Building a GraphQL application

GraphQL is a mechanism that provides a platform-agnostic CRUD transaction across the applications without specifying the actual database connectivity details and database dialects. This mechanism is model-centric or data-centric and focuses on the data the users want to fetch through the backend API implementations.

Our microservice designed the Flask sub-application to be a GraphQL application with the following HTTP GET endpoint that creates the GraphQL UI explorer:

```
from ariadne.explorer import ExplorerGraphiQL

… … … … … …
flask_sub_app = create_app_sub("../config_dev_sub.toml")
CORS(flask_sub_app)
explorer_html = ExplorerGraphiQL().html(None)

@flask_sub_app.route("/graphql", methods=["GET"])
def graphql_explorer():
    return explorer_html, 200
```

Our solution used the `Ariadne` module because it is updated and can integrate with Flask 3.x components. *Figure 12.8* shows the actual GraphQL Explorer after accessing the given `/flask/graphql` endpoint.

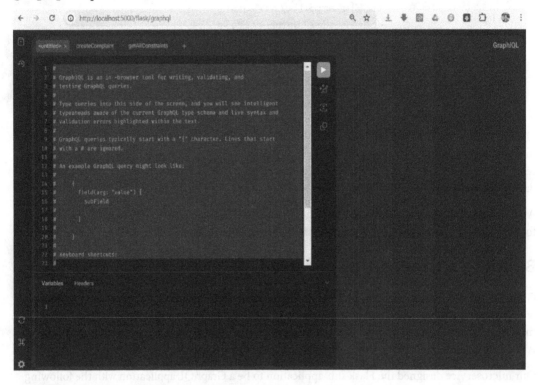

Figure 12.8 – The Ariadne GraphQL Explorer

Next, build the crucial GraphQL schema configuration, the `schema.graphql` file, which creates all the GraphQL model classes derived from the ORM models, request data, and response objects of the supposed API endpoints. The following is a snapshot of the `schema.graphql` file used by our Flask sub-application:

```
schema {
    query: Query
    mutation: Mutation
}

# These are the GraphQL model classes
type Complainant {
    id: ID!
    firstname: String!
    lastname: String!
```

```
    middlename: String!
    email: String!
    date_registered: String!
}

type Complaint {
    id: ID!
    ticketId: String!
    catid: Int!
    complainantId: Int!
    ctype: Int!
}
```

The language used to implement the schema.graphql file is the **Schema Definition Language (SDL)**, and the first portion is the declaration of the operation types needed by the GraphQL REST services – namely, Mutation and Query. Then, what follows are the definitions of GraphQL *object types*, the building blocks of GraphQL that represent the records that the REST services will fetch from or persist in the data repository. In the given definition file, the GraphQL transactions will focus on utilizing Complainant, Complaint, and their related model classes to manage the feedback sub-module of the *Online Library Management System*.

Each model class consists of Int, Float, Boolean, ID, or any custom scalar object type. GraphQL also allows model classes to have enum and list ([ ]) field types. The scalar or multi-valued fields can be nullable or non-nullable (!). So far, the given model classes all consist of non-nullable scalar fields. By the way, the octothorpe or hashtag (#) sign is the comment symbol of the SDL.

After building the model classes, the next step is to define the Query and Mutation operations with their parameters and return types.

```
# The GraphQL operations
type Query {
    listAllComplainants: ComplainantListResult!
    listAllComplaints: ComplaintListResult!
    listAllCategories: CategoryListResult!
    listAllComplaintTypes: ComplaintTypeListResult!
}

type Mutation {
    createCategory(name: String!): CategoryInsertResult!
    createComplaintType(name: String!): ComplaintTypeInsertResult!
    createComplainant(input: ComplainantInput!):
ComplainantInsertResult!
    createComplaint(input: ComplaintInput!): ComplaintInsertResult!
}
```

Our Flask sub-application focuses on the persistence and retrieval of feedback about the Online Library's processes. Its Query operations involve retrieving the complaints (listAllComplaints), complainants (listAllComplainants), and the category (listAllCategories) and complaint type (listAllComplaintTypes) lookups. On the other hand, the Mutation operations involve adding complaints (createComplaint), complainants (createComplainant), complaint categories (createCategory), and complaint types (createComplaintType) to the database. createCategory and createComplaintType have their respective String parameter name, but the other mutators use input types to organize and manage their lengthy parameter list. Here are the implementations of the ComplaintInput and ComplainantInput types:

```
# These are the input types
input ComplainantInput {
    firstname: String!
    lastname: String!
    middlename: String!
    email: String!
    date_registered: String!
}

input ComplaintInput {
    ticketId: String!
    complainantId: Int!
    catid: Int!
    ctype: Int!
}
```

Aside from input types, Query and Mutation operators need result types to manage the response of GraphQL's REST service executions. Here are some of the result types used by our Query and Mutation operations:

```
# These are the result types
type ComplainantInsertResult {
    success: Boolean!
    errors: [String]
    model: Complainant!
}

type ComplaintInsertResult {
    success: Boolean!
    errors: [String]
    model: Complaint!
}
... ... ... ... ... ...
type ComplainantListResult {
```

```
    success: Boolean!
    errors: [String]
    complainants: [Complainant]!
}

type ComplaintListResult {
    success: Boolean!
    errors: [String]
    complaints: [Complaint]!
}
```

Now, all these object types, input types, and result types build GraphQL resolvers that implement these `Query` and `Mutation` operations. A *GraphQL resolver* connects the application's repository and data layer to the GraphQL architecture. Although GraphQL can provide auto-generated resolver implementations, it is still practical to implement a custom resolver for each operation to capture the needed requirements, especially if the operations involve complex constraints and scenarios. The following snippet from `modules_sub_flask/resolvers/complainant_repo.py` implements the resolvers of our defined `Query` and `Mutation` operations:

```python
from ariadne import QueryType, MutationType
from typing import List, Any, Dict
from modules_sub_flask.models.db import Complainant
from sqlalchemy.orm import Session

query = QueryType()
mutation = MutationType()

class ComplainantResolver:
    def __init__(self, sess:Session):
        self.sess = sess

    @mutation.field('complainant')
    def insert_complainant(self, obj, info, input) -> bool:
        try:
            complainant = Complainant(**input)
            self.sess.add(complainant)
            self.sess.commit()
            payload = {
                "success": True,
                "model": complainant
            }
        except Exception as e:
            print(e)
            payload = {
```

```
                    "success": False,
                    "errors": [f"Complainant … not found"]
            }
        return payload
        … … … … … …
```

The `insert_complainant()` transaction accepts the input from the GraphQL dashboard and saves the data to the database, while the following `select_all_complainant()` retrieves all the records from the database and renders them as a list of complainant records to the GraphQL dashboard:

```
def select_all_complainant(self, obj, info) -> List[Any]:
    complainants = self.sess.query(Complainant).all()
    try:
        records = [todo.to_json() for todo in complainants]
        print(records)
        payload = {
            "success": True,
            "complainants": records
        }
    except Exception as e:
        print(e)
        payload = {
            "success": False,
            "errors": [str("Empty records")]
        }
    … … … … … …
```

The `ariadne` module has `QueryType` and `MutationType` that map GraphQL components such as *input types*. The `MutationType` object, for instance, maps the `ComplainantInput` type to the `input` parameter of the `insert_complainant()` method.

Our GraphQL provider looks like a repository class, but it can also be a service type as long as it meets the requirements of the `Query` and `Mutation` functions defined in the `schema.graphql` definition file.

Now, the mapping of each resolver function to its respective HTTP request function in `schema.graphql` always happens in `main.py`. The following snippet in `main_sub_flask.py` performs mapping of these two GraphQL components:

```
… … … … … …
from ariadne import load_schema_from_path, make_executable_schema, \
    graphql_sync, snake_case_fallback_resolvers, ObjectType
from ariadne.explorer import ExplorerGraphiQL
from modules_sub_flask.resolvers.complainant_repo import
ComplainantResolver
```

```
from modules_sub_flask.resolvers.complaint_repo import
ComplaintResolver
from modules_sub_flask.models.config import db_session


… … … … … …
complainant_repo = ComplainantResolver(db_session)
category_repo = CategoryResolver(db_session)
… … … … … …

query = ObjectType("Query")
query.set_field("listAllComplainants", complainant_repo.select_all_
complainant)
query.set_field("listAllComplaints", complaint_repo.select_all_
complaint)
… … … … … …
mutation = ObjectType("Mutation")
mutation.set_field("createComplainant", complainant_repo.insert_
complainant)
mutation.set_field("createComplaint", complaint_repo.insert_complaint)
… … … … … …

type_defs = load_schema_from_path("./schema.graphql")
schema = make_executable_schema(
    type_defs, query, mutation,
)
… … … … … …
```

main_sub_flask.py loads all the components from schema.graphql and maps all its operations to the repository and model layers of the mounted application. It is recommended to place the schema definition file in the main project directory for easy access to the file. *Figure 12.9* shows the sequence of operations needed to run the createComplainant mutator.

Figure 12.9 – Syntax for running a GraphQL mutator

And *Figure 12.10* shows how to run the `listAllComplainants` query operation.

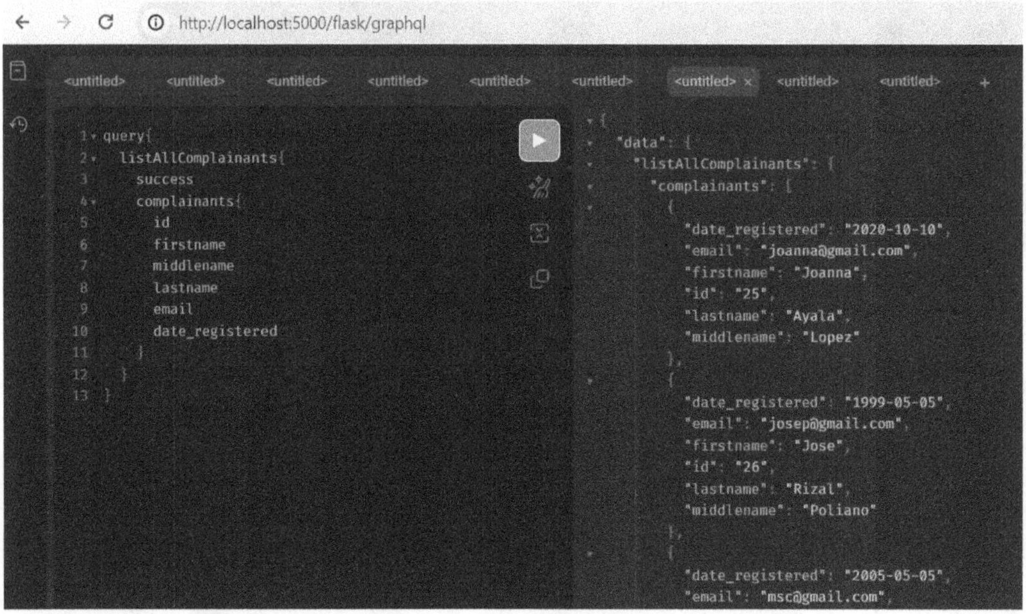

Figure 12.10 – Syntax for running a GraphQL query operator

There are other libraries Flask can integrate to implement the GraphQL architecture, but they need to be up to date to support Flask 3.x.

Summary

Flexibility, adaptability, extensibility, and maintainability are the best adjectives that fully describe Flask as a Python framework.

Previous chapters have proven Flask to be a simple, minimalist, and Pythonic framework that can build API and web applications with fewer configurations and setups. Its vast support helps us build applications that manage workflows and perform scientific calculations and visualization using plots, graphs, and charts. Although a WSGI application at the core, it can implement asynchronous API and view functions with `async` services and repository transactions.

Flask has *Flask-SQLAlchemy, Flask-WTF, Flask-Session, Flask-CORS*, and *Flask-Login* that can lessen the cost and time of development. Other than that, stable and up-to-date extensions are available to help a Flask application secure its internals, run on an HTTPS platform, and protect its form handling from **Cross-Site Request Forgery (CSRF)** problems. On the other hand, Flask can use SQLAlchemy, Pony, or Peewee to manage data persistency and protect applications from SQL injection. Also, the framework can can manage NoSQL data using MongoDB, Neo4j, Redis, and CouchBase databases.

Flask can also build WebSocket and SSE using standard and `asyncio` platforms.

This last chapter has added, to Flask's long list of capabilities and strengths, the ability to connect to various Python frameworks and to provide interfaces and services to applications outside the Python environment.

Aside from managing project modules using Blueprints, Flask can use Werkzeug's **DispatcherMiddleware** to dispatch requests to other mounted WSGI applications such as Django and Flask sub-applications and compatible ASGI applications, such as FastAPI. This mechanism shows Flask's interoperability feature, which can lead to building microservices. On the other hand, Flask can help provide services to Flutter apps, React web UIs, and GraphQL Explorer to run platform-agnostic query transactions.

Hopefully, this book showcased Flask's strengths as a web and API framework cover to cover and also helped discover some of its downsides along the way. Flask 3.x is a lightweight Python framework that can offer many things in building enterprise-grade small-, middle-, and hopefully large-scale applications.

This book has led us on a long journey of learning, understanding, and hands-on experience about Flask 3's core and new asynchronous features. I hope this reference book has provided the ideas and solutions that may help create the necessary features, deliverables, or systems for your business requirements, software designs, or daily goals and targets. Thank you very much for choosing this book as your companion for knowledge. And do not forget to share your Flask experiences with others because mastering something starts with sharing what you learned.

Index

`packtpub.com`

Subscribe to our online digital library for full access to over 7,000 books and videos, as well as industry leading tools to help you plan your personal development and advance your career. For more information, please visit our website.

Why subscribe?

- Spend less time learning and more time coding with practical eBooks and Videos from over 4,000 industry professionals

- Improve your learning with Skill Plans built especially for you

- Get a free eBook or video every month

- Fully searchable for easy access to vital information

- Copy and paste, print, and bookmark content

Did you know that Packt offers eBook versions of every book published, with PDF and ePub files available? You can upgrade to the eBook version at `packtpub.com` and as a print book customer, you are entitled to a discount on the eBook copy. Get in touch with us at `customercare@packtpub.com` for more details.

At `www.packtpub.com`, you can also read a collection of free technical articles, sign up for a range of free newsletters, and receive exclusive discounts and offers on Packt books and eBooks.

Other Books You May Enjoy

If you enjoyed this book, you may be interested in these other books by Packt:

Full-Stack Flask and React

Olatunde Adedeji

ISBN: 978-1-80324-844-8

- Explore the fundamentals of React for building user interfaces.
- Understand how to use JSX to render React components.
- Handle data and integrate third-party libraries and APIs into React applications.
- Secure your Flask application with user authentication and authorization.
- Discover how to use Flask RESTful API to build backend services with React frontend.
- Build modular and scalable Flask applications using blueprints

Flask Framework Cookbook

Shalabh Aggarwal

ISBN: 978-1-80461-110-4

- Explore advanced templating and data modeling techniques.
- Discover effective debugging, logging, and error-handling techniques in Flask.
- Work with different types of databases, including RDBMS and NoSQL.
- Integrate Flask with different technologies such as Redis, Sentry, and Datadog.
- Deploy and package Flask applications with Docker and Kubernetes.
- Integrate GPT with your Flask application to build future-ready platforms.
- Implement continuous integration and continuous deployment (CI/CD) to ensure efficient and consistent updates to your Flask web applications.

Packt is searching for authors like you

If you're interested in becoming an author for Packt, please visit authors.packtpub.com and apply today. We have worked with thousands of developers and tech professionals, just like you, to help them share their insight with the global tech community. You can make a general application, apply for a specific hot topic that we are recruiting an author for, or submit your own idea.

Share Your Thoughts

Now you've finished *Mastering Flask Web and API Development*, we'd love to hear your thoughts! Scan the QR code below to go straight to the Amazon review page for this book and share your feedback or leave a review on the site that you purchased it from.

https://packt.link/r/1-837-63322-3

Your review is important to us and the tech community and will help us make sure we're delivering excellent quality content.

Download a free PDF copy of this book

Thanks for purchasing this book!

Do you like to read on the go but are unable to carry your print books everywhere?

Is your e-book purchase not compatible with the device of your choice?

Don't worry!, Now with every Packt book, you get a DRM-free PDF version of that book at no cost.

Read anywhere, any place, on any device. Search, copy, and paste code from your favorite technical books directly into your application.

The perks don't stop there, you can get exclusive access to discounts, newsletters, and great free content in your inbox daily

Follow these simple steps to get the benefits:

1. Scan the QR code or visit the following link:

https://packt.link/free-ebook/9781837633227

2. Submit your proof of purchase.
3. That's it! We'll send your free PDF and other benefits to your email directly.

www.ingramcontent.com/pod-product-compliance
Lightning Source LLC
Chambersburg PA
CBHW060642060326
40690CB00020B/4482